Parasitology

Symposia of the British Society for Parasitology Volume 37

Veterinary Parasitology – recent developments in immunology, epidemiology and control

EDITED BY

RICHARD J. MARTIN and HENK D. F. H. SCHALLIG

CO-ORDINATING EDITOR

L. H. CHAPPELL

CAMBRIDGE
UNIVERSITY PRESS

Contents

List of contributions   vii

Preface   S1

Simulating *Dictyocaulus viviparus* infection in calves: the parasitic phase   S3
Summary   S3
Introduction   S3
Description of the basic model   S4
  General characteristics   S4
  Life cycle of *Dictyocaulus viviparus*   S4
  Development time of juvenile parasitic stages   S4
  Using the model for experimental infections   S5
Parameter functions and estimates   S6
  Establishment   S6
  Mortality rate of juvenile stages   S9
  Mortality rate of adult worms   S10
  Fecundity   S11
Preliminary model results   S11
  Initial estimates for parameters and constants   S11
  Comparison with experimental data   S12
Acknowledgement   S14
References   S14

Vaccine development and diagnostics of *Dictyocaulus viviparus*   S17
Summary   S17
Introduction   S17
Sources of protective antigens in *D. viviparus*   S18
Rational approach to vaccine design   S18
Immunodiagnosis of *D. viviparus*   S20
Conclusions   S21
Acknowledgements   S21
References   S21

The immune response and the evaluation of acquired immunity against gastrointestinal nematodes in cattle: a review   S25
Summary   S25
Introduction   S25
The immune response against gastrointestinal nematodes   S26
  Antigen presentation and recruitment of different T cell subpopulations   S26

Effector responses   S27
  Immunoglobulins   S27
    IgE   S27
    IgA   S27
    IgG   S28
  Eosinophils   S29
  Mucosal mast cells and globule leucocytes   S29
  Mucus, gut motility and the enteric nervous system   S31
  Immunomodulation   S32
Evaluating the acquired immunity of cattle against gastrointestinal nematodes   S32
  Immunological parameters   S32
    Immunoglobulins   S32
    Mucosal mast cell and globule leucocyte counts   S33
  Parasitological parameters   S33
    Faecal egg counts   S33
    Worm counts   S34
    Worm length   S34
    Number of eggs per female worm   S34
    Vulval flap development   S34
Discussion   S34
References   S35

Development of vaccines against gastrointestinal nematodes   S43
Summary   S43
Introduction   S43
Vaccine attributes and efficacy   S44
Epidemiological and immunological considerations   S44
Vaccination with whole worm material and radiation-attenuated parasites   S45
Vaccination with parasite antigens   S45
  Biochemical fractionation of complex parasite extracts   S46
  Isolation of parasite molecules essential for parasite maintenance within the host   S46
  Antigen selection using antibodies or cells from immune animals   S48
  Isolation of specific antigens from parasite organs that may serve as targets for antibody or immune effector cells   S50
Recombinant protein production   S52
Functional genomics and antigen selection   S53
Mucosal immune responses and antigen delivery   S53
DNA vaccines   S54
Concluding comments   S54
References   S55

Immunological responses of sheep to
Haemonchus contortus S63
Summary S63
Introduction S63
Antibody S63
  Ovine immunoglobulins S63
  IgA and IgG S64
  IgE S64
Lymphocyte proliferation responses S65
Eosinophils and mast cells S65
  Mast cell responses S65
  Eosinophils S66
Cytokines S66
  Th$_1$/Th$_2$ responses S66
  Cytokines in nematode-infected ruminants S66
The possible causes of unresponsiveness S67
  Unresponsiveness of young lambs against
    infectious diseases S67
  Unresponsiveness against gastrointestinal
    nematodes S68
Immunity induced by vaccination S68
  Natural antigens S69
  Hidden antigens S69
Some concluding remarks S69
Acknowledgements S69
References S69

Impact of nutrition on the
  pathophysiology of bovine
  trypanosomiasis S73
Summary S73
Introduction S73
Background S73
The disease S74
Role of nutrition S74
Trypanosomiasis and parasitic
  gastrointestinal infections S74
  Feed and water intakes S74
  Effects on feed utilisation S75
Recent studies with trypanotolerant N'Dama
  cattle S75
  Anaemia and parasitaemia S75
  Fever S76
  Voluntary feed intake S76
  Liveweight S77
  Feeding behaviour S77
  Interactions between infection, nutrition
    and milk yield S77
Studies in small ruminants S78
  Parasitaemia S78
  Bodyweight S78
  Packed red cell volume (PCV) S79
  Recovery from anaemia and erythropoietic
    responses S79
  Serum lipid fractions S80
  Albumin metabolism S80
Influence of trypanosome infections on the
  digestive physiology of sheep S80
  Feed intake S81

  Bodyweight gain S81
  Apparent digestibility S81
  Mean retention time of the roughage in
    the digestive tract S81
  Nitrogen retention S82
Conclusions S82
References S83

Electrophysiological investigation of
  anthelmintic resistance S87
Summary S87
Introduction S87
Properties of ligand-gated ion-channels S88
The patch-clamp technique for recording
  opening of levamisole receptor ion-
  channels S90
Comparison of properties of levamisole
  receptors in sensitive (SENS) and
  resistant (LEVR) isolates S90
  Number of active levamisole receptors is
    less in the resistant isolate S90
  The mean open-time is less in the
    resistant isolate S90
  Channel-conductance histograms show
    heterogeneity of receptors S90
Two-micropipette current-clamp technique S92
  Ascaris pharyngeal muscle S92
  Body muscle contains nicotinic receptors
    and GABA receptors revealed with
    current clamp S92
Conclusion S93
References S93

The development of anthelmintic
  resistance in sheep nematodes S95
Summary S95
Introduction S95
Prevalence of anthelmintic resistance in UK
  and Europe S95
Selection of anthelmintic resistance S95
  Reversion S99
  Detection of resistance S101
  Management of resistance S102
References S103

Value of present diagnostic methods for
  gastrointestinal nematode infections
  in ruminants S109
Summary S109
Introduction S109
Herd health monitoring S109
Clinical diagnosis S110
Faecal examination S110
  Faecal egg counts S110
  Faecal larval cultures S111
Pasture larval counts S111

Contents

Post-mortem examination   S112
  Tracer worm counts   S112
  Sentinel worm counts   S113
Mucous membranes/red blood cell values   S113
Blood pepsinogen values   S113
Blood gastrin values   S113
Serology   S113
DNA technology   S114
Towards herd health monitoring of
  gastrointestinal nematode infections   S116
Conclusions   S117
Acknowledgements   S117
References   S117

**Prospects for controlling animal
  parasitic nematodes by predacious
  micro fungi**   S121
Summary   S121
Introduction   S121
Biological control   S122
Endoparasitic fungi as biological control
  agents   S123
Biological control of nematode eggs by fungi   S124
Predacious fungi as biological control agents   S124
Potential for implementation of biological
  control   S127
Conclusion   S127
References   S128

***Onchocerca ochengi* infections in cattle
  as a model for human
  onchocerciasis: recent
  developments**   S133
Summary   S133
Introduction   S133

Biology, parasitology and infection   S133
  Experimental infections   S133
  O. ochengi infections do not cause disease   S134
  Experimental field exposure reveals the
    kinetics of nodule acquisition and
    insusceptible individuals   S136
  Drugs freely penetrate the nodule and
    adult male worm   S136
  O. ochengi contains *Wolbachia*-like bacteria   S136
Attributes as a model system   S137
  High degree of genetic and antigenetic
    similarity with *O. volvulus*   S137
  Multiple infection allows sequential
    evaluations of worm status   S137
Chemotherapy and chemoprophylaxis   S137
  Methods for drug efficacy studies   S137
  The search for macrofilaricidal agents   S137
  Avermectins/milbemycins and their action
    against adult *Onchocerca*   S138
  Oxytetracycline is macrofilaricidal   S139
  Avermectins are prophylactic against *O.
    ochengi*   S139
Immunology and immunity   S139
  Immune responses are down-regulated at
    patency in experimental infections   S139
  Putative-immune (PI) cattle are
    significantly protected against heavy
    challenge infections but radically drug-
    cured cattle are not   S140
  Immunization with irradiated L_3 larvae
    induces significant protection   S140
Conclusions and future research   S140
Acknowledgements   S141
References   S141

List of contributions

1 *Preface by* RICHARD MARTIN   S1

2 Simulating *Dictyocaulus viviparus*
infection in calves: the parasitic phase *by*
H. W. PLOEGER *and* M. EYSKER   S3

3 Vaccine development and diagnostics of
Dictyocaulus viviparus by J. B. MᶜKEAND   S17

4 The immune response and the evaluation
of acquired immunity against
gastrointestinal nematodes in cattle: a
review *by* E. CLAEREBOUT *and*
J. VERCRUYSSE   S25

5 Development of vaccines against
gastrointestinal nematodes *by* D. KNOX   S43

6 Immunological responses of sheep to
Haemonchus contortus by
H. D. F. H. SCHALLIG   S63

7 Impact of nutrition on the
pathophysiology of bovine
trypanosomiasis *by* P. H. HOLMES,

E. KATUNGUKA-RWAKISHAYA, J. J. BENISON,
G. J. WASSINK *and* J. J. PARKINS   S73

8 Electrophysiological investigation of
anthelmintic resistance *by* R. J. MARTIN
and A. P. ROBERTSON   S87

9 The development of anthelmintic
resistance in sheep nematodes *by*
F. JACKSON *and* R. L. COOP   S95

10 Value of present diagnostic methods for
gastrointestinal nematode infections in
ruminants *by* M. EYSKER *and*
H. W. PLOEGER   S109

11 Prospects for controlling animal parasitic
nematodes by predacious micro fungi *by*
M. LARSEN   S121

12 *Onchocerca ochengi* infections in cattle as
a model for human onchocerciasis: recent
developments *by* A. J. TREES, S. P. GRAHAM,
A. RENZ, A. E. BIANCO *and* V. TANYA   S133

S1

Symposia of the British Society for Parasitology

VOLUME 37

Veterinary Parasitology – recent developments in immunology, epidemiology and control

EDITED BY RICHARD J. MARTIN AND HENK D. F. H. SCHALLIG

CO-ORDINATING EDITOR L. H. CHAPPELL

Preface

There have been important developments in the field of veterinary parasitology over the last few years. This symposium was called to collect individuals together, who have made significant contributions to their field of study, to present and summarize their work.

I would like to pause for a moment before introducing the Symposium in this preface to comment on the sad loss of Professor Peter Nansen, a particularly eminent Danish scientist who developed our field of study. I, like many others, remember him with affection. He was a very helpful colleague and outstanding leader of the Danish Centre for Experimental Parasitology, Royal Veterinary and Agricultural University, Frederiksberg C. We are all saddened by his death and will continue to carry our memories of him with us.

In the symposium Henk Schallig and I wanted to show how epidemiology is now an important contributor to veterinary parasitology so we were fortunate to get Harm Ploeger to contribute a paper on *Dictyocaulus viviparus*. We also wanted to review progress in the development of vaccines for the control of parasites: for this we thank Jackie McKeand and David Knox for their contributions. Immunology is also a major component of the science of veterinary parasitology, and we thank Jozef Vercruysse for this contribution. I thank Henk for his review of *H. contortus* immunology. Protozoa

are also very significant parasites of domestic animals and Ely Katunguka-Rwakishaya delivered a paper for Peter Holmes which reviewed the importance of nutrition for the survival of animals infected with bovine trypanosomiasis. I have reviewed electrophysiological studies of anthelmintic resistance, and Frank Jackson and Bob Coop have reviewed anthelmintic resistance in sheep. Developments in the diagnosis of ruminant gastrointestinal worms were reviewed by Marten Eysker. Two reviews, on the recent developments on control, were given: Michael Larsen, on predacious microfungi and Sandy Trees on the chemotherapy of onchocerciasis.

The meeting was supported by Pfizer Central Research, Fort Dodge and the British Society for Parasitology. We acknowledge this support gratefully. Both Pfizer and Fort Dodge have continued to support the discipline of veterinary parasitology.

A joy of meetings of veterinary parasitologists is the camaraderie that exists. It is not found to the same level in other life science subjects. I carry good memories of the meeting and thank both the audience and the contributors for the success of the meeting. Time will tell how useful the publications prove to be. Henk and I hope that you will find them to be a useful summary of recent developments in our field.

RICHARD J. MARTIN
December 1999

Simulating *Dictyocaulus viviparus* infection in calves: the parasitic phase

H. W. PLOEGER* *and* M. EYSKER

Division of Parasitology and Tropical Veterinary Medicine, Department of Infectious Diseases and Immunology, Faculty of Veterinary Medicine, Utrecht University, P.O. Box 80.165, 3508 TD Utrecht, The Netherlands

SUMMARY

A model simulating *Dictyocaulus viviparus* infection in calves is described. The present paper only deals with the parasitic phase of the life-cycle. Descriptions are given for establishment, development rate of juvenile stages, mortality rates of both juvenile and adult stages, and fecundity. Literature data were used to to develop parameter functions and to estimate initial values for constants. Development of acquired immunity, defined as the proportional ability of the host to reduce the number of parasite individuals in some stage or about to move into a next stage, against establishment (protection) or affecting mortality rates of juvenile or adult parasite stages has been included. The effect of immunity on one parameter or process is viewed as distinctly separate from the effect on another. Preliminary comparisons between model prediction and observations gives encouraging results, indicating that the model simulates experimental *D. viviparus* infection in calves reasonably well. Some quantitative discrepancies between prediction and observation make clear however, that not all parts of the model are accurate. Further experimentation is needed to re-evaluate current model description and to improve model simulation.

Key words: *Dictyocaulus viviparus*, simulation, model.

INTRODUCTION

It has been agreed for a long time that the epidemiology of lungworm (*Dictyocaulus viviparus*) infection in calves is unpredictable, certainly in comparison with that of gastrointestinal nematode infections. Then why try to create a simulation model for this parasite of the bovine lung?

First, though an effective vaccine has been available for many decades, lungworm infections continue to be a matter of concern (Eysker, 1989, 1994). Not only in calves, but nowadays also in older cattle (David, 1993, 1997; cf. comments of F. H. M. Borgsteede and J. Vercruysse, 1997, in 'Discussion: general session panel', Veterinary Parasitology **72**, 525–537). This undoubtedly is linked to all efforts to control gastrointestinal nematode infections. Vaccine-stimulated immunity requires a natural booster to become fully effective (Eysker, 1994). Most modern anthelmintics are active against a broad range of parasites, including *D. viviparus*. These drugs are extensively applied. Efforts are currently being undertaken in The Netherlands to reduce the use of anthelmintic drugs. In addition, farmers and veterinarians are being made more aware of the fact that control of gastrointestinal nematode infections begins with an assessment of the effects of applied pasture management. Such efforts may also influence the occurrence of lungworm infections. This alone is sufficient reason to increase

our understanding of lungworm epidemiology and the development of immunity in relation to exposure and what the effects are of various pasture management and parasite control strategies. Modelling can greatly assist this process.

Second, modelling the dynamics of any kind of system presumes at least some degree of predictability, or reproducibility, in that system. The predictability in epidemiology for several gastrointestinal nematodes of both cattle and sheep has led to mathematical models which produced useful results (Callinan *et al.* 1982; Grenfell, Smith & Anderson, 1987; Leathwick, Barlow & Vlassoff, 1992; Smith, 1994; Smith & Grenfell, 1994). The seeming unpredictability of lungworm infections may lead to the idea that there is some fundamental difference between *D. viviparus* and gastrointestinal nematodes such as *Ostertagia ostertagi*. However, *D. viviparus* is a nematode and a member of the same family to which the gastrointestinal nematodes belong. It may well be that there is no fundamental difference in the life-cycle or epidemiology, but that it is just a matter of parameter values as has been found for other trichostrongylid nematode species (Smith, 1994). Attempting to create a model may help to elucidate such a fundamental difference, if indeed it exists. Both field and experimental observations indicate that it may be possible to capture at least parts of lungworm epidemiology in a simulation model. Eysker *et al.* (1993, 1994) showed in field trials that the epidemiology of lungworm infection becomes predictable once the start of the initial infection is known. It also has been shown that

* Corresponding author: Tel: +31 30 253 1223. Fax: +31 30 254 0784. E-mail: h.ploeger@vet.uu.nl

different dose levels result in correspondingly different patterns of faecal larval shedding (Boon, Kloosterman & Breukink, 1984).

The present paper first describes the basic model simulating the life-cycle dynamics of *D. viviparus*, and subsequently focuses on the parasitic phase in the life-cycle. Data from literature, in so far as they were available, were used to estimate parameters for establishment, parasite mortality and fecundity. Some preliminary comparisons of model outcomes with experimental data will be shown. The free-living phase will be dealt with elsewhere.

DESCRIPTION OF THE BASIC MODEL

General characteristics

The type of model (1) is based on the *a priori* assumption that any difference between *D. viviparus* and other members of the trichostrongylid family is quantitative rather than qualitative in nature, and (2) is determined by the ultimate goal, i.e. to simulate a variety of control strategies, including pasture management, for use on commercial farms. Consequently, we started to build the simulation model along similar lines to the one created for gastrointestinal nematodes and today known as PARABAN©-Merial (Smith, Jacobsen & Guerrero, 1995). Our model, basically, is a series of difference equations describing the temporal parasite population dynamics, taking account of effects of host responses as well as of some weather conditions on those dynamics. In the initial model building a deterministic approach is followed. The model runs from day to day.

Life-cycle of Dictyocaulus viviparus

The description of the life-cycle can be reduced to what happens to four parasite life-stages (Fig. 1). For the infective larval stage two phases are distinguished, one phase within dung pats and one on grass. This distinction is made because of possible differences in mortality rates dictated by differences in environment. This gives five difference equations:

$$L_{3,t+1} = e^{-\mu_{L3,t}} \cdot L_{3,t} + m_t \cdot L_{3,t} - \beta_t \cdot L_{3,t}. \tag{1}$$

$$L_{t+1} = e^{-\mu_{L,t}} \cdot L_t + \epsilon_{t-mt} \cdot \frac{\beta_{t-mt}}{H_{t-mt}} \cdot L_{3,t-mt} \tag{2}$$

$$- \delta_{L,t} \cdot L_{t-(dL-mt)}.$$

$$Ad_{t+1} = e^{-\mu_{Ad,t}} \cdot Ad_t + \delta_{L,t} \cdot L_{t-(dL-mt)}. \tag{3}$$

$$L_{1,t+1} = \zeta_t \cdot \tfrac{1}{2} \cdot Ad_{t-dE}. \tag{4}$$

$$fL_{3,t+1} = e^{-\mu_{fL3,t}} \cdot fL_{3,t} + \delta_{L1,t} \cdot L_{1,t-dL1,t} \cdot H_{t-dL1,t} \tag{5}$$

$$- m_t \cdot fL_{3,t}.$$

Tables 1 and 2 present the symbolic notation for all variables, parameters and constants, along with their definitions. Notice that in Eqn (2) some time has been allowed for ingested larvae before they enter the juvenile stage. This specifically refers to the time required for migration from the intestine to the lungs, and in effect covers the period between ingestion of L_3 and the first time parasite stages can accurately be recovered from the lungs.

Several assumptions have been made to simplify the initial modelling phase. The ones of relevance for the parasitic phase are: (1) all hosts are equal, so that group mean data represent what happens in an average host; (2) the adult worm population consists of equal numbers of male and female worms; and (3) parasitic juvenile stages do not become inhibited at some point during development. Arrested development appears to relate in particular to the carrier state as a means of surviving the winter period in temperate climatic regions (Eysker, 1994). Although carriers, usually older animals, are important as a source for primary infections in calves (Saatkamp, Eysker & Verhoeff, 1994), the phenomenon of arrested development itself plays only a minor role in the pattern of infection during the first grazing season.

Development time of juvenile parasitic stages

Within the model described in Eqns (1–5) switches are made between what happens in an individual host to numbers of infective larvae per unit of area. It is obvious that the description given above is intended for simulating natural infections, with or without additional artificial infection. It is also clear from the above equations that on each day the number of parasite individuals entering a certain stage are treated as cohorts. Development from one stage to another stage has not yet been considered to be subject to some biologically-determined temporal distribution. In the case of simulating natural infections with daily ingestion of infective larvae, this may not present a problem or give immediately obvious aberrant observations. However, testing (parts of) the model can initially be best done using data from experimental infections. Then, particularly with single infection or widely-spaced trickle infections, the absence of temporal distributions does present a problem. It will not be possible to reproduce the initial increase in parasite individuals in e.g. the adult worm population or to reproduce temporal faecal larval excretion patterns. Further, a difficulty arises whenever host immunity influences the rate of development of certain life stages thereby increasing the period in which individuals in one stage appear as individuals in the next stage.

To overcome the problem outlined above, a final aspect of the basic model specifically subjects the development time of juvenile parasite stages in a host to a temporal distribution function. Here, the delayed gamma distribution is used as was applied by Young *et al.* (1980*a*, *b*) for modelling the hatching

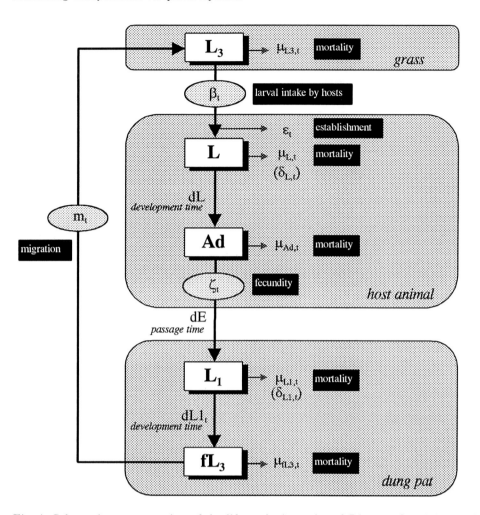

Fig. 1. Schematic representation of the life-cycle dynamics of *Dictyocaulus viviparus* showing the parasite life-stages, major (rate) parameters and processes modelled. Definitions and symbolic notation used are given in Tables 1 and 2.

times of *Ostertagia circumcincta* eggs and the development times of the free-living stages of *O. ostertagi*. The cumulative form of the distribution is:

if $t > dL_{min}$ then
$$F(t;1,\lambda,dL_{min}) = 1 - [1 + \lambda(t - dL_{min})]e^{-\lambda(t - dL_{min})}. \quad (6)$$

Eqn (6) is a three-parameter function. Two are included as parameters in symbolic notation. The third is assigned the value 1, because the parameter represents the probability that individual larvae will develop and move into the next stage. The probability of surviving the juvenile stage in the host is addressed with the parameters $\mu_{L,t}$ and $\delta_{L,t}$. The cumulative probability distribution is shown in Fig. 2, along with the distribution of proportions of developed larvae moving into the adult stage.

The above still does not consider the possibility of host responses influencing the rate of development. Michel *et al.* (1965) observed that acquired immunity slows down the rate at which juvenile stages develop. This part of acquired immunity seems to be linked to arrested development, which is for the moment not considered. Hence, in the initial modelling phase it is

assumed that the development rate of juvenile stages cannot be influenced by host responses. The inclusion of Eqn (6), however, facilitates an easy implementation of any effect of host immunity on the rate of development at later stages.

Using the model for experimental infections

In the present paper attention is restricted to the parasitic phase under experimental conditions. Therefore the description given above can be simplified. For experimental infections only Eqns (2–4) are relevant with one change in Eqn (2). The term β_t/H_t can be set to unity and L_{3t} has to be viewed as the number of infective larvae artificially given to calves at some time t.

Usually, 1st stage larvae are counted per amount of faeces. Hence, Eqn (4) is followed by recalculating L_{1t} into number of L_1 excreted by the host per gram of faeces (LPG). Daily faeces production by calves is estimated as described by Smith, Grenfell & Anderson (1987).

Table 1. Definitions and symbolic notation of variables and parameters

t	Time in days

Variables

Ad_t	Number of adult worms in host
$fL_{3,t}$	Number of infective larvae in dung pats (on 1 ha pasture)
H_t	Stocking density (number of calves per hectare pasture)
$I_{P,t}$	Proportional protection afforded by acquired immunity against the establishment of newly ingested L_3s
$I_{\mu Ad,t}$	Immunity (as a proportion) influencing the rate of loss of adult worms
$I_{\mu L,t}$	Immunity (as a proportion) influencing the rate of loss of developing juvenile stages
$iL_{3,t}$	Number of ingested infective larvae on day t
L_t	Number of juvenile stages in one host
$L_{1,t}$	Number of first-stage larvae produced in one host (day^{-1})
$L_{3,t}$	Number of infective larvae (on 1 ha pasture)
LPG_t	Number of excreted first-stage larvae per gram of faeces
WB_t	Actual worm burden (total number of juvenile stages and adult worms)

Parameters

β_t	Removal of larvae through ingestion (grazing) by calves
$dL_{1,t}$	Average duration parasites stay in the first larval stage or, more accurately, stay in the pre-infective larval stages (days)
$\delta_{L,t}$	Proportion of juvenile stages surviving that stage to enter the adult parasite stage
$\delta_{I,1,t}$	Proportion of first-stage larvae appearing as 1st stage larvae dL_1 days ago that survives to enter the next larval parasite stage, in this case the infective stage
ϵ_t	Proportion of ingested larvae that establishes in the host
m_t	Rate of migration of infective larvae from dung pats to herbage (larva^{-1} day^{-1})
$\mu_{L3,t}$	Mortality rate of infective larvae on herbage (larva^{-1} day^{-1})
$\mu_{L,t}$	Mortality rate of juvenile stages in a host (larva^{-1} day^{-1})
$\mu_{Ad,t}$	Mortality rate of adult worms (worm^{-1} day^{-1})
$\mu_{fL3,t}$	Mortality rate of third-stage larvae, as well as of pre-infective larvae, in dung pats (larva^{-1} day^{-1})
$\Theta_{wb,t}$	Step function with Θ being 0 if WB > 0, otherwise Θ is 1
ζ_t	Fecundity, i.e. the number of eggs produced by female worms (female worm^{-1} day^{-1})

PARAMETER FUNCTIONS AND ESTIMATES

The parameters that will be discussed are establishment, mortality rates of juvenile and adult stages, and fecundity. The passage time (dE) of eggs and larvae hatched from these eggs through the alimentary tract will not be discussed. For convenience it is assumed constant, being equal to the average passage time of food. Values for the parameters involved in Eqn (6) were given in Fig. 2 (see also Table 2) and will not be discussed further.

A strong immunity rapidly develops against *D. viviparus*. As a consequence, the population dynamics of the parasite stages inside the host will be for a large part under the control of the immune system of the host. Hence, it is appropriate to first make some general remarks on how acquired immunity and its effects on the parasite population dynamics are viewed.

Acquired immunity may influence some or all of the following processes: establishment of ingested larvae, rate of development to the adult stage including arrestment of development, mortality during development as well as in the adult stage, and fecundity of female worms. Acquired immunity will be described in terms of its effect on the aforementioned processes, with the initial exclusion of an effect on the rate of larval development. No reference will be made to specific components of humoral or cell-mediated immune responses. Hence, immunity or some unidentified component of it is here defined as the proportional ability of the host to reduce the number of parasite individuals in some stage or about to move into a next stage. The effect of immunity on one parameter or process is viewed as distinctly separate from the effect on another, following the suggestion of Michel *et al.* (1965).

Age-resistance is neglected here. Michel *et al.* (1965), although referring to some studies indicating the possible existence of age-related immunity, presented data which showed that calves at least up to an age of two years do not show an increase of resistance to infection. The only two effects observed were a reduced growth rate of parasite individuals in the early stages of (re-)infection and a tendency for an increased proportion of the parasites to become arrested in their development. These phenomena were ascribed to an increased ability of older hosts to mount stronger relevant immune responses at an earlier point in time. However, these were relatively minor effects and became apparent after the calves were older than 1 year.

Establishment

Michel *et al.* (1965) defined establishment rate as the proportion of ingested infective larvae that reached the lungs, moulted into juvenile stages and was

Table 2. *Definitions, symbolic notation and initial estimates of constants*

Ad_{max}	Number of adult worms above which the rate of development of immunity against the survival of adult worms is maximal	1000
dE	Average duration before produced eggs appear as L_1 in the faeces (days)	3
dI_P	Timelag after first infection before protection starts to decrease (days)	—*
dL	Average time required before ingested L_3 become adult (days)	28
dL_{min}	The minimum development time required before ingested L_3 become adult (days)	20
ϵ_0	The basic proportion of L_3s that, after ingestion, establishes as juvenile larvae and survives at least up to day 10 after ingestion in a fully susceptible host	0·3
γ_P	The rate of loss of protection against the establishment of reinfection (day^{-1})	0·011
i	Number of days ingested larvae contribute to the development of protective immunity ($i = 1 \ldots n$)	3
$iL3_{max}$	Number of ingested L_3 above which the rate of development of protective immunity (I_P) is maximal	1000
λ	A parameter, together with dL_{min}, determining the average development time (dL), which is given by $dL_{min} + 2/\lambda$	0·25
L_{max}	Number of juvenile stages above which the rate of development of immunity against the survival of those juvenile stages is maximal	1000
mt	Time required for larvae to migrate from the intestine to the lungs (days)	10
$\mu_{L,0}$	Basic rate of loss of juvenile stages during development to maturity in fully susceptible calves (day^{-1})	0·0148
$\mu_{Ad,0}$	Basic rate of loss of adult worms in fully susceptible calves (day^{-1})	0·0148
τ	Timelag before effective protective immunity develops after ingestion of infective larvae (days)	8
T	Timelag before a present worm burden, either juvenile or adult, results in immunity effectively acting against the survival of immature or adult stages (days)	21
v_P	Rate of development of protective immunity I_P (day^{-1})	0·25
$\overline{\omega}_{Ad}$	Rate of development of acquired immunity acting on the survival of adult worms (day^{-1})	0·60
$\overline{\omega}_L$	Rate of development of acquired immunity acting on the survival of developing immature stages (day^{-1})	0·20
$\overline{\omega}_P$	Rate of reinforcement of protective immunity (day^{-1})	0·035
WB_{max}	Worm burden above which reinforcement of protective immunity is maximal	1000
ζ	Fecundity, i.e. the number of eggs produced by female worms (female worm^{-1} day^{-1})	11 340†

* This constant is only used in Eqn (8) and not in the simulation model as such.
† Initially, fecundity is kept constant.

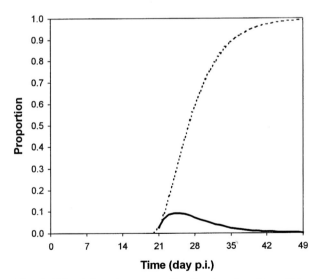

Fig. 2. The delayed gamma distribution (see Eqn (6)) for the development time of juvenile parasite stages in a host. The solid line gives the daily proportion of parasite individuals becoming mature. The dashed line represents the corresponding cumulative distribution. Parameter estimates used here were $dL_{min} = 20$ and $\lambda = 0.25$, meaning that the average development time was set at 28 days.

present at 10 days p.i. Here, we follow this definition because it is one of the earliest times for fairly accurate parasite collection and counting. It implicitly includes any mortality of parasite individuals during larval penetration of the intestinal wall and subsequent migration to the lungs.

After a primary infection the proportional establishment is assumed to lie around some basic biological value (ϵ_0), a value not yet influenced by host responses. Not many data are available to estimate the value for ϵ_0. Michel (1962) infected 29 calves between 3 and 9 months of age with 35 000–40 000 L_3, and subsequently necropsied these animals 10 days later. The average establishment was 21·2 % (range 10·9–32·1 %). Michel *et al.* (1965) infected 12 calves 6–21 months of age with 66 L_3 kg^{-1} liveweight; 9–11 days p.i. they observed an average establishment of 31·0 % (range 20·7–42·2 %). In both studies no effect of host age was seen. Finally, Cornwell & Jones (1970*a*) conducted two studies with calves 8 weeks of age. In the first, eight calves were infected with 1000 L_3 and necropsied either 14 or 28 days p.i. At 14 days p.i. an average establishment of 15·3 % (range: 13·4–17·0 %) was

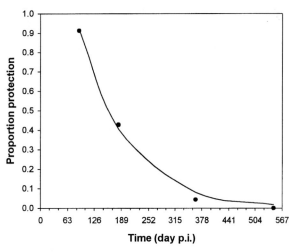

Fig. 3. Proportion protection against establishment of a severe challenge infection (circles), given after varying time-intervals following two initial immunizing infection doses (data from Michel et al. 1965). The two immunizing doses were given at days 0 and 28. The solid line is a fitted exponential function to the data (for estimates see text). One data-point from a challenge at 27 months is omitted from the graph because the level of protection was the same as the level at 18 months p.i.

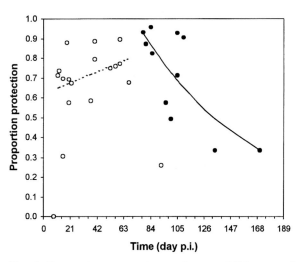

Fig. 4. Proportion protection against establishment of a severe challenge infection ($35\,000$–$40\,000$ L_3), given after varying time-intervals following an initial immunizing dose of 3200 to 3500 L3 (data from Michel, 1962). The open circles denote calves which were patent, while the closed circles denote calves which stopped excreting larvae. The dashed line is a linear fit to the data between days 11 and 67 p.i., with a slope of 0·0026 (s.e. 0·0018). The solid line is an exponential fit to the data represented by the closed circles. Parameter estimates of this fit are given in the text.

observed, while at 28 days 25·8 % (range: 9·4–41·4 %) of the larval dose was recovered. In the second experiment, six calves received 4000 L_3 of which on average 27·3 % (range 17·1–41·1 %) was recovered 8–12 days p.i. Based on the above primary infection data, the establishment rate ϵ_0 can be roughly

estimated to be 0·3. It should be noted that most of the above data originate from experiments in which calves were given extremely high larval doses. In the present model it is assumed that the rate of establishment in fully susceptible calves does not depend on the level of the larval dose.

Once the host starts to respond to infection, the rate of establishment depends on the level of host immunity specifically directed against establishment of newly ingested larvae (Michel, 1962; Michel et al. 1965). The immunity against establishment of super- or reinfections is called protection. In equation form this can be written as:

$$\epsilon_t = \epsilon_0(1 - I_{P,t}). \tag{7}$$

Michel et al. (1965) estimated $I_{P,t}$ after an immunizing infection with 1100 L_3 twice, given with an interval of 4 weeks. At 3, 6, 12, 18 and 27 months after the first immunizing dose calves were challenged with 66 L_3 kg^{-1}. The observed average protection in time is plotted in Fig. 3. To the observations an exponential function can be fitted of the form:

$$I_{P,t} = e^{-\gamma_P(t - dI_P)}, \quad \text{if } t - dI_P \geqslant 0. \tag{8}$$

The estimates for γ_P and dI_P are 0·00898 (s.e. 0·00081) and 81·8 (s.e. 4·3). The latter suggested that protection afforded by the immunizing infection starts to decline 82 days after the first dose was given. Necropsy data of calves during the course of the immunizing infection suggested that protection was already high at the time of the second immunizing dose. Establishment of the second dose was virtually zero as judged by the absence of large numbers of developing stages 8–18 days after that second dose. The estimate of 82 days roughly coincided with the end of the patent period, which suggests that the decline in protection starts after the worm burden has been lost.

From the data of Michel et al. (1965) it is not clear when protection actually begins to develop. It did appear that an infection dose of 1100 L_3 was sufficient to cause a very high level of protection being evident 28 days p.i. as mentioned above. It is also not clear whether or not protection is dependent upon infection dose, or if protection may wane in the presence of a remaining (small) worm burden. Earlier, Michel (1962) resolved some of these questions, notably the development of protection early during a severe primary infection. Using 3200–3500 L_3 in a single immunizing dose he found the data presented in Fig. 4. Michel (1962) suggested that there were three phases. The first being rapid development of protection between days 8 and 11 p.i. Such a rapid development was earlier noted by Michel & Parfitt (1956). The second being a phase of consolidation and a slow reinforcement of protection lasting until patency ends around days 70–80 p.i. And finally, a phase in which protection wanes in the absence of both an existing worm burden and reinfection. Fitting Eqn (8) to the third phase of declining protection now resulted in estimates for γ_{p} and $\mathrm{dI_p}$ of 0·01105 (s.e. 0·00391) and 70·8 (s.e. 10·3). These estimates come close to those for the data of Michel *et al.* (1965). The difference in estimates for $\mathrm{dI_p}$ may be explained by the fact that Michel (1962) used one immunizing dose, whereas Michel *et al.* (1965) used two immunizing doses 4 weeks apart. Noteworthy is that again loss of protection apparently started after calves were no longer patent (see Fig. 4). No effect of immunizing dose was investigated, although Michel mentioned that 'protection resulting from a smaller dose of larvae is either substantially less or develops less rapidly'. A relationship between initial dose level and subsequent course of infection was noted earlier by Michel & Parfitt (1956).

It seems likely that the magnitude of protection is influenced by three main forces: (1) a force stimulating the development of protection to some level acting rapidly and strongly and probably in a dose-dependent manner; (2) a force subsequently maintaining and reinforcing developed protection so that it steadily, and at a low rate, increases in magnitude; and (3) a force reducing the magnitude of protection. It appears reasonable that the first largely depends on the number of ingested infective larvae (i.e. exposure to infection), and that the second force depends on the presence of an adult worm burden. The third force may be considered as a continuous breakdown of protection related to loss of entities conferring that protection and thus also occurring in the presence of reinfection and worm burdens. The breakdown of protection becomes apparent only after worm burdens have been lost in the absence of reinfection. Therefore, it is assumed that loss of protection occurs irrespective of reinfection or worm burdens, and that this loss in the presence of a worm burden is counterbalanced by the second force mentioned above. Note that, as a consequence, it is assumed that there is no immunological memory concerning protection against establishment. Eqn (8) now can be revised and extended to:

$$\mathrm{I_{P,t}} = \left(\frac{\sum_{i=1}^{n} \mathrm{iL_{3,t-\tau-i}}}{\mathrm{iL_{3max}}}\right) v_{\mathrm{P}}(1 - \mathrm{I_{P,t-1}})$$
$$+ \left(\frac{\mathrm{WB_{t-1}}}{\mathrm{WB_{max}}}\right) \overline{\omega}_{\mathrm{P}}(1 - \mathrm{I_{P,t-1}})$$
$$+ (1 - \gamma_{\mathrm{P}}\Theta_{\mathrm{wb,t}}) \, \mathrm{I_{P,t-1}}. \qquad (9)$$

The ratios in Eqn (9) can never exceed unity. This implies that for development of protection at infection levels exceeding $\mathrm{iL_{3max}}$ and $\mathrm{WB_{max}}$ the dependency on infection level or dose will be lost. This does not disagree with the intuitive expectation that with increasing infection levels additional increases contribute disproportionately less to stimulating development of protection. The term

$(1 - \mathrm{I_{P,t-1}})$ ensures that any increase in protection is proportional to the distance left between current level of protection and the maximum possible level of protection being 1.

Although Eqn (9) is not a very elegant equation, it does allow experimental falsification. For instance, the first part can be tested in experiments studying the relationship between truncated primary infections and protective immunity at different times after infection. Noteworthy in this respect is the finding of an immunodominant antigen on the $\mathrm{L_3}$ sheath surface coat by Gilleard, Duncan & Tait (1995 *a, b*). This finding prompted Taylor *et al.* (1999) to investigate the development of immunity to a challenge infection in calves following immunizing trickle infections while being treated with a long-acting endectocide. They found that the endectocide did not prevent development of immunity against a challenge infection, indicating that incoming infective larvae were capable of eliciting a strong immune reaction. Further, a comparison is possible with what happens in vaccinated animals. In fact, X-irradiated larvae have been found to reach the lungs at the same rate as normal infective larvae, but their further development to the adult stage is hindered. Still, vaccination with these irradiated larvae produces a high, but not 100%, level of protection against challenge infection (Poynter *et al.* 1960), supporting the relevancy of the first part of Eqn (9). Similar findings were reported by Cornwell & Jones (1970 *b, c*).

Mortality rate of juvenile stages

Apart from an acquired immunity becoming effective against establishment of reinfection(s), acquired immunity can affect the mortality rate of developing larvae (Michel *et al.* 1965). The latter is discussed here under the general heading of loss of parasite individuals between days 10 and 30 after infection, i.e. more or less between becoming established at the predilection site and becoming mature.

Again assuming that there is no age-dependent immunity involved, the worm count data from Michel *et al.* (1965) can be used to obtain an estimate of both the normal rate of loss of parasite individuals (i.e. after primary infection) as the effect of developed immunity on this rate of loss. Using the data of the calves up to an age of 18 months, the average proportion of parasites surviving the period between days 10 and 30 p.i., and having developed (i.e. being > 3 mm according to Michel *et al.* 1965), was 0·743. It is here assumed that those individuals being smaller than 3 mm at day 30 were arrested and did not suffer any mortality. Consequently this fraction was left out in the calculations. Then assuming an exponential decay function, this implies a daily rate of loss ($\mu_{\mathrm{L,t}}$) of 0·0148 in susceptible, previously parasite-naïve calves. In previously infected calves only a proportion of, on average, 0·0407 of the parasites present at day 10 and developing, survived up to day 30. This was found after high challenge infections from 3 months p.i. onwards in calves primary infected with 1100 L_3 twice. This means a daily rate of loss of 0·1601, more than 10 times that seen in susceptible calves. A very interesting set of data was provided by the vaccinated calves in the same experiment. In these animals, the average rate of loss was similar to what was observed in the susceptible control animals (proportion survived being 0·771 leading to a daily loss rate of 0·0130). Since X-irradiated larvae do penetrate and migrate to the lungs but do not develop further after becoming established, the immunity involved here cannot be related to the intake of infective larvae. Apparently, established larvae have to develop beyond the early 5th stage before the immunity of relevance here can develop.

The data of Michel *et al.* (1965) indicate that this particular aspect of acquired immunity does not wane in time, as opposed to protection against establishment. This appears to be supported by data from Jarrett *et al.* (1959). These authors infected calves with 2500 L_3 and subsequently challenged these animals after 160 (with 4500 L_3) and 310 days (with 13 000 L_3). At autopsy 32 days after the second challenge very few worms were recovered, on average 0·14 % of the challenge dose with all worms being smaller than usual. The same authors, using 25 doses of 300 L_3 over a period of 62 days, found calves to be highly resistant to a challenge infection of 15 000 L_3 155 days after the last immunizing dose. Only 0·16 % of the challenge dose could be recovered 30 days after the challenge was given. If all aspects of acquired immunity were to wane with time, much higher worm burdens would have been expected.

The proportion of established parasite individuals surviving the development period and becoming adult can be described by:

$$e^{-\mu_{L,t}(dL-mt)} = (1 - I_{\mu L,t})\, e^{-\mu_{L,0}(dL-mt)}. \tag{10}$$

The left-hand part of Eqn (10) is equal to the parameter $\delta_{L,t}$ in Eqns (2) and (3). From Eqn (10) $\mu_{L,t}$ can be calculated as a function of $I_{\mu L,t}$. All that remains to be done is to define how $I_{\mu L,t}$ depends upon the juvenile worm burden. Thus:

$$I_{\mu L,t} = I_{\mu L,t-1} + (1 - I_{\mu L,t-1})\frac{L_{t-1-T}}{L_{max}}\,\overline{\omega}_L. \tag{11}$$

As in Eqn (9), the ratio in Eqn (11) cannot exceed unity. A further constraint to Eqn (11) is that $I_{\mu L,t}$ is not allowed to exceed a value of 0·9453 to prevent the mortality rate from increasing above 0·16, which was the value found in immune calves (Michel *et al.* 1965). Such a restriction is further needed to allow for the presence of carrier animals in later stages of model development.

Mortality rate of adult worms

No specific data were found concerning the mortality rate of adult worms. An initial estimate for this, therefore, has to be made based on the pattern of faecal larval excretion after primary infections. As far as it is possible to find a typical pattern of infection, represented by the L_1 excretion in the faeces, it is shown in Fig. 5 for four high single primary infection doses. These four trials in Fig. 5 are certainly not the only ones available in the literature, but appear to represent the general temporal pattern of LPG fairly well. The day of peak LPG and length of the period of larval excretion (i.e. the temporal pattern as such) appear to be fairly consistent between trials. The peak LPG occurs somewhere between days 30 and 40 p.i. and the duration of the patent period is approximately 30–40 days.

Then, it is first assumed that the adult worm mortality can be described with the general exponential decay function. Second, it is assumed that all parasite individuals have matured by the time that the peak LPG is observed. Third, it is assumed that the end of the patent period coincides with the time that virtually all worms have been lost. Consequently, the worm population present at the time of the peak LPG occurrence (say at day 37 p.i.) has died off some 20 days later (day 57 p.i.). A mortality rate of 0·2303 worm^{-1} day^{-1} would result in a reduction of the worm population to 1 % within 20 days. A rate of 0·4605 would result in a reduction to 0·01 %. These estimates are very high and far higher than those estimated for developing immature stages, even in immune host animals.

Faced with a lack of relevant data, it seems reasonable to assume that the mortality rate of adult worms can be described in a similar way as was done for the juvenile stages. Further, it is reasonable to assume that if there is an adult population, this population will always be subject to some mortality. However, it seems unlikely that a young adult population, e.g. the worms responsible for the initial increase in LPG, is subject to the high mortality rate of 0·23 or higher as estimated above. Therefore, the mortality rate of adult worms will vary in time. The equations then become:

$$\mu_{Ad,t} = \mu_{Ad,0} - \frac{\ln(1 - I_{\mu Ad,t})}{20}. \tag{12}$$

$$I_{\mu Ad,t} = I_{\mu Ad,t-1} + (1 - I_{\mu Ad,t-1})\frac{Ad_{t-1-T}}{Ad_{max}}\,\overline{\omega}_{Ad}. \tag{13}$$

The number 20 in Eqn (12) substitutes the term $(dL - mt)$ in Eqn (10). Originally the latter was set at 20 because data referred to observations between days 10 and 30 p.i. for the juvenile stages. The estimate for the parameter $\overline{\omega}_{Ad}$ will be higher than for its counterpart $\overline{\omega}_L$. In contrast with the mortality rate

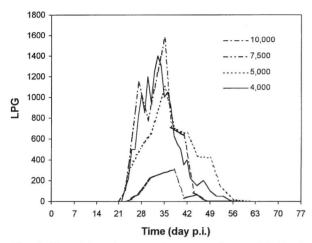

Fig. 5. Faecal larval output (per gram faeces, LPG) after a single primary infection. Data originate from Jarrett *et al.* (1957) (4000 L$_3$), Cornwell & Jones (1970*b*) (7500 and 10000 L$_3$ doses), and Ploeger *et al.* (1990) (5000 L$_3$).

for the juvenile stages, the restriction that this rate cannot exceed 0·16 will be neglected in the case of an adult worm population.

Fecundity

Here, fecundity (ζ) is defined as the total daily number of first-stage larvae which can be produced by one female worm and which appear in the excreted faeces. At first, fecundity has to be estimated from primary infection trials in which the host presumably has not mounted immune responses affecting infection, or at least the reproductive capacity of the worms. Estimates can be derived in several ways. First, it can be derived from trials in which calves have been necropsied relatively early during the primary infection and for which both worm counts and LPG data are available on the same day p.i. Second, estimates can be made using the peak LPG during a primary infection assuming several things: establishment of 30%, a male:female ratio of 1:1, a mortality rate of 0·0148 day^{-1} under the additional assumption that the peak LPG occurs shortly after the last larvae have matured, and finally that the calves usually used in trials excrete some 8 kg faeces per day. The latter assumption is not very accurate. Smith *et al.* (1987) used a function to describe faeces production in dependence of calf liveweight assuming certain types of food. However, in many instances no accurate weight data are available. Therefore, the above given assumption concerning daily faeces production is made simply for convenience. These assumptions together will not provide for a highly accurate estimation of fecundity. However, the estimate will be sufficiently accurate to be used as a provisional first estimate in the model. Table 3 presents some estimates on fecundity obtained from various trials.

A few things can be noted with respect to the

estimates based on the necropsy data in Table 3. First, Kloosterman, Frankena & Ploeger (1989), Kloosterman, Ploeger & Frankena (1990) and Kloosterman (unpublished data) found that the male:female ratio at day 35 does not differ significantly from 1:1. This was found irrespective of the size of the worm burden which varied among the experimental calves. Second, the data from Kloosterman & Frankena (1988) and Kloosterman *et al.* (1989) showed an extremely low establishment, with < 1% and about 5% of the larval dose recovered at necropsy. In the other two trials from Kloosterman and co-workers (unpublished) some 20% of the larval dose was recovered at necropsy, which conforms better to what is usually observed. Third, in two instances (Kloosterman *et al.* 1989, 1990) a strongly significant non-linear positive relationship was reported between worm count and LPG on day 35 p.i., implying that fecundity per worm increased with increasing worm burden. This contrasts to the observation that fecundity of *O. ostertagi* is negatively related to worm burden (Smith *et al.* 1987). The unpublished data of Kloosterman did not support the existence of the relationship between fecundity and worm burden reported by Kloosterman *et al.* (1989, 1990). Therefore, it is felt that data are insufficient to implement any kind of relationship between fecundity and size of worm burden into the model.

Until further studies are performed, fecundity will be assumed to be a constant irrespective of size of worm burden and also irrespective of the host's experience with infection or level of immunity. The initial approximation of fecundity will be based on the mean of the values listed in Table 3, with the exclusion of the values from Kloosterman & Frankena (1988) and Kloosterman *et al.* (1989) because of the apparently extremely low establishment of infection in these trials.

PRELIMINARY MODEL RESULTS

Initial estimates for parameters and constants

For several parameters and constants initial estimates have already been given above. Estimates for the remaining parameters were obtained by systematically changing parameter values until reasonably satisfactory fits were obtained with available data from the literature, in particular with those from Michel (1962). Table 2 lists the initial estimates for all constants, as used for the preliminary comparisons with experimental data below.

It is emphasised that, at present, estimates were sought which would result in model predictions mimicking observations in a qualitatively satisfactory way, rather than be quantitatively as accurate as possible. Because of a lack of relevant data several aspects in the present model may not be accurate descriptions of what really happens. Hence, the

Table 3. *Estimates on fecundity from single primary infection experiments*

Study	L_3 dose	Day of peak LPG or of necropsy	Peak LPG or LPG on day of necropsy	Estimated or counted no. of female worms	Estimated fecundity (female^{-1} day^{-1})
Based on peak LPG					
Jarrett *et al.* (1957)	4000	33	1400	446	25112
Jarrett *et al.* (1959)	2500	50	200	217	7377
Cornwell (1962)	5000	28	237	601	3155
Michel *et al.* (1965)*	1100	42	130	107	16200
Cornwell & Jones (1970*b*)	7500	38	306	777	3151
Cornwell & Jones (1970*b*)	10000	35	1569	1083	11590
Ploeger *et al.* (1990)	5000	35	1105	542	16310
Eysker *et al.* (1995)‡	20	38	4·53†	2	18133
Based on necropsy data					
Kloosterman & Frankena (1988)	6000	35	2·55§	46·9	435
Kloosterman *et al.* (1989)	5000	35	‖	‖	1142
Kloosterman *et al.* (1990)	5000	35	326 ‖	422 ‖	6187
Kloosterman, unpublished data	5000	35	417	539	6189

* Note that here no consideration is given to the efficiency of the larvae counting technique, except in this case because authors specifically mentioned that the recovery rate of their technique was 60% on average.
† In some instances larval counts were made per 10 to 30 g of faeces, resulting in LPG values having decimals.
‡ Individual LPG data of the animals were published by Eysker (1997).
§ LPG value here is geometric group mean.
‖ These authors found a strongly significant correlation across experimental groups between worm count and LPG on the day of necropsy, which in one case was used here to calculate fecundity (Kloosterman *et al.* 1989) and in the other case was at first used as an indication that all data could be pooled together irrespective of experimental group.

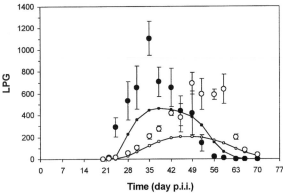

Fig. 6. Comparison between predicted (lines) and observed (circles) LPG. Observations represent the arithmetic means (± S.E.). Dataset from Ploeger *et al.* (1990). Experimental designs (4·5 month old calves were used): Open circles – 4 calves infected with 400 L_3 three times per week for 4 weeks; Closed circles – 4 calves infected with a single primary dose of 5000 L_3. The latter is included here for comparative reasons, although these data were also used during the modelling process.

present model must be viewed as a first attempt to describe the population dynamics of *D. viviparus*, in order to produce hypotheses for experimental validation which subsequently can be used for improving the model both qualitatively and quantitatively.

Comparison with experimental data

Several experimental datasets were used. With one exception, none of these was used in the initial model building. These datasets comprise different artificial

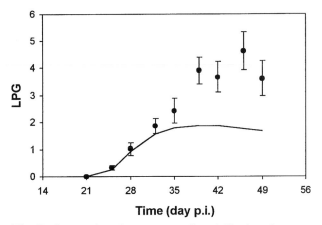

Fig. 7. Comparison between predicted (line) and observed (circles) LPG. Observations represent the arithmetic means (± S.E.). Dataset from Eysker (1997). Experimental design (3–4 month old calves were used): 16 calves (12 in 1991 and 4 in 1994) infected with a single primary dose of 20 L_3. Animals were subsequently turned out on pasture as seeder calves. Larval excretion during the first 7 weeks was assumed to be solely due to the primary infection dose. Similar data from another 4 calves, infected in 1992 (Eysker, 1997), were used in the initial modelling process.

infection regimens, including single primary infection with different larval dose levels, trickle infections and a primary infection followed by a challenge infection. The resulting LPG patterns are presented in Figs 6–9. The most relevant details concerning experimental designs are given in the respective figure headings, together with the origin of the dataset.

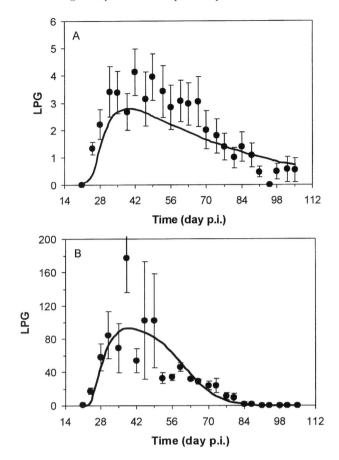

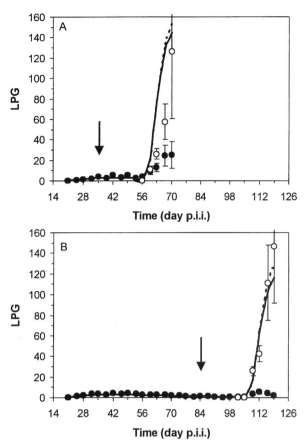

Fig. 8. Comparison between predicted (lines) and observed (circles) LPG. Observations represent the arithmetic means ($\pm$ s.e.). Dataset from Eysker & Ploeger (unpublished). Experimental designs (in all cases 3–4 month old calves were used): (A) 9 calves infected with a single primary dose of 30 L_3; (B) 3 calves infected with a single primary dose of 1000 L_3. The latter were followed until faecal larval shedding no longer occurred. The calves primary infected with 30 L_3 were challenged on day 84 (see Fig. 9B).

Fig. 9. Comparison between predicted (lines) and observed (circles) LPG. Observations represent the arithmetic means ($\pm$ s.e.). Dataset from Eysker & Ploeger (unpublished). Experimental designs (in all cases 3–4 month old calves were used); (A) 6 calves infected with a single primary dose of 30 L_3 and challenged with 2000 L_3 on day 35 (closed circles and solid line) compared with 6 calves receiving only 2000 L_3 at day 35 (open circles and dashed line); (B) 9 calves infected with a single primary dose of 30 L_3 and challenged with 2000 L_3 on day 84 (closed circles and solid line) compared with 9 calves receiving only 2000 L_3 at day 84 (open circles and dashed line). The arrow in both graphs indicates the time of the challenge infection of 2000 L_3.

Generally, the results in Figs 6–9 show that the patterns of infection as exemplified by L_1 excretion in the faeces were mimicked to varying degrees of accuracy by the model predictions. For some of the single primary infection experiments model prediction underestimated LPG values compared to the observed values. Part of the reason for this discrepancy may be found in the estimate for fecundity, which may depend on parasite strains involved. From Table 3 it was apparent that estimates varied substantially between studies. In contrast, model prediction for the single primary infection with 1000 L_3 (Fig. 8B) followed the average observed LPGs quite closely. The large variability within as well as between consecutive sampling days in this experiment may have been the result of small group size and the fact that these animals unexpectedly experienced clinical disease complicated with secondary bacterial infections. Model prediction following a single primary infection with 2000 L_3 over-estimated the initial increase in observed LPGs in one case (Fig. 9A), but not in another group of calves (Fig. 9B). The results of the single primary infection experiments depicted in Figs 6–9 support a strong dose-dependent larval excretion. One of the most interesting results was obtained with the groups primary infected with 30 L_3 and challenged with 2000 L_3 35 or 84 days later (Fig. 9). In the model it was assumed that a low primary infection would not result in strong (protective) immunity. Hence, the similar prediction for both the 'primary infected and challenged' and 'challenged only' groups in Figs 9A and 9B. However, observations clearly show that even a dose level as low as 30 L_3 induces a strong immunity. Moreover, larval excretion after a challenge on day 84 (Fig. 9B) was lower compared to that after a challenge on day 35 (Fig. 9A). This difference cannot be explained by a difference in faecal production between both groups of calves. Hence, the data support that duration of infection also plays a role. It has yet to be determined which aspects of immunity are particularly important here. However, it is reminded that the second part of Eqn (9) on protection implicitly implicates a role for duration of infection. Thus, it appears that the noted discrepancy between model prediction and observation is quantitative rather than qualitative in nature.

The discrepancies between prediction and observation are interesting. They illustrate how the model can be used to test the adequacy of the prevailing assumptions about the development of immunity to *D. viviparus*. Further comparisons between model output and the preliminary data shown in Figs 8 and 9 (data collected specifically for the purpose of validation) and with data from experiments to follow, will be used to re-evaluate the current model architecture and improve simulation.

ACKNOWLEDGEMENTS

Prof. A. W. C. A. Cornelissen is thanked for his support for this work and for critically reading of the manuscript. Dr A. Vermeulen from Intervet, Boxmeer, The Netherlands, is thanked for kindly providing the lungworm larvae for the experiments described in Figs 8 and 9.

REFERENCES

BOON, J. H., KLOOSTERMAN, A. & BREUKINK, M. (1984). Parasitological, serological and clinical effects of continuous graded levels of *Dictyocaulus viviparus* inoculations in calves. *Veterinary Parasitology* **16**, 261–272.

CALLINAN, A. P. L., MORLEY, F. H. W., ARUNDEL, J. H. & WHITE, D. H. (1982). A model of the life cycle of sheep nematodes and the epidemiology of nematodiasis in sheep. *Agricultural Systems* **9**, 199–225.

CORNWELL, R. L. (1962). Production of immunity to *Dictyocaulus viviparus* by the parenteral administration of fourth stage larvae. *Journal of Comparative Pathology* **72**, 181–189.

CORNWELL, R. L. & JONES, R. M. (1970a). Immunity to *Dictyocaulus viviparus* in calves. I. Exposure of larvae to triethylene melamine (TEM) and preliminary vaccination experiments. *Research in Veterinary Science* **11**, 553–559.

CORNWELL, R. L. & JONES, R. M. (1970b). Immunity to *Dictyocaulus viviparus* in calves. II. Vaccination with normal larvae or larvae attenuated by triethylene melamine (TEM) or X-irradiation and exposure to experimental challenge. *Research in Veterinary Science* **11**, 560–568.

CORNWELL, R. L. & JONES, R. M. (1970c). Immunity to *Dictyocaulus viviparus* in calves. III. Vaccination with larvae attenuated by triethylene melamine (TEM) or X-irradiation and protection against field challenge. *Research in Veterinary Science* **11**, 569–574.

DAVID, G. P. (1993). Increased prevalence of husk. *Veterinary Record* **133**, 627.

DAVID, G. P. (1997). Survey on lungworm in adult cattle. *Veterinary Record* **141**, 343–344.

EYSKER, M. (1989). The epidemiology of lungworm infections in cattle. In *The Veterinary Annual* 29 (ed. Grunsell, C. S. G., Hill, F. W. G. & Raw, M. E.), pp. 69–72. London, Wright.

EYSKER, M. (1994). Epidemiologie en bestrijding van longworminfecties bij het rund. *Tijdschrift voor Diergeneeskunde* **119**, 322–325.

EYSKER, M. (1997). The sensitivity of the Baermann method for the diagnosis of primary *Dictyocaulus viviparus* infections in calves. *Veterinary Parasitology* **69**, 89–93.

EYSKER, M., BOERSEMA, J. H., CORNELISSEN, J. B. W. J., KOOYMAN, F. N. J., DE LEEUW, W. A. & SAATKAMP, H. W. (1994). An experimental field study on the build up of lungworm infections in cattle. *The Veterinary Quarterly* **16**, 144–147.

EYSKER, M., BOERSEMA, J. H., CORNELISSEN, J. B. W. J. & KOOYMAN, F. N. J. (1995). Efficacy of Michel's 'dose and move' system against *Dictyocaulus viviparus* infections in cattle using moxidectin as anthelmintic. *Veterinary Parasitology* **58**, 49–60.

EYSKER, M., SAATKAMP, H. W. & KLOOSTERMAN, A. (1993). Infection build-up and development of immunity in calves following primary *Dictyocaulus viviparus* infections of different levels at the beginning or in the middle of the grazing season. *Veterinary Parasitology* **49**, 243–254.

GILLEARD, J. S., DUNCAN, J. L. & TAIT, A. (1995a). *Dictyocaulus viviparus*: Surface antigens of the L3 cuticle and sheath. *Experimental Parasitology* **80**, 441–453.

GILLEARD, J. S., DUNCAN, J. L. & TAIT, A. (1995b). An immunodominant antigen on the *Dictyocaulus viviparus* L3 sheath surface coat and a related molecule in other strongylid nematodes. *Parasitology* **111**, 193–200.

GRENFELL, B. T., SMITH, G. & ANDERSON, R. M. (1987). A mathematical model of the population biology of *Ostertagia ostertagi* in calves and yearlings. *Parasitology* **95**, 389–406.

JARRET, W. F. H., JENNINGS, F. W., McINTYRE, W. I. M., MULLIGAN, W., THOMAS, B. A. C. & URQUHART, G. M. (1959). Immunological studies on *Dictyocaulus viviparus* infection: The immunity resulting from experimental infection. *Immunology* **2**, 252–261.

JARRETT, W. F. H., McINTYRE, W. I. M., JENNINGS, F. W. & MULLIGAN, W. (1957). The natural history of parasite bronchitis with notes on prophylaxis and treatment. *Veterinary Record* **69**, 1329–1336.

KLOOSTERMAN, A. & FRANKENA, K. (1988). Interactions between lungworms and gastrointestinal worms in calves. *Veterinary Parasitology* **26**, 305–320.

KLOOSTERMAN, A., FRANKENA, K. & PLOEGER, H. W. (1989). Increased establishment of lungworms (*Dictyocaulus viviparus*) in calves after previous infections with gastrointestinal nematodes (*Ostertagia ostertagi* and *Cooperia oncophora*). *Veterinary Parasitology* **33**, 155–163.

KLOOSTERMAN, A., PLOEGER, H. W. & FRANKENA, K. (1990). Increased establishment of lungworms after exposure to a combined infection of *Ostertagia ostertagi* and *Cooperia oncophora*. *Veterinary Parasitology* **36**, 117–122.

LEATHWICK, D. M., BARLOW, N. D. & VLASSOFF, A. (1992). A model for nematodiasis in New Zealand lambs. *International Journal for Parasitology* **22**, 789–799.

MICHEL, J. F. (1962). Studies on resistance to *Dictyocaulus* infection. IV. The rate of acquisition of protective immunity in infection of *D. viviparus*. *Journal of Comparative Pathology* **72**, 281–285.

MICHEL, J. F., MacKENZIE, A., BRACEWELL, C. D., CORNWELL, R. L., ELLIOT, J., HEBERT, C. N., HOLMAN, H. H. & SINCLAIR, I. J. B. (1965). Duration of the acquired resistance of calves to infection with *Dictyocaulus viviparus*. *Research in Veterinary Science* **6**, 344–395.

MICHEL, J. F. & PARFITT, J. W. (1956). An experimental study of the epidemiology of parasitic bronchitis in calves. *Veterinary Record* **68**, 706–710.

PLOEGER, H. W., HOOGEVEEN, J. C., KLOOSTERMAN, A. & VAN DEN BRINK, R. (1990). Angiotensin-converting-enzyme activity in serum from calves infected continuously or with a single dose of *Dictyocaulus viviparus* infective larvae. *Veterinary Parasitology* **37**, 237–241.

POYNTER, D., JONES, B. V., NELSON, A. M. R., PEACOCK, R., ROBINSON, J., SILVERMAN, P. H. & TERRY, R. J. (1960). Recent experiences with vaccination. *Veterinary Record* **72**, 1078–1086.

SAATKAMP, H. W., EYSKER, M. & VERHOEFF, J. (1994). Study on the causes of outbreaks of lungworm disease on commercial dairy farms in The Netherlands. *Veterinary Parasitology* **53**, 253–261.

SMITH, G. (1994). Population biology of the parasitic phase of trichostrongylid nematode parasites of cattle and sheep. *International Journal for Parasitology* **24**, 167–178.

SMITH, G. & GRENFELL, B. T. (1994). Modelling of parasite populations: gastrointestinal nematode models. *Veterinary Parasitology* **54**, 127–143.

SMITH, G., GRENFELL, B. T. & ANDERSON, R. M. (1987). The regulation of *Ostertagia ostertagi* populations in calves: density-dependent control of fecundity. *Parasitology* **95**, 373–388.

SMITH, G., JACOBSEN, J. & GUERRERO, J. (1995). PARABAN: a case study of how macroparasitic models can be useful at farm level. *Proceedings of the Annual Meeting of the Dutch Society for Veterinary Epidemiology and Economics*, pp. 9–24. Lelystad: The Netherlands.

TAYLOR, S. M., KENNY, J., EDGAR, H. W., MALLON, T. R., CANAVAN, A. & McCARVILL, E. (1999). Introduction of immunity to *Dictyocaulus viviparus* while under treatment with endectocides. *Abstract Proceedings 17th W. A. A. V. P.*, 15–19 August, Copenhagen: Denmark.

YOUNG, R. R., ANDERSON, N., OVEREND, D., TWEEDIE, R. L., MALAFANT, R. L. & PRESTON, G. A. N. (1980a). The effect of temperature on times to hatching of eggs of the nematode *Ostertagia circumcincta*. *Parasitology* **81**, 477–491.

YOUNG, R. R., NICHOLSON, R. M., TWEEDIE, R. L. & SCHUH, H. J. (1980b). Quantitative modelling and prediction of development times of the free-living stages of *Ostertagia ostertagi* under controlled and field conditions. *Parasitology* **81**, 493–505.

Vaccine development and diagnostics of *Dictyocaulus viviparus*

J. B. McKEAND*

Department of Veterinary Clinical Science and Animal Husbandry, Faculty of Veterinary Science, University of Liverpool, Leahurst, South Wirral CH64 7TE, UK

SUMMARY

Parasitic bronchitis is a serious disease of cattle and is caused by the nematode, *Dictyocaulus viviparus*. For over 30 years, a radiation-attenuated larval vaccine has been used for prevention of this disease. This vaccine has been used with considerable success in the UK and parts of Western Europe, however, it has several disadvantages. It has a short shelf-life and the vaccine has to be produced annually necessitating the use of donor calves. Following vaccination, calves must receive further boosting from natural challenge to maintain protective immunity. Sales of the irradiated larval vaccine have decreased dramatically since the 1970s. This is thought to be due to increased reliance of farmers on anthelmintic programmes to control lungworm infection. It is possible that, under certain circumstances, these programmes do not allow sufficient parasite exposure to stimulate protective immunity to further *Dictyocaulus* challenge. This is borne out by the recent documented increase in the number of outbreaks of parasitic bronchitis in the UK. A stable vaccine against *D. viviparus* that is capable of stimulating a more prolonged immunity would be beneficial. Recent research has been directed at identification and isolation of components thought to be involved in parasite survival in the host and examination of their potential as vaccine candidates. One of these components is acetylcholinesterase (AChE), an enzyme secreted by adult worms. This review describes the development of the secreted AChE as a vaccine candidate, as well as documenting recent developments in the immunodiagnosis of *D. viviparus*.

Key words: *Dictyocaulus viviparus*, vaccine development, immunodiagnosis, acetylcholinesterase.

INTRODUCTION

Dictyocaulus viviparus is a trichostrongylid nematode whose adult stages inhabit the main stem bronchi and tracheae of cattle. This parasite causes a severe, sometimes fatal, bronchopneumonia, the most common clinical manifestations being coughing, respiratory distress and weight loss. Since the 1950s, a vaccine composed of irradiated larvae (Dictol or Huskvac, Intervet) has been available for the control of parasitic bronchitis (husk). This vaccine stimulates a strong protective immunity and, following its commercial introduction, was very successful in reducing the number of outbreaks of disease. Traditionally, husk was a disease seen in calves during the latter half of the first grazing season, however, reports of lungworm in adult cattle have increased dramatically over the last decade (David, 1993, 1996, 1997; Robinson, Jackson & Sarchet, 1993; Williams, 1996; Veterinary Investigation Diagnosis Analysis (VIDA), 1997, Central Veterinary Laboratory, Weybridge). Figures from VIDA show that between 1992 and 1994 there were a total of 538 recorded outbreaks compared with 232 in the preceding 3 years. Since 1994, there has been a further steady increase in the number of reported annual outbreaks, peaking at 543 in 1997 which

represented a 64% increase on the figures quoted for 1996 (see Fig. 1). The numbers that are recorded by VIDA are thought to be a vast underestimate of the actual prevalence as many clinical cases are treated by farmers without veterinary consultation or without involvement of the veterinary investigation services. The increase in recorded diagnoses of parasitic bronchitis is thought to reflect the increasing commercial involvement and awareness of the disease, as well as the availability of an ELISA for detection of specific antibody. Nevertheless, the rise in outbreaks probably reflects the growing tendency of farmers to replace vaccination with strategic treatments using highly effective anthelmintics (Connan, 1993; Mawhinney, 1996). The sales of Huskvac have dropped considerably over the last two decades with approximately 750 000 animals being vaccinated per annum in the mid 1970s, falling to 250 000 cattle per annum in the 1990s (I. Mawhinney, personal communication). Whilst most modern anthelmintics have excellent efficacy against *D. viviparus*, their repeated use may result in reduced exposure to the parasite antigens causing animals to remain susceptible to disease in later life (Urquhart *et al.* 1981; Vercruysse *et al.* 1987). Even if farmers do vaccinate, but subsequently use repeated anthelmintic treatments, animals will have reduced exposure to further natural challenge and immunity will wane leading to susceptibility to disease. Parasitic bronchitis in adult cattle is potentially serious

* Tel: 0151 794 6089, Fax: 0161 794 6034;
E-mail: J. B. Matthews@liv.ac.uk

Parasitology (2000), **120**, S17–S23. Printed in the United Kingdom   © 2000 Cambridge University Press

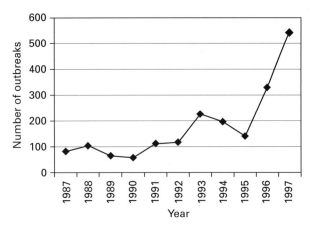

Fig. 1. Recorded outbreaks of *Dictyocaulus viviparus* (1987–1997). Data taken from VIDA analysis III (1998), Central Veterinary Laboratory, Weybridge.

and can have high morbidity with significant economic consequences through costs of treatment and reductions in milk yield, fertility and body weight (Woolley, 1997).

The irradiated larval vaccine has several disadvantages. For example, it is relatively unstable and therefore not suitable for extensive application abroad. Furthermore, the *D. viviparus* larvae must be produced annually, requiring the sacrifice of many donor calves. As the vaccine does not confer a parasitologically-sterile immunity, animals can continue to act as carriers (Hendriks & van Vliet, 1980; Urquhart, 1985). Thus, on farms with a history of parasitic bronchitis, each new batch of calves must be vaccinated. Furthermore, as mentioned above, in the absence of subsequent field challenge (as in the case of animals treated long term with anthelmintics), immunity induced by the vaccine falls dramatically leading to susceptibility to disease (Michel & McKenzie, 1965). A recombinant vaccine against *D. viviparus* that is more stable and would provide longer lasting immunity would be beneficial.

SOURCES OF PROTECTIVE ANTIGENS IN
D. VIVIPARUS

Several studies have been performed to investigate the sources of putative protective antigens in *D. viviparus*. Many of the earlier studies provided indirect evidence that immune responses mounted against antigens expressed by the later developmental stages such as fourth stage larvae (L_4), fifth stage larvae (L_5) or adult worms are involved in protective immunity. For example, when immune cattle were re-infected, it was observed that larvae penetrated the lungs where further development was terminated from day 11 p.i. onwards (Michel, 1969). Furthermore, in studies with irradiated larvae, it appeared that the worms had to invade the lungs for protection to be induced (Cornwell, 1960; Poynter *et*

al. 1960). The precise antigens that stimulate immunity to *D. viviparus* have not been identified but molecules released by adult worms may play an important role. When adult excretory/secretory (ES) products were characterized by immunoprecipitation followed by SDS–PAGE, 18 of the 20 polypeptides present in the ES products were seen to be recognized by infected, immune calves (Britton *et al.* 1993*a*). Subsequently, different antigen fractions from several life cycle stages of *D. viviparus* were compared for their ability to induce protection in the laboratory model host, the guinea pig. In one study where animals were immunized in the context of Freund's complete adjuvant, those groups that received somatic extracts of adult worms or third stage larvae (L_3) were not protected significantly against challenge (McKeand *et al.* 1995*a*). In contrast, animals that received adult ES products were significantly protected against re-infection when compared with challenge controls (Mann–Whitney U-test, $P < 0.05$). To examine the protective role of antibody, naive guinea pigs were passively immunized with sera from animals that had been immunized previously with adult ES products or from guinea pigs exposed to experimental L_3 infection (McKeand *et al.* 1995*a*). The serum recipients in both groups were challenged with *D. viviparus* L_3 1 hour after passive transfer and their worm burdens examined at day 6 p.i. The numbers of lung larvae in both sera recipient groups were significantly lower than in guinea pigs that had received normal sera (Mann–Whitney U-test, $P < 0.05$) suggesting that, in the laboratory host at least, antibody-mediated mechanisms contribute to immune protection in both L_3 infection and adult ES immunization. These results corroborated with earlier passive transfer studies in calves where antibody was shown to play a protective role in immunity to L_3 challenge (Jarrett *et al.* 1955). These results argue that measurement of antibody responses to *D. viviparus* is pertinent to the definition of protective antigens.

RATIONAL APPROACH TO VACCINE DESIGN

This involves identification of molecules that are potentially critical to parasite survival in the host. One strategy for reducing nematode survival *in vivo* is to induce immune responses to enzymatic components that may assist the worm in invasion of host tissue, feeding, replication or evasion of host immunity. Often, these components are found on the parasite surface or in their ES products. Antibody responses to the surface antigens of *D. viviparus* have been characterized in detail (Britton *et al.* 1993*b*; McKeand *et al.* 1996; Scott, McKeand & Devaney, 1996), however, the relationship of these responses to protective immunity is unclear. The antibody responses to the adult surface, for example, may have little relevance to immunity as it has been shown that these stages readily shed surface-bound antibody when maintained at 37 °C (McKeand & Kennedy, 1995).

The ES products of adult *D. viviparus* have been shown to contain enzyme activities of different classes. These include acetylcholinesterases (McKeand *et al.* 1994*a*), proteinases (Britton *et al.* 1992) and superoxide dismutases (Britton, Knox & Kennedy, 1994). Five migratory isoforms of AChE were shown to be present in *D. viviparus* adult ES products by gel electrophoresis followed by specific enzyme staining (McKeand *et al.* 1994*a*). These AChEs were common to adult ES products and adult worm somatic extracts, however, chemical assay showed the AChE activity to be 200 times more abundant (per unit protein) in the ES products. Expression of these AChE isoforms appeared to be regulated in that these enzymes were not detected in somatic extracts of L_3 stages. All five isoforms of the adult AChEs were shown to be immunogenic in that they were recognized by antibody from infected calves, indicating that these enzymes are released *in vivo* (McKeand *et al.* 1994*a*).

Why would a parasitic nematode expend the energy to release AChE? These enzymes have been ascribed a role in immunomodulation via the hydrolysis of host acetylcholine (ACh) which, in addition to its role in neurotransmission, has been found to enhance the functions of several types of host immune effector cell. Lymphocyte activation, mast cell degranulation and neutrophil-mediated antibody-dependent cytotoxicity have all been shown to be enhanced in the presence of ACh (reviewed by Rhoads, 1984). This is of relevance to *D. viviparus* infection because, in infected animals, increased numbers of neutrophils (Jarrett & Sharp, 1963), mast cells (H. R. P. Miller, personal communication) and lymphocytes (Jarrett & Sharp, 1963) are observed in the lungs. Thus, it could be postulated that adult *D. viviparus* release AChE to hydrolyse ACh and ablate its effect on these types of effector cell within the immediate environment of the worm. The theory that AChE secretion by gastrointestinal nematodes leads to a so-called biochemical holdfast by reducing intestinal contractions has not been substantiated. Although many intestinal nematodes release AChE (for example, Pritchard *et al.* 1991; Blackburn & Selkirk, 1992), these enzymes have now been shown to be secreted by non gut-dwelling parasites such as *Brugia malayi* (Rathaur *et al.* 1987) and the bird lungworm, *Syngamus trachea* (Riga *et al.* 1995).

In an attempt to investigate the role of *D. viviparus* AChE in protective immunity, adult ES fractions were enriched for AChE activity by electroelution of protein from the appropriate region of poly-acrylamide gels and the fractions used to immunize guinea pigs (McKeand *et al.* 1995*b*). The AChE-enriched fraction produced significant levels of protection after challenge compared with the adjuvant control group (Mann–Whitney *U*-test, $P < 0·05$) and stimulated high levels of AChE-specific antibody. In a separate study, adult ES products were used to immunize two inbred strains of guinea pigs as opposed to the outbred strain used in the experiments described above (McKeand *et al.* 1994*b*). The results showed that the two inbred strains differed in their susceptibility to challenge infection following ES immunization and also in their antibody responses to *D. viviparus* antigens. Interestingly, when antibody binding to *D. viviparus* AChE was compared between the two strains, it was observed that strain 13 guinea pigs, which were significantly protected against re-infection, recognized more isoforms of AChE than strain 2 guinea pigs, which were not protected. It is not yet known, however, whether antibodies specific to *D. viviparus* AChE can passively confer immunity to challenge.

It has proved impossible to purify sufficient native AChE from *D. viviparus* adult ES products to perform meaningful immunization studies in either guinea pigs or calves. Recently, an AChE-encoding cDNA has been isolated from *D. viviparus*. Initially, an AChE-encoding fragment of 365 bp was generated from adult worm RNA by reverse transcriptase polymerase chain reaction. This was used to screen an adult *D. viviparus* cDNA expression library. Several cDNA clones were isolated which sequencing showed to be different lengths of the same gene. The longest cDNA was 1·7 kbp and this showed between 50 and 60 % identity to other AChE genes (J. B. McKeand, unpublished observations). At the protein level, this sequence was 550 amino acids in length and presented 45 % identity with the *Nippostrongylus brasiliensis* acetylcholinesterase B precursor (over 510 amino acids); 50 % identity with the *C. elegans ace-2* product (over 395 amino acids) and 35 % identity with the human cholinesterase precursor (over 506 amino acids). The *D. viviparus* AChE sequence contains conserved positions of the components of the catalytic triad. Part of this cDNA has been subcloned and expressed in a bacterial expression system to produce a polypeptide of approximately 55 kDa. This recombinant molecule is recognized by sera from guinea pigs immunized with the AChE-enriched ES fraction as well as by sera from immune, infected calves (J. B. McKeand, unpublished observations). Moreover, Western blotting experiments have shown that rabbit serum raised against the recombinant AChE recognized a component of approximately 55 kDa in adult ES products. This recombinant protein was recently assessed in an immunization study in calves and, unfortunately, did not induce significant levels of protective immunity when compared with challenge control animals (J. B. McKeand, unpublished data).

Immunization studies using AChE from other parasitic nematodes have also produced equivocal results. In an earlier study, guinea pigs were immunized with AChE-enriched *Trichostrongylus colubriformis* fractions and were not protected against challenge (Rothwell & Merritt, 1975). In a later study, purified secretory AChE from *T. colubriformis* was used to immunize sheep against infections with the homologous parasite, in addition to *Haemonchus contortus* and *Cooperia oncophora* (Griffiths & Pritchard, 1994). A low degree of cross-species protection was achieved with an average reduction in worm burden of all species of 31 % but there were no consistent reductions in faecal egg counts and modest increases in AChE-specific antibody.

As mentioned above, SOD activity has also been detected in *D. viviparus* adult ES products. An ES-specific copper/zinc-dependent SOD was found to be released in large quantities by adult worms (Britton *et al.* 1994). The antigenicity of this SOD isoform was demonstrated by reduction of enzyme activity following incubation of adult ES products with IgG antibody purified from the sera of infected or vaccinated calves (Britton *et al.* 1994). The high level of SOD released by adult *D. viviparus* may be a reflection of the oxygen-rich pulmonary environment of this parasite and antibody inhibition of this SOD may be an important target of protective immunity. Attempts are now under way to clone and express the gene encoding this SOD for use in further immunological studies.

Proteinase activities have also been identified in *D. viviparus* adult ES products (Britton *et al.* 1992). Serine-, cysteine- and metalloproteinases were identified in several stages with secreted materials being more active against protein substrates per unit protein than the somatic extracts. Again, antigenicity of the parasite proteinases was demonstrated by inhibition of enzyme activity with Protein G-purified serum IgG antibody from infected and vaccinated calves. Antibody response to these proteinases may limit parasite-mediated tissue damage thus limiting pathology, as well as reducing worm survival. No further work has been performed on the *D. viviparus* proteinases.

IMMUNODIAGNOSIS OF *D. VIVIPARUS*

Serodiagnosis of parasitic bronchitis by ELISA has been evaluated in naturally infected, experimentally infected and vaccinated animals. These studies have been carried out to aid diagnosis of infection in the field or to provide information on prevalence. Positive ELISA titres appear to be a satisfactory indicator of recent herd exposure, however, they are rather inaccurate in determining the immune status of individual animals (Bos & Beekman, 1985). Furthermore, seroprevalence rates do not always reflect actual outbreaks of clinical disease. For example, in one study where 75 % of herds tested were found to have positive titres to *D. viviparus* only 15 % had clinical husk, although 51 % of the farms had experienced husk in the past (Boon, Kloosterman & Van Der Lende, 1984). Furthermore, all of the current ELISAs used diagnostically incorporate antigen preparations from adult worms (for example, the ELISA used by the UK veterinary investigation service), so that larval invasion following vaccination or during pre-patent infection is not detected. Thus, it is difficult to assess vaccinated animals that have not been exposed to subsequent pasture challenge. For example, in experimentally infected cattle, the response to adult worm antigens was delayed when responses were studied in vaccinated animals which had a slow uptake of infective larvae (Bos, Beekman-Boneschanscher & Boon, 1986). To overcome this, an L_4 stage ELISA has been used in some experimental studies to examine responses in vaccinated calves (Mawhinney, 1997), however, these are not generally available to veterinarians in practice.

Much of the recent work concerned with immuno-diagnosis of dictyocaulosis has been performed in the Netherlands and Germany. De Leeuw & Cornelissen (1991) identified a specific 17 kDa antigen for use in diagnosis by comparing somatic extracts of adult worms, ES antigens of adult worms and somatic antigens of L_3 in an indirect ELISA. Species-specificity was examined using sera from calves with mono-infections of heterologous helminth species. When adult worm somatic antigens were analysed by Western blotting, a 17 kDa protein was identified that did not react with the heterologous sera. This was isolated by ultrafiltration and anion chromatography and compared with whole somatic antigen in an indirect ELISA. Extinction values measured in both assays correlated well and the protein has been developed for use in diagnostic ELISA. Subsequently, three different ELISAs were compared for sensitivity, specificity and sero-conversion after primary infection (de Leeuw & Cornelissen, 1993). These assays were an indirect ELISA using crude somatic adult antigen; an indirect ELISA containing purified antigens isolated from adult worm somatic antigens and a competition ELISA incorporating purified antigen in combination with *D. viviparus*-specific monoclonal antibodies. The specificity of the competition and the purified antigen ELISA was 97 %, whereas the specificity of the crude antigen ELISA was 67 %. Sensitivities of the purified antigen, the competition and the crude antigen ELISAs were 97, 73 and 99 %, respectively. All three assays detected sero-conversion 4–6 weeks p.i. although none detected seroconversion in recently vaccinated calves. In another study, a *D. viviparus*-specific ELISA and IHA were compared for sensitivity, specificity, time of seroconversion and persistence of antibody responses (Cornelissen, Borgsteede & van Milligen, 1997). Specificity of both tests was very high, however, the sensitivity of the ELISA (100 %) was far superior (IHA 78·1 %) and detected antibodies earlier in infection, although vaccinates were not seropositive. Seroprevalence using this ELISA was determined in a field study where 48·6 % animals were found to be positive and when the ELISA was used in five different laboratories, the repeatability and reproducibility were promising enough to introduce this as the routine test in The Netherlands. The sequence of the purified antigen used in this ELISA has not been characterized but is likely to be a native equivalent to the recombinant major sperm protein (MSP) mentioned below. This ELISA is available commercially as a kit (Ceditest, Institute for Animal Science and Health ID-DLO, The Netherlands) and has now been used widely in The Netherlands, Germany, Belgium, France and Sweden (F. H. M. Borgsteede, personal communication).

In separate studies, an adult worm antigen of diagnostic potential was detected by Western blotting of crude adult worm antigen with sera from infected and vaccinated cattle (Schnieder, 1992). A *D. viviparus*-specific region around 18 kDa was identified, isolated and a lambda ZAP II cDNA library screened with rabbit antiserum to the antigen. Of the clones identified, the one with the highest expression yields was expressed as a glutathione S-transferase protein (DvGST3-14) and also, after cleavage with thrombin, as pure recombinant parasite protein (Dv3-14). An immunoblot dipstick test was developed which provided results within 90 min of blood sampling and was found to detect infections with more than 99 % specificity and sensitivity by 30–85 days p.i. (Schnieder, 1993*a*). Subsequently, this antigen was shown to be encoded by a gene fragment with homology to MSP from *Ascaris suum*, *C. elegans* and *Onchocerca volvulus* (Schnieder, 1993*b*). The recombinant antigen had now been subcloned in a *Drosophila* expression system and has been shown to work well in the dipstick format (T. Schnieder, personal communication). This test is offered commercially but the response so far in Germany has been disappointing (T Schnieder, personal communication).

CONCLUSIONS

Thus, despite the presence of the irradiated larval vaccine for nearly 40 years, there is still much research interest in the immunology of *D. viviparus* infection. Partly, this reflects the re-emergence of husk as an important disease in cattle in wet, temperate climates but also because an improved vaccine is sought that would promote a longer-term immunity in the absence of pasture challenge. In addition, *D. viviparus* provides a rather unique system in which to study immune responses to a parasitic nematode, as unlike the situation with most ruminant helminths, exposure to this parasite results in a relatively strong and rapid immunity. Attempts have been made to define the precise protective responses and antigens involved in dictyocaulosis but these still elude us. In terms of the recombinant AChE, it is likely that a molecule more similar to the native enzyme, than the form described here, may induce better protection and, to this end, the AChE clone will be sub-cloned into a baculovirus system. *D. viviparus* serology too has its drawbacks, with a lack of larval specific diagnostic tests available that can identify the immune status of an individual animal.

ACKNOWLEDGEMENTS

The unpublished AChE work cited in this review was supported by the BBSRC. The author is grateful to Fred Borgsteede, Graham David, Ian Mawhinney, Keith Matthews and Thomas Schnieder for their helpful comments in the preparation of this manuscript and Amanda Davidson for much of the technical work relating to the AChE work.

REFERENCES

BLACKBURN, C. C. & SELKIRK, M. E. (1992). Characterisation of the secretory acetylcholinesterases from adult *Nippostrongylus brasiliensis*. *Molecular and Biochemical Parasitology* **53**, 79–88.

BOON, J. H., KLOOSTERMAN, A. & VAN DER LENDE, T. (1984). The incidence of *Dictyocaulus viviparus* infections in cattle in the Netherlands. II. Survey of sera collected in the field. *Veterinary Quarterly* **6**, 13–17.

BOS, H. J. & BEEKMAN, J. (1985). Serodiagnosis of lungworm infection calves using ELISA. *Developments in Biological Standardization* **62**, 45–52.

BOS, H. J., BEEKMAN-BONESCHANSCHER, J. & BOON, J. H. (1986). Use of ELISA to assess lungworm infection in calves. *Veterinary Record* **119**, 153–156.

BRITTON, C. B., CANTO, G. J., URQUHART, G. M. & KENNEDY, M. W. (1993*a*). Characterisation of excretory secretory products of adult *Dictyocaulus viviparus* and the antibody response to them. *Parasite Immunology* **15**, 163–174.

BRITTON, C. B., CANTO, G. J., URQUHART, G. M. & KENNEDY, M. W. (1993*b*). Stage-specific antigens of the cattle lungworm, *Dictyocaulus viviparus*. *Parasite Immunology* **15**, 625–634.

BRITTON, C. B., KNOX, D. P., CANTO, G. J., URQUHART, G. M. & KENNEDY, M. W. (1992). The secreted and somatic proteinases of the bovine lungworm *Dictyocaulus viviparus* and their inhibition by antibody from infected and vaccinated animals. *Parasitology* **105**, 325–333.

BRITTON, C., KNOX, D. P. & KENNEDY, M. W. (1994). Superoxide dismutase (SOD) activity of *Dictyocaulus viviparus* and its inhibition by antibody from infected and vaccinated bovine hosts. *Parasitology* **109**, 257–263.

CONNAN, R. M. (1993). Calfhood vaccination for Dictyocaulosis. *Veterinary Record* **133**, 554–555.

CORNELISSEN, J. B., BORGSTEEDE, F. H. & VAN MILLIGEN, F. J. (1997). Evaluation of an ELISA for the routine diagnosis of *Dictyocaulus viviparus* infections in cattle. *Veterinary Parasitology* **70**, 153–164.

CORNWELL, R. L. (1960). The complement fixing antibody response of calves to *Dictyocaulus viviparus*. III. Vaccinated calves exposed to challenge. *Journal of Comparative Pathology* **70**, 499–513.

DAVID, G. P. (1993). Increased prevalence of husk. *Veterinary Record* **135**, 627.

DAVID, G. P. (1996). An epidemiological study of husk in adult cows in 32 UK herds (preliminary findings). *Cattle Practice* **5**, 295–297.

DAVID, G. P. (1997). Survey on lungworm in adult cattle. *Veterinary Record* **141**, 343–344.

DE LEEUW, W. A. & CORNELISSEN, J. B. (1993). Comparison of three enzyme immunoassays for diagnosis of *Dictyocaulus viviparus* infection. *Veterinary Parasitology* **49**, 229–241.

DE LEEUW, W. A. & CORNELISSEN, J. B. (1991). Identification and isolation of a specific antigen with diagnostic potential from *Dictyocaulus viviparus*. *Veterinary Parasitology* **39**, 137–147.

GRIFFITHS, G. & PRITCHARD, D. I. (1994). Vaccination against gastrointestinal nematodes of sheep using purified secretory acetylcholinesterase from *Trichostrongylus colubriformis* – an initial pilot study. *Parasite Immunology* **16**, 507–510.

HENDRIKS, J. & VAN VLIET, G. (1980). The value of lungworm vaccination in calves. *Tijdschr Diergeneeskd* **105**, 764–770.

JARRETT, W. F. H., JENNINGS, F. W., McINTYRE, W. I. M., MULLIGAN, W. & URQUHART, G. M. (1955). Immunological studies on *Dictyocaulus viviparus* infection. Passive immunisation. *Veterinary Record* **67**, 291–296.

JARRETT, W. F. H. & SHARP, N. C. C. (1963). Vaccination against parasitic disease: reactions in vaccinated and immune hosts in *Dictyocaulus viviparus* infection. *Journal of Parasitology* **49**, 177–189.

McKEAND, J. B., DUNCAN, J. L., URQUHART, G. M. & KENNEDY, M. W. (1996). Isotype-specific antibody responses to the surface exposed antigens of adult and larval stages of *Dictyocaulus viviparus* in infected and vaccinated calves. *Veterinary Parasitology* **61**, 287–295.

McKEAND, J. B. & KENNEDY, M. W. (1995). Shedding of surface bound antibody by adult *Dictyocaulus viviparus*. *International Journal for Parasitology* **25**, 1255–1258.

McKEAND, J. B., KNOX, D. P., DUNCAN, J. L. & KENNEDY, M. W. (1994a). The immunogenicity of the acetycholinesterases of the cattle lungworm, *Dictyocaulus viviparus*. *International Journal for Parasitology* **24**, 501–510.

McKEAND, J. B., KNOX, D. P., DUNCAN, J. L. & KENNEDY, M. W. (1994b). Genetic control of the antibody repertoire against excretory/secretory products and acetylcholinesterases of *Dictyocaulus viviparus*. *Parasite Immunology* **16**, 251–260.

McKEAND, J. B., KNOX, D. P., DUNCAN, J. L. & KENNEDY, M. W. (1995a). Protective immunisation of guinea pigs against *Dictyocaulus viviparus* using excretory secretory products of adult parasites. *International Journal for Parasitology* **26**, 95–104.

McKEAND, J. B., KNOX, D. P., DUNCAN, J. L. & KENNEDY, M. W. (1995b). Immunisation of guinea pigs against *Dictyocaulus viviparus* using adult ES products enriched for acetylcholinesterases. *International Journal for Parasitology* **25**, 829–837.

MAWHINNEY, I. C. (1996). Control of lungworm. *Veterinary Record* **138**, 263.

MAWHINNEY, I. C. (1997). A survey of lungworm serology and vaccination in first season and second season grazing cattle. *Cattle Practice* **5**, 323–325.

MICHEL, J. (1969). The epidemiology and control of some nematode infections of grazing animals. *Advances in Parasitology* **7**, 211–282.

MICHEL, J. & McKENZIE, A. (1965). Duration of acquired resistance of calves to infection with *Dictyocaulus viviparus*. *Research in Veterinary Science* **6**, 344–395.

POYNTER, D., JONES, B. V., NELSON, A. M. R., PEACOCK, R., SILVERMAN, P. H. & TERRY, R. J. (1960). Recent experiences with vaccination. *Veterinary Record* **72**, 1078–1090.

PRITCHARD, D. I., LEGGETT, K. V., ROGAN, M. T., McKEAN, P. G. & BROWN, A. (1991). *Necator americanus* secretory acetycholinesterase and its purification from excretory-secretory products by affinity chromatography. *Parasite Immunology* **13**, 187–199.

RATHAUR, S., ROBERTSON, B. D., SELKIRK, M. E. & MAIZELS, R. M. (1987). Secretory acetylcholinesterases from *Brugia malayi* adult and microfilarial parasites. *Molecular and Biochemical Parasitology* **26**, 257–265.

RHOADS, M. L. (1984). Secretory cholinesterase of nematodes: possible functions in the host-parasite relationship. *Tropical Veterinarian* **2**, 3–10.

RIGA, E., PERRY, R. N., BARRETT, J. & JOHNSTON, M. R. (1995). Biochemical analyses on single amphidial glands, excretory-secretory gland cells, pharyngeal glands and their secretions from the avian nematode *Syngamus trachea*. *International Journal for Parasitology* **25**, 1151–1158.

ROBINSON, T. C., JACKSON, E. R. & SARCHET, R. W. (1993). Husk in heifers. *Veterinary Record* **132**, 143.

ROTHWELL, T. L. & MERRITT, G. C. (1975). Vaccination against the nematode *Trichostrongylus colubriformis*. II. Attempts to protect guinea-pigs with worm acetylcholinesterase. *International Journal for Parasitology* **5**, 453–460.

SCHNIEDER, T. (1992). *Dictyocaulus viviparus*: isolation and characterization of a recombinant antigen with potential for immunodiagnosis. *International Journal for Parasitology* **22**, 933–938.

SCHNIEDER, T. (1993a). A dipstick immunoassay using a recombinant antigen for the rapid diagnosis of bovine Dictyocaulosis. *Research in Veterinary Science* **54**, 278–282.

SCHNIEDER, T. (1993b). The diagnostic antigen encoded by gene fragment Dv3–14: a major sperm protein of *Dictyocaulus viviparus*. *International Journal for Parasitology* **23**, 383–389.

SCOTT, C. A., McKEAND, J. B. & DEVANEY, E. (1996). A longitudinal study of local and peripheral isotype/subclass antibodies in *Dictyocaulus viviparus*-

infected calves. *Veterinary Immunology and Immunopathology* **53**, 235–247.

URQUHART, G. M. (1985). Field experience with the bovine lungworm vaccine. *Development of Biological Standards* **62**, 109–112.

URQUHART, G. M., JARRETT, W. F., BAIRDEN, K. & BONAZZI, E. F. A. (1981). Control of parasitic bronchitis in calves: vaccination or treatment? *Veterinary Record* **108**, 180–182.

VERCRUYSSE, J., DORNY, P., BERGHEN, P. & FRANKENA, K. (1987). Use of an oxfendazole pulse release bolus in the control of parasitic gastroenteritis and parasitic bronchitis in first season grazing claves. *Veterinary Record* **121**, 297–300.

WILLIAMS, P. C. (1996). Control of lungworm. *Veterinary Record* **138**, 263.

WOOLLEY, H. (1997). The economic impact of husk in dairy cattle. *Cattle Practice* **5**, 315–317.

The immune response and the evaluation of acquired immunity against gastrointestinal nematodes in cattle: a review

E. CLAEREBOUT* *and* J. VERCRUYSSE

Department of Parasitology, Faculty of Veterinary Medicine, University of Gent, Salisburylaan 133, 9820 Merelbeke, Belgium

SUMMARY

The present review discusses the immune responses to gastrointestinal nematodes in cattle and the different immunological and parasitological parameters used to assess acquired immunity. Measuring acquired immunity to gastrointestinal nematodes in cattle (e.g. for the evaluation of candidate parasite vaccines) is hampered by the limited understanding of bovine immune responses against gastrointestinal parasites. In this paper the available data on protective immunity against gastrointestinal nematodes, and especially *Ostertagia ostertagi*, in cattle are compared with the current knowledge of protective immune responses against gastrointestinal nematodes in rodent models and small ruminants. In contrast to the immune response in mice, which is controlled by T helper 2 (Th$_2$) lymphocytes and results in mast cell- or goblet cell-mediated expulsion of adult worms, bovine immune responses to *O. ostertagi* do not show a clear Th$_2$ cytokine profile, nor do they result in rapid expulsion of the parasite. The first manifestation of immunity to *O. ostertagi* in calves is a reduction of worm fecundity, possibly regulated by the local IgA response. Worm numbers are only reduced after a prolonged period of host–parasite contact, and there are indications that *O. ostertagi* actively suppresses the host's immune response. Until the mechanisms of protective immunity against *O. ostertagi* are revealed, the use of immunological parameters to estimate acquired immunity in cattle is based on their correlation with parasitological parameters and on extrapolation from rodent and small ruminant models. Assessing the resistance of calves against a challenge infection by means of parasitological parameters is probably still the most accurate way to measure acquired immunity against gastrointestinal nematodes.

Key words: Gastrointestinal nematodes, *Ostertagia ostertagi*, cattle, immunity, rodent models.

INTRODUCTION

Parasitic gastro-enteritis (PGE) is a major source of production losses in animal husbandry. In temperate climate regions the most important cause of PGE in cattle is infection with the abomasal nematode *Ostertagia ostertagi*. Concurrent infections with other gastrointestinal (GI) nematodes are common, especially with the intestinal helminth *Cooperia oncophora*. At present, control of GI nematode infections relies heavily on the use of anthelmintics (Vercruysse & Dorny, 1999). However, increasing consumer concerns regarding drug residues and the threat of emerging anthelmintic resistance in cattle will force veterinarians to adopt alternative control strategies in the future (Vercruysse & Dorny, 1999).

Vaccination may become a feasible control alternative (Emery, McClure & Wagland, 1993). The aim of a parasite vaccine is to induce long-lasting protective immunity. Consequently, both the efficacy

and the duration of protection provided by a candidate vaccine should be assessed, as opposed to its efficacy alone (Anderson, 1998). A drawback for the development and evaluation of candidate parasite vaccines in cattle is the limited knowledge on protective immune responses against GI nematodes. As a result, specific immunologic parameters for acquired immunity against *Ostertagia* and *Cooperia* are lacking.

The first part of this paper presents an overview of the immune responses in ruminants against GI nematodes, in particular the immune responses in cattle against *O. ostertagi* and *C. oncophora*. In the second part, different immunological and parasitological parameters to estimate acquired immunity against GI nematodes in cattle are evaluated. Because much of the present understanding of protective immunity against GI nematodes comes from laboratory animal models, this review will refer to the immune response against GI nematodes in rodents as a model for GI nematode infections in ruminants. Nevertheless, care must be taken in extrapolation from one host–parasite system to another. Differences between host species, nematode species, and in the localization of particular parasites within the host gut have important consequences for the interaction between the parasite and the host immune

* Corresponding author: Department of Parasitology, Faculty of Veterinary Medicine, University of Gent, Salisburylaan 133, 9820 Merelbeke, Belgium, Tel: +32 9 264 73 93. Fax: +32 9 264 74 96. E-mail: Edwin.Claerebout@rug.ac.be

Parasitology (2000), **120**, S25–S42.   Printed in the United Kingdom   © 2000 Cambridge University Press

system. Compared to many other GI nematodes, protective immune responses against *O. ostertagi* are weak and require a prolonged exposure period before they are discernible (Gasbarre, 1997).

THE IMMUNE RESPONSE AGAINST GASTROINTESTINAL NEMATODES

Antigen presentation and recruitment of different T cell subpopulations

Different cell types are thought to be involved in the recognition, processing and presentation of antigens in the gut: epithelial cells, dendritic cells, macrophages and B cells (Owen, 1994). Although there is no proof of the capacity of any of these cell types to present nematode antigens in the gut *in vivo* (Miller, 1996a), there is evidence that soon after infection of calves with *O. ostertagi*, worm antigens are presented to the host in the abomasal lymph nodes (Gasbarre, 1997). However, information on the nature of *O. ostertagi* antigens is limited (Canals & Gasbarre, 1990; Keith *et al.* 1990; Mansour *et al.* 1990; Hilderson *et al.* 1993b), and it is not known which *Ostertagia* antigens elicit a protective immune response (Hilderson *et al.* 1995a).

After presentation of the parasite antigens to T lymphocytes, the T cells further regulate the host response against the GI nematodes. The importance of T lymphocytes for protection against GI nematodes has been demonstrated in laboratory animals infected with *Trichinella spiralis* (Grencis, Riedlinger & Wakelin, 1985), *Nippostrongylus brasiliensis* (Katona, Urban & Finkelman, 1988), *Heligmosomoides polygyrus* (Urban, Katona & Finkelman, 1991a), *Trichuris muris* (Koyama, Tamauchi & Ito, 1995) and *Strongyloides stercoralis* (Rotman *et al.* 1997). In each of these model systems, the T cell subset to which this protection has been ascribed is the CD4-bearing T helper cell (Grencis *et al.* 1985; Katona *et al.* 1988; Urban *et al.* 1991a; Koyama *et al.* 1995; Rotman *et al.* 1997). In mice, CD4+ T cells can be segregated into T helper 1 (Th$_1$) and T helper 2 (Th$_2$) cell subsets, based on the cytokines they secrete (Mosmann *et al.* 1986). Th$_1$ cells secrete interferon (IFN)-γ, interleukin (IL)-2 and lymphotoxin (LT)-α, whereas Th$_2$ cells secrete, among others, IL-4, IL-5, IL-9, IL-10 and IL-13. Products of one subset negatively regulate the other subset (Mosmann & Coffman, 1989). Whether the Th$_1$ or the Th$_2$ subset gains dominance in the immune response depends not only on a number of host factors, such as the antigen presenting cell type, co-stimulatory molecules and cytokine environment, but also on the nature and the dose of the parasite antigens (Grencis, 1996; Constant & Bottomly, 1997). The phenomena that are typically seen during nematode infection, such as eosinophilia, mucosal mastocytosis amd IgE responses, are controlled

mainly by Th$_2$ cytokines (Finkelman *et al.* 1991). A Th$_2$ type response in the gut has been associated with expulsion of *T. muris* (Else & Grencis, 1991; Else, Hültner & Grencis, 1992), *N. brasiliensis* (Urban *et al.* 1992), *T. spiralis* (Grencis, Hültner & Else, 1991) and *H. polygyrus* (Urban *et al.* 1992; Svetic *et al.* 1993). Finally, *in vivo* studies have demonstrated the importance of specific cytokines in controlling expulsion of *T. muris* (Else *et al.* 1994; Bancroft *et al.* 1997; Bancroft, McKenzie & Grencis, 1998), *N. brasiliensis* (Finkelman *et al.* 1994; McKenzie *et al.* 1998), *H. polygyrus* (Urban *et al.* 1991b, 1995), *T. spiralis* Finkelman *et al.* 1997) and *S. stercoralis* (Rotman *et al.* 1997). Overall, IL-4 and IL-13 promote protective immunity against gastrointestinal nematodes, while IL-12 and IFN-γ promote the survival of the parasites (reviewed by Finkelman *et al.* 1997 and Else & Finkelman, 1998).

In sheep, lymphocytes have also been shown to be important in immunity against GI nematodes. Transfer of gastric lymph lymphocytes from lambs, immunized against *Haemonchus contortus* (Smith *et al.* 1984a) or *Teladorsagia circumcincta* (Smith *et al.* 1986) to their genetically identical uninfected twins, transferred protection against a homologous challenge infection (Smith *et al.* 1984a, 1986). Subsequently, Gill, Watson & Brandon (1993b) demonstrated the crucial importance of T helper cells in the protective immune response against *H. contortus*. *In vivo* depletion of CD4+ T cells abrogated immunity to *Haemonchus* in genetically resistant sheep. More recently, *in vivo* depletion of CD4+ T cells partially abrogated immunity to *H. contortus* induced by gut membrane immunisation (Karanu *et al.* 1997). There is, as yet, no proof of the existence of two distinct subsets of T helper cells in ruminants. However, despite the lack of data on Th$_1$/Th$_2$ bias in ruminants, there are many features of nematode infection in ruminants which would be considered a Th$_2$ type of response if they had occurred in the mouse. Eosinophilia and mucosal mast cell hyperplasia are often observed during GI nematode infection in ruminants (Miller, 1996a). Unlike in rodents, the role of specific cytokines in regulating immune responses against GI nematode infections has not yet been determined in ruminants. However, McClure *et al.* (1995) demonstrated that prolonged administration to sheep of a monoclonal antibody to IFN-γ resulted in significantly increased protection against *Trichostrongylus colubriformis*. These latter results suggest that IFN-γ negatively influences a protective response against *T. colubriformis*.

The importance of a specific set of T cells, or specific cytokines, for the protective immune response of cattle against *O. ostertagi* has not yet been demonstrated. During primary infection with *O. ostertagi* a strong increase in number of both parasite-specific and non-specific lymphocytes has been observed in the abomasal lymph nodes (Gasbarre,

1986), and in the abomasal mucosa (Almeria *et al.* 1997*a*). Percentages of T cells bearing the α-β T cell receptor were decreased, while the percentages of B lymphocytes and γ-δ bearing T cells were elevated (Gasbarre, 1994; Almeria *et al.* 1997*a*). In naturally infected cattle, Baker, Scott & Gershwin (1993) also described increased percentages of B lymphocytes in the abomasal lymph. Cytokine profiles during a primary *O. ostertagi* infection showed a less restricted Th_1/Th_2 profile than that observed in murine models. In both the abomasal lymph nodes (Canals *et al.* 1997) and in lymphocytes isolated from the mucosa (Almeria *et al.* 1997*b*). IL-4, IL-10 and IFN-γ mRNA levels were increased after a primary infection. However, the observed changes in T cell ratios and cytokine expression during primary infection were not related to protective immunity against *Ostertagia*. In immunized animals exhibiting protection against reinfection with *O. ostertagi*, a shift to higher percentages of B lymphocytes characteristic of a primary infection was not observed (Almeria *et al.* 1998). Protected animals showed a decreased IL-4 response upon challenge when compared to non-immune animals. Moreover, IL-4 mRNA levels were significantly correlated with the number of worms, suggesting that IL-4 levels may not be associated with the protective mechanisms (Almeria *et al.* 1998). This interpretation is in conflict with the hypothesis developed using murine models, which assumes that protection against GI nematodes requires Th_2 activation with increased levels of IL-4. However, the mechanism of rapid expulsion of adult worms which occurs in most murine models involving GI nematodes has not been shown in *O. ostertagi*, and neither bovine T cell clones nor *in vivo* responses seem to have a clear Th_1/Th_2 cytokine profile (Brown *et al.* 1994; Almeria *et al.* 1997*b*, 1998; Canals *et al.* 1997).

EFFECTOR RESPONSES

Immunoglobulins

A direct effector role of antibodies in the immune response against GI nematodes has not been demonstrated. However, in mice there are numerous reports of passive transfer of immunity against parasitic nematodes by serum or immunoglobulin fractions (Miller, 1984; Finkelman *et al.* 1997). Passive transfer of immunity by immunoglobulins has not been successful in ruminants (Adams, Merritt & Cripps, 1980; Kloosterman, Benedictus & Aghina, 1980).

IgE. Intestinal nematode infections are often associated with the development of an IgE-mediated immediate (Type-1) hypersensitivity response in the gut. These allergic reactions, including IgE-mediated activation of mucosal mast cells, are generally considered to have a direct protective function (Jarrett & Miller, 1982; Miller, 1996*b*). Depletion of IgE antibodies in rats reduced the resistance of the animals to infection with *T. spiralis* (Dessein *et al.* 1981). Transfer of IgE from *T. spiralis* immune rats resulted in expulsion of *T. spiralis* L_1 larvae from the recipients (Ahmad, Wang & Bell, 1991). In contrast, anti-IgE treatment had no effect on protective immunity against *S. stercoralis* L_3 in mice (Brigandi *et al.* 1996).

In sheep, an elevation of circulating total IgE and/or parasite-specific IgE antibodies has been demonstrated during infections with *H. contortus* (Kooyman *et al.* 1997), *T. colubriformis* (Shaw, Gatehouse & McNeill, 1988) and *T. circumcincta* (Huntley *et al.* 1998*a, b*). In sheep repeatedly infected with *H. contortus*, a negative relationship between worm burdens and total serum IgE levels was described, but not between worm burdens and *H. contortus*-specific serum IgE levels (Kooyman *et al.* 1997). The total IgE response was thought to be suggestive of the involvement of IgE in the anti-*Haemonchus* immune mechanisms (Kooyman *et al.* 1997). This conclusion is in contradiction with the hypothesis that non-specific IgE may actually favour the survival of the parasites, by competing with parasite-specific IgE for IgE receptors on mast cells, basophils, eosinophils and/or T cells (Jarrett & Miller, 1982; Pritchard, 1993). Shaw *et al.* (1998) and Huntley *et al.* (1998*b*) observed an alteration in the proportion of total to parasite-specific IgE during the development of immunity against *T. colubriformis* and *T. circumcincta*, respectively. During a primary response, IgE antibodies to parasite antigens were low, despite elevated total IgE levels (Huntley *et al.* 1998*b*; Shaw *et al.* 1998). In contrast, there was a pronounced parasite-specific IgE response after challenge of previously infected animals (Huntley *et al.* 1998*b*; Shaw *et al.* 1998). The generation of relatively high levels of specific IgE may be important in the activation of a mast cell-mediated effector response (Huntley *et al.* 1998*b*).

Only a limited number of studies have investigated the IgE responses in cattle during *O. ostertagi* infections (Thatcher, Gershwin & Baker, 1989; Baker & Gershwin, 1992, 1993). The results of these studies were conflicting. In one study (Baker & Gershwin, 1992) total and *Ostertagia*-specific IgE levels followed the seasonal pattern of parasitism as estimated by faecal egg counts. In contrast, Thatcher *et al.* (1989) and Baker & Gershwin (1993) reported that serum and lymph IgE responses were negatively correlated with numbers of *O. ostertagi* in naturally infected calves.

IgA. In rodents, a correlation between IgA levels and immunity against GI nematodes has been used to suggest a role for this isotype in the protective response (Wedrychowicz, Maclean & Holmes, 1984;

Almond & Parkhouse, 1986). In one study, passive transfer of immunity to *T. muris* in mice by monoclonal IgA antibodies has been demonstrated (Roach *et al.* 1991).

Challenge infections with *T. colubriformis* in immune sheep resulted in an increased number of IgA-containing cells in the *lamina propria* of the small intestine (Adams, Merritt & Cripps, 1980). Charley-Poulain, Luffau & Pery (1984) observed a close temporal relationship between the rise in local anti-*H. contortus* IgA antibodies and the self-cure reaction. Adult sheep that were immunized against *H. contortus* (Duncan, Smith & Dargie, 1978; Smith *et al.* 1984*a*) or *T. circumcincta* (Smith *et al.* 1983, 1984*b*, 1985, 1986) showed a strong increase in IgA-containing cells and/or IgA in the gastric lymph (Smith *et al.* 1983, 1984*a*, *b*, 1985, 1986) or abomasal mucosa (Duncan *et al.* 1978; Smith *et al.* 1984*b*) after a homologous challenge infection. In lambs, the same immunization procedures produced a lower level of protection, together with a lower IgA response (Duncan *et al.* 1978; Smith *et al.* 1985). Transfer of immunity to *H. contortus* (Smith *et al.* 1984*a*) and *T. circumcincta* (Smith *et al.* 1986) by lymphocytes also resulted in a transfer of a local IgA response (Smith *et al.* 1984*a*, 1986). Sheep that were genetically resistant to *H. contortus* had higher anti-*Haemonchus* IgA levels in their serum and faecal extracts (Gill *et al.* 1993*a*), and more IgA antibody-containing cells in the abomasal mucosa (Gill *et al.* 1994) than random-bred sheep. Furthermore, serum and faecal IgA responses were negatively related with faecal egg counts in both genotypes (Gill *et al.* 1993*a*).

The mode of action of IgA in the protective immune response has not been defined. Roach *et al.* (1991) showed that anti-*T. muris* IgA had the direct capacity to expel the parasite. They suggested that the protection resulted from antibody binding of excretion/secretion (ES) antigens with a possible role in tissue penetration or feeding. Smith *et al.* (1985) and Stear *et al.* (1995*b*) reported a negative correlation between local IgA levels and worm length in *T. circumcincta*-infected sheep. There was no negative association between IgA levels and worm numbers. Results from an earlier experiment (Smith *et al.* 1984*b*) has also indicated that worm loss in immune sheep occurred before the IgA response developed. IgA-mediated suppression of worm development (Smith *et al.* 1985; Stear *et al.* 1995*b*) and worm fecundity (Stear *et al.* 1995*b*) were suggested as possible mechanisms.

Ostertagia-specific IgA antibodies increase in the serum of calves that are artificially infected (Canals & Gasbarre, 1990) or naturally infected (Gasbarre *et al.* 1993*a*; Gasbarre, Leighton & Davies, 1993*b*) with *O. ostertagi*. Serum IgA levels are low, compared to IgG (Hilderson *et al.* 1993*b*) and may be a spill-over of IgA secreted into the gastrointestinal lumen.

Increased numbers of IgA antibody-containing cells in the gut mucosa have been reported during primary and secondary *O. ostertagi* and *C. oncophora* infection (Frankena, 1987). The role of IgA responses in immunity to *O. ostertagi* in calves has yet to be resolved. Christensen (1991) and Gasbarre *et al.* (1993*a*) suggested that anti-*Ostertagia* IgA levels in serum may be indicative of the presence of developing larvae and/or adult worms. Alternatively, serum IgA levels could reflect the degree of sensitisation of the animal: a high level of specific serum IgA might indicate a high level of resistance to re-infection (Christensen, 1991). Claerebout *et al.* (1999*b*) observed a significant correlation between the mucosal IgA response and resistance to a challenge infection in calves that were previously infected with *O. ostertagi*. Parasite-specific IgA antibodies in the abomasal mucus were significantly and negatively related with faecal egg counts and the number of eggs per female worm, suggesting that local IgA responses in the abomasum are associated with a reduction in worm fecundity (Claerebout *et al.* 1999*b*).

IgG. In mice, IgG$_1$ has been shown to be the principal protective factor against *H. polygyrus* in immune serum (Pritchard *et al.* 1983; Williams & Behnke, 1983). Challenge infections with *T. colubriformis* in immune sheep resulted in an increased number of IgG$_1$-containing plasma cells in the *lamina propria* of the small intestine, while the numbers of IgG$_2$-containing cells were not affected (Adams *et al.* 1980). High serum IgG levels have been associated with resistance to *H. contortus* in immunized adult sheep, while low serum IgG levels in immunized lambs were related to unresponsiveness to *Haemonchus* (Duncan *et al.* 1978). In sheep that were genetically resistant to *H. contortus*, serum and faecal IgG$_1$ levels were significantly higher than in random-bred sheep (Gill *et al.* 1993*a*). Resistant sheep also had significantly higher numbers of IgG$_1$ antibody-containing cells in the abomasal mucosa (Gill *et al.* 1994). IgG$_2$ and IgM levels and the numbers of IgG$_2$- and IgM-containing cells were not significantly different between genetically resistant and random-bred sheep (Gill *et al.* 1993*a*, 1994). In contrast, Yong *et al.* (1991) reported that genetic resistance to *T. circumcincta* was associated with high levels of anti-parasite IgA and IgG$_2$ antibodies in the abomasal mucosa and that there was no difference in IgG$_1$ levels. Gill *et al.* (1993*a*) suggested that resistance against these two parasites may occur via different immune effector mechanisms. An interaction between nematode surface antigens, parasite-specific IgG and mucin glycoproteins could be responsible for physical entrapment of *T. spiralis* and *N. brasiliensis* in superficial mucus in rats (Miller, 1987). Immune serum from goats infected with *T. colubriformis* has been shown to suppress helminth feeding *in vitro*, while serum from uninfected goats does not (Bottjer, Klesius & Bone, 1985). Feeding inhibition of immune serum was associated with IgG$_1$ isotype (Bottjer *et al.* 1985). Incubation of *T. colubriformis* in immune serum IgG$_1$ also depressed egg production by the worm (Bone & Klesius, 1985).

Frankena (1987) observed an increase in the number of IgG$_2$ antibody-containing cells in the abomasal mucosa of 2 calves during primary and secondary *O. ostertagi* infections, while IgG$_1$-containing cells were absent. Similar observations were made in the small intestinal mucosa during *C. oncophora* infection (Frankena, 1987). These results are in agreement with the findings of Yong *et al.* (1991) in *T. circumcincta* infected sheep. Serum IgG$_2$ levels have been correlated with protection of calves against *Oesophagostomum radiatum* (Gasbarre & Canals, 1989). However, IgG$_1$ is the predominant immunoglobulin in serum from calves during artificial (Canals & Gasbarre, 1990; Hilderson *et al.* 1993 *a, b*) or natural (Christensen, 1991; Gasbarre *et al.* 1993 *a*) *O. ostertagi* infections. Serum IgG$_2$ responses against *O. ostertagi* require more time and are lower, compared to IgG$_1$ (Christensen, 1991; Gasbarre *et al.* 1993 *a*). It has been suggested that *Ostertagia*-specific IgG$_1$ antibodies may be an indication of the presence of infection, whereas IgG$_2$ responses may be correlated with a protective immune response (Christensen, 1991). Total IgG levels in serum have been related to acquired immunity to *O. ostertagi* and *C. oncophora* (e.g. Kloosterman, Albers & Van Den Brink, 1978, 1984; Frankena, 1987). Animals with higher IgG titres had fewer and shorter worms with less ova per female, and more female worms with reduced vulval flaps (Kloosterman *et al.* 1984).

Eosinophils

Blood eosinophilia and increased numbers of eosinophils in the parasitised gastrointestinal mucosa are typically seen during helminth infections (Rothwell, 1989). The proliferation and differentiation of eosinophils is promoted by the Th$_2$ cell cytokine IL-5 (Korenaga & Tada, 1994). The ability of eosinophils to kill a variety of parasites *in vitro* (Butterworth, 1984), including larvae from the GI nematode *T. spiralis* (Gransmuller *et al.* 1987), has led to the suggestion that eosinophils are antiparasite effector cells. However, evidence for an *in vivo* effector function of eosinophils against GI nematodes is less convincing. Results from a number of studies in rodents have suggested that, although eosinophils may be involved in protection, they have no direct anti-worm effect (Rothwell, 1989). More recently, administration of either anti-IL-5 or anti-IL-5 receptor monoclonal antibodies has been shown to block the development of eosinophilia in mice, but

not the development of resistance to *N. brasiliensis* (Coffman *et al.* 1989), *H. polygyrus* (Urban *et al.* 1991 *b*), *Strongyloides venezuelensis* (Korenaga *et al.* 1991), *T. spiralis* (Herndon & Kayes, 1992) or *T. muris* (Betts & Else, 1999). In contrast, Rotman *et al.* (1997) reported that treatment with anti-IL-5 monoclonal antibody of mice immunized against *S. stercoralis* reduced their eosinophil response and resistance to a challenge infection.

In sheep, eosinophilia has been associated with resistance to *T. colubriformis* (Dawkins, Windon & Eagleson, 1989; Buddle *et al.* 1992; Rothwell *et al.* 1993), *T. circumcincta* (Stear *et al.* 1995 *a*) and *Nematodirus battus* infections (Winter, Wright & Lee, 1997). Patterson *et al.* (1996) found significantly higher numbers of eosinophils in blood, abomasal and jejunal mucosae of goats that were resistant to *T. circumcincta* and *T. colubriformis*, compared to susceptible goats. In contrast, Gill (1991) and Pernthaner *et al.* (1995) observed no relationship between eosinophilia and resistance to *H. contortus* and *T. colubriformis*, respectively. Nevertheless, it has been demonstrated that eosinophils can immobilize the L$_3$ stage of *T. colubriformis* and *H. contortus* in *in vitro* cultures in the presence of specific antiparasite antibodies (Jonas, Stankiewicz & Rabel, 1995; Rainbird, MacMillan & Meeusen, 1998). Eosinophils obtained from mammary washes of sheep primed by repeated infusion of *H. contortus* larvae were more effective than eosinophils obtained after a single infusion of parasite extract in *Fasciola hepatica* infected ewes, suggesting the former were activated *in vivo* (Rainbird *et al.* 1998).

In cattle, *O. ostertagi* infection has been associated with accumulation of eosinophils in the abomasal mucosa (e.g. Ritchie *et al.* 1966; Snider *et al.* 1988; Wiggin & Gibbs, 1989, 1990). However, an effector role of eosinophils against *Ostertagia* has not been established. Washburn (1984) showed that eosinophils bind to L$_3$ larvae of *O. ostertagi*, but it is not known whether the larvae are damaged. It has been demonstrated that *O. ostertagi* L$_3$ larvae are capable of attracting eosinophils *in vitro* (Washburn & Klesius, 1984) and *in vivo* (Klesius, Haynes & Cross, 1985). These data suggest that *O. ostertagi* itself is directly responsible for the eosinophil accumulation in the abomasum of infected cattle.

Mucosal mast cells and globule leucocytes

In rodents, expulsion of GI nematodes from the host has been associated with infiltration of mast cells in the gut mucosa (Woodbury *et al.* 1984). Originally, the hypothesis of mucosal mast cell involvement in worm expulsion was based on the temporal correlation between the kinetics of mucosal mastocytosis and that of worm loss (e.g. Rothwell & Dineen, 1972; Nawa & Miller, 1979). The demonstration that mucosal mast cell proteinases are secreted in serum and local intestinal secretions during expulsion of GI nematodes in rats (Woodbury *et al.* 1984) and mice (Huntley *et al.* 1990) indicated that mucosal mast cells were functionally active during worm rejection. In abomasal tissue from sheep that were immunized against *H. contortus*, the numbers of mucosal mast cells and the concentrations of sheep mast cell proteinase (SMCP) were raised when compared to those in non-infected abomasa (Huntley *et al.* 1987). After larval challenge, a significant release of SMCP occurred in sera and gastric lymph of sheep that were immune against *H. contortus* and *T. circumcincta*, respectively (Huntley *et al.* 1987), and in gut contents of sheep immunized against *T. colubriformis* (Jones *et al.* 1994). In addition, mucosal mast cells isolated from immunized sheep released SMCP when they were incubated with parasitic antigen *in vitro* (Jones, Huntley & Emery, 1992; Bendixsen, Emery & Jones, 1995). Recent studies have presented further evidence of a functional role of mucosal mast cells in the protective immune response against GI nematodes in mice. In rodents, mastocytosis is controlled by a variety of Th_2 type cytokines, including IL-3, IL-9 and stem cell factor (SCF) (Miller, 1996 *a*, *b*; Else & Finkelman, 1998). Administration of IL-3 resulted in mucosal mast cell-mediated expulsion of *Strongyloides ratti* (Nawa *et al.* 1994) and *T. spiralis* (Korenaga, Abe & Hashiguchi, 1996) in mice. IL-9 transgenic mice, which constitutively overexpress IL-9, exhibited an enhanced Th_2 response to *T. spiralis* infection (including mastocytosis) and expelled their worms extremely rapidly compared with normal wild-type mice (Faulkner *et al.* 1997). Treatment of *T. spiralis*-infected mice with a monoclonal antibody against the mast cell receptor of SCF (c kit) depleted the intestinal mucosal mast cell population and abrogated the protective response (Grencis *et al.* 1993; Faulkner *et al.* 1997). In contrast, there is evidence that mucosal mast cells are not essential for the expulsion of *N. brasiliensis* and *T. muris* from mice. Mast cell deficient W/W mice were still able to expel a *N. brasiliensis* infection (Über, Roth & Levy, 1980; Crowle & Reed, 1981). Treatment of mice with an anti-IL-3 monoclonal antibody (Betts & Else, 1999) or a combination of anti-IL-3 and anti-IL-4 monoclonal antibodies (Madden *et al.* 1991) depressed the mucosal mast cell response, but did not alter the expulsion of *T. muris* (Betts & Else, 1999) or *N. brasiliensis* (Madden *et al.* 1991).

Different effector mechanisms have been suggested for mucosal mast cells in the expulsion of GI nematodes (Miller, 1996 *b*). Mediators released from mast cell granules, such as histamine, 5-hydroxytryptamine, proteinases, prostaglandins and leukotrienes, may have a direct anti-parasitic effect (e.g. Douch *et al.* 1983), and/or may increase mucosal permeability (King & Miller, 1984;

Scudamore *et al.* 1995), facilitating transport of serum antibodies and complement into the gut lumen. Alternatively (or additionally), a mast cell-mediated epithelial chloride ion secretory response may help to flush the worms from the mucosal surface (Baird & O'Malley, 1993). Mast cells are also a source of cytokines (Galli, Gordon & Wershil, 1991), and may influence the T helper cell response or up-regulate IgA production (Ramsay *et al.* 1994).

In rats (Murray, Miller & Jarrett, 1968) and in sheep (Huntley, Newlands & Miller, 1984) mucosal mast cells that are stimulated with antigen migrate to the epithelial surface, and release their contents, resulting in globule leucocytes (Huntley, 1992). Accumulation of globule leucocytes in the gut mucosa of sheep has been associated with rejection of the intestinal dwelling nematodes *T. colubriformis* (e.g. Dineen, Gregg & Lascelles, 1978; Dineen & Windon, 1980; Douch *et al.* 1986) and *N. battus* (Winter *et al.* 1997) and the abomasal-dwelling nematodes *H. contortus* (Gamble & Zajac, 1992) and *T. circumcincta* (Smith *et al.* 1984 *b*; Stear *et al.* 1995 *b*). The absence of a negative relationship between the number of mucosal mast cells and worm numbers (Gamble & Zajac, 1992; Stear *et al.* 1995 *b*) suggests that the critical step in reducing worm numbers is the degranulation of mast cells to produce globule leucocytes (Stear *et al.* 1995 *b*).

Accumulation of globule leucocytes has been described in cattle infected with *O. ostertagi* (e.g. Ritchie *et al.* 1966; Snider *et al.* 1981, 1988; Wiggin & Gibbs, 1987, 1989, 1990; Hilderson *et al.* 1995 *b*, Claerebout *et al.* 1996, 1998 *a*, *b*) or *C. oncophora* (Armour *et al.* 1987). While a significant association between the presence of globule leucocytes in the gut mucosa and a protective immune response against *O. ostertagi* has been observed (Claerebout *et al.* 1998 *a*), in two other studies no such correlation could be demonstrated (Claerebout *et al.* 1996, 1998 *b*). This inconsistency may be partly due to the fact that histological examination of the abomasal mucosa only gives a snapshot at a given moment of the dynamic process of mast cell stimulation, maturation, migration and degranulation (Rothwell, 1989). However, differences in the functional activity of the mucosal mast cells/globule leucocytes between ruminant species can not be excluded. Goats had substantially larger numbers of globule leucocytes than sheep after they were challenged with *T. circumcincta* and *Trichostrongylus vitrinus*, despite a less effective immune expulsion of the parasites in the goats (Huntley *et al.* 1995). Patterson *et al.* (1996) observed no significant difference in the numbers of globule leucocytes or mucosal mast cells between goats that were resistant or susceptible to *T. circumcincta* and *T. vitrinus*. Furthermore, effector mechanisms may be different for different GI nematode species, as shown in rodents (Nawa *et al.* 1994).

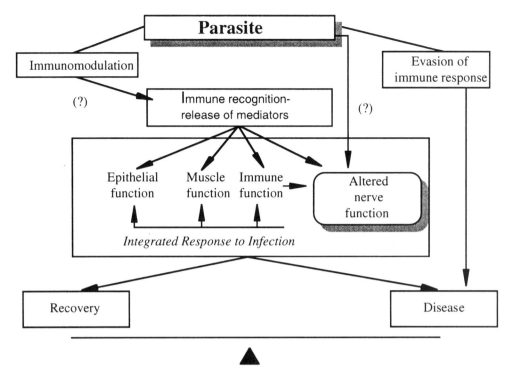

Fig. 1. Hypothetical scheme of how parasite factors and various host cells/mediators can interact to determine the balance between the development of disease and recovery or prevention of infection (from McKay & Fairweather, 1997).

Mucus, gut motility and the enteric nervous system

Increased mucus production, increased fluidity in the gut lumen and increased motility of the gut are associated with the host's response to GI nematode infection (Miller, 1996*b*).

Intestinal mucus is involved in protection against *N. brasiliensis* and *T. spiralis* in rats (reviewed by Miller, 1987 and Nawa *et al.* 1994). Expulsion of *N. brasiliensis* is associated with goblet cell hyperplasia (Nawa *et al.* 1994) and with changes in the physicochemical properties of the mucins produced and secreted by the goblet cells (Ishikawa, Horii & Nawa, 1993; Ishikawa *et al.* 1995). These changes are thought to be at least in part under the control of a Th$_2$ type response (Ishikawa, Wakelin & Mahida, 1997). In guinea pigs infected with *T. colubriformis*, the numbers of goblet cells and the proportion of sulphomucins in these cells increased significantly earlier in high responder animals, compared to low responder guinea pigs (Manjili *et al.* 1998). Sheep that were immunized against *H. contortus* showed significant reductions in neutral mucin at the mucosal surface of the abomasum, and increased quantities of acidic mucin (Newlands, Miller & Jackson, 1990). Treatment with the corticosteroid dexamethasone caused depletion of mucus and abrogated the protective response in challenged sheep (Newlands *et al.* 1990). Douch *et al.* (1983) reported that mucus from the GI tract of sheep resistant to nematode infection affected larval motility *in vitro*. Intestinal mucus from sheep immunized against *T. colubriformis* inhibited the motility of *H. contortus*, *Nematodirus spathiger* and *T. circumcincta*. The secretion of compounds with larval paralysing activity has been attributed to mucosal mast cells (Douch, Morum & Rabel, 1996). Abomasal mucus from calves infected with *O. ostertagi* also showed larval migration inhibition (LMI) activity *in vitro* (Claerebout *et al.* 1999*a*). Although the LMI capacity of the abomasal mucus was very variable, the highest paralysing activity was consistently observed in mucus from previously immunized calves.

Another function that contributes to host protection against GI nematodes is fluid secretion, led by epithelial ion transport (Baird & O'Malley, 1993). *In vitro* antigen challenge of isolated intestinal mucosae from previously parasitised animals evoked an ion transport response, whereas such responses did not occur in tissues from uninfected control animals. Secretory diarrhoea, a consequence of chloride secretion, may flush the GI tract of its parasites and/or may produce an environment that is unfavourable for the parasite (Baird & O'Malley, 1993).

In rodents there is convincing evidence of the involvement of the intestinal motor system in host defence against GI nematode infection (reviewed by Vallance & Collins, 1998). Infections with *T. spiralis* evoke a heightened intestinal muscle function in rats and mice. In mice which expel *T. spiralis* quickly ('responders') the maximal longitudinal muscle tension was significantly higher compared to 'non-responders'. The muscle contractility appears to be modulated by the immune system. Studies in mice indicated that CD4+ T cells play an important role in the initiation as well as in the persistence of altered muscle function. The T cell mediators active in altering intestinal motility still need to be defined, but results from recent studies suggest that IL-4 may be involved (Vallance & Collins, 1998).

Increased water secretion and peristalsis, and the physiology of immune cells (e.g. mast cells) are possibly regulated by the enteric nervous system (McClure & Emery, 1994; McKay & Fairweather, 1997). Alterations in the distribution of nerves, in the levels of neurochemicals and in neuronal function have been observed following GI infections in rodents (reviewed by McKay & Fairweather, 1997). Fig. 1 presents a hypothetical scheme of the integrated intestinal response to nematode infection.

Immunomodulation

GI nematodes may survive by modulating or evading the host immune system (Fig. 1). The ability of *H. polygyrus* to cause chronic infections in mice has been attributed to immunomodulatory capacities of the adult worms (reviewed by Behnke, 1987). Other nematodes, such as *T. spiralis*, *N. brasiliensis* and *T. muris* have greatly prolonged survival in mice concurrently harbouring *H. polygyrus* (Behnke, 1987). The exact mechanism of the immuno-suppression is still unknown. Mesenteric lymph node lymphocytes from mice harbouring either *H. polygyrus* alone or *H. polygyrus* and *T. spiralis* secreted lower levels of IL-9 and IL-10 when stimulated *in vitro*, compared to lymphocytes from mice infected with *T. spiralis* alone. This down-regulation of *in vitro* IL-9 and IL-10 responses during *H. polygyrus* infection correlated with poor mastocytosis and chronic survival of adult worms (Behnke *et al.* 1993). Those mouse strains that did eventually expel *H. polygyrus* were able to maintain a sustained IL-9 and IL-10 response compared with those strains that did not expel (Behnke *et al.* 1993). Seemingly adult *H. polygyrus* secrete an immuno-modulatory factor that counteracts the Th_2-driven immune responses of the host (Telford *et al.* 1998). Grencis & Entwistle (1997) showed production of an IFN-γ-like molecule by *T. muris* in mice. Although the functional activity of the molecule *in vivo* has not yet been determined, production of an IFN-γ homologue by the worm might be a mechanism to interfere with the regulation of the host immune response (Grencis & Entwistle, 1997).

Adult cows that have spent several seasons on infected pastures still harbour low numbers of *O. ostertagi*, indicating that sterilizing immune responses against this parasite are uncommon. Possible reasons for this apparent lack of protective immune responses have been reviewed by Klesius

(1993) and Gasbarre (1997). The reduced ability of peripheral blood lymphocytes from *O. ostertagi*-infected cattle to respond to mitogens (Klesius *et al.* 1984; Cross, Klesius & Haynes, 1986; Snider *et al.* 1986; Wiggin & Gibbs, 1990) indicated that *Ostertagia* infection can exhibit non-specific sup-pressive effects on the host immune response. In contrast, *Ostertagia* L_3 antigen has induced pro-liferation of bovine lymphocytes *in vitro* (Klesius *et al.* 1984; Wiggin & Gibbs, 1989, 1990; De Marez *et al.* 1997). However, the lymphocyte responses to *O. ostertagi* L_3 antigen were suppressed during the prepatent period of an *Ostertagia* infection (Klesius *et al.* 1984). Moreover, proliferative responses induced by L_3 antigen extract were suppressed by L_4 and adult antigens (De Marez *et al.* 1997). Antigen-specific immunosuppression was also suggested as a possible explanation for a decrease in the frequency of *Ostertagia*-specific peripheral blood lymphocytes and abomasal lymph node lymphocytes of calves during *Ostertagia* infection (Gasbarre, 1986, 1994). The suppression possibly involves activation of macrophages and/or dendritic cells, which then act as suppressor cells (T. De Marez, unpublished observations). Alternatively, it has been postulated that a unique subset of helper T cells (Th_3 cells) can suppress both Th_1 and Th_2 responses by secretion of TGF-β (Chen *et al.* 1994). Decreased levels of different cytokines, including IL4, IFN-γ and TGF-β were observed in immunized calves after a challenge infection with *O. ostertagi* (Almeria *et al.* 1998). One possible explanation could be that, once induced, TGF-β suppressed all cytokine responses including itself by autoregulation (Almeria *et al.* 1998). Finally, polyclonal activation of lymphocytes with specificities different from *Ostertagia* has also been suggested as an alternative mechanism of immunosuppression (Gasbarre, 1994, 1997).

EVALUATING THE ACQUIRED IMMUNITY OF CATTLE AGAINST GASTROINTESTINAL NEMATODES

Immunological parameters

Immunoglobulins. Perhaps the most widely used im-munological marker for acquired immunity against GI nematodes in cattle is the level of parasite-specific IgG antibodies in serum. Total serum IgG levels mainly mirror an IgG_1 response, since IgG_1 is the predominant immunoglobulin in serum from *Ostertagia*- and *Cooperia*-infected calves (Canals & Gasbarre, 1990; Christensen, 1991; Gasbarre *et al.* 1993*a*; Hilderson *et al.* 1993*a, b*). Although it has been suggested that IgG_2 could be more important in the immune response to GI nematodes (Gasbarre & Canals, 1989; Christensen, 1991), total IgG levels have been used successfully to measure acquired immunity to *O. ostertagi* and *C. oncophora* (e.g. Kloosterman *et al.* 1984; Frankena, 1987). However, the IgG response is not only dependent on the acquired immunity of the calves, but also on the antigenic stimulation, i.e. the level of exposure to infection. Within exposure levels, IgG levels correlate with the immunological responsiveness of individual animals (Kloosterman *et al.* 1978, Kloosterman, Parmentier & Ploeger, 1992).

In enzyme-linked immunosorbent assays (ELISA) for detection of parasite-specific IgG, crude larval or adult worm extract is generally used as an antigen source (e.g. Keus, Kloosterman & Van Den Brink, 1981; Klesius, Washburn & Haynes, 1986; Gasbarre *et al.* 1993 *a*, *b*; Hilderson *et al.* 1993 *a*). However, protective immunity may be correlated with the recognition of specific epitopes, and these responses could be hidden in the response against the crude antigens used (Gasbarre *et al.* 1993 *b*). Recently, the use of purified or recombinant low molecular weight *C. oncophora* antigens has shown potential for studies on immune-mediated resistance to *Cooperia* (Van Diemen *et al.* 1997).

Although parasite-specific IgA antibodies can be detected in serum from calves infected with *Ostertagia* (Canals & Gasbarre, 1990; Gasbarre *et al.* 1993 *a*, *b*), serum IgA levels are low, compared to IgG (Hilderson *et al.* 1993 *b*). In addition, serum IgA levels are only moderately correlated with IgA responses in the abomasal mucus (Sinski *et al.* 1995). Since IgA antibodies are mainly secreted locally in the mucosa (Bienenstock & Befus, 1980), determination of IgA levels in the serum may not be the best method to evaluate acquired immunity against GI nematodes.

A negative relationship between serum IgE and *Ostertagia* worm burdens has been described (Baker & Gershwin, 1993), but a test for the detection of bovine IgE is currently not available.

Mucosal mast cell and globule leucocyte counts. Although there are numerous indications that mucosal mast cells are effector cells in the immune response against GI nematodes, caution is needed when mucosal mast cell counts are used as indicators of acquired immunity. The most important problem with mast cell counts is that they do not reflect the level of mast cell activity in the mucosa, because mast cells lose their identifying granules during function (Rothwell, 1989). Therefore, globule leucocyte counts may provide a more accurate reflection of mast cell activity (Stear *et al.* 1995 *b*). A direct and reliable method of monitoring mucosal mast cell activity *in vivo* is to quantify the release of chymotrypsin-like proteases from the mast cell granules (Miller, 1996 *a*). Immuno-assays to measure mast cell proteases have demonstrated the secretion of these enzymes in serum and intestinal secretions during expulsion of GI nematodes in rodents (Woodbury *et al.* 1984; Huntley *et al.* 1990) and sheep (Huntley *et al.* 1987).

In sheep, resistance against *T. colubriformis* has been associated with the presence of larval paralysing activity in the GI mucus (e.g. Douch *et al.* 1983, 1984, 1986; Stankiewicz *et al.* 1993; Jones *et al.* 1994). In some studies, the larval paralysing activity was correlated with the presence of globule leucocytes (Douch *et al.* 1986; Stankiewicz *et al.* 1993). In addition, mucosal mast cells/globule leucocytes isolated and purified from immunized sheep secreted compounds having larval migration inhibition (LMI) activity when cultured with homologous nematode larvae or antigens (Douch *et al.* 1996). LMI activity has also been detected in faeces and ileal digesta of immunized sheep (Douch *et al.* 1983; Kimambo & MacRae, 1988; Jones *et al.* 1994), and determination of the level of LMI activity in faeces has been proposed as a useful indicator of the resistance status of sheep undergoing challenge infection (Douch *et al.* 1983). In contract, no relationship between LMI activity in abomasal mucus and resistance against *H. contortus* was observed by Gamble & Zajac (1992) despite the presence of significantly higher numbers of globule leucocytes in the abomasal mucosa of resistant lambs.

In calves infected with *O. ostertagi*, both abomasal mucus and serum exhibited larval paralysing activity (Claerebout *et al.* 1999 *a*). Sera from immunized animals showed significantly higher LMI capacity after a challenge infection, compared to previously uninfected calves, and serum LMI activity was significantly negatively correlated with *Ostertagia* worm counts.

Parasitological parameters

Acquired immunity has the potential to regulate the establishment, development, fecundity and survival of GI nematodes (Quinnell & Keymer, 1990). The different processes of the parasite life cycle are not necessarily regulated by the same immune effector mechanisms (Kloosterman *et al.* 1978; Stear *et al.* 1995 *b*), nor are they affected simultaneously. The first manifestation of acquired immunity to GI nematodes in ruminants is usually a decreasing egg output. Decreasing egg counts and stunting of growth are followed by retardation and arrestment of development, adult worm loss and, finally, resistance to establishment of ingested larvae (Vercruysse, Hilderson & Claerebout, 1994). These expressions of acquired immunity can be assessed by parasitological parameters, such as faecal egg counts, worm counts, worm length, number of eggs *in utero* and vulval flap development (Klesius, 1988).

Faecal egg counts. This is the only parasitological parameter that can be obtained regularly from the same animal during a GI parasitic nematode infection. Animals that were immunized against *C.*

oncophora (Frankena, 1987), *O. ostertagi* (Christensen *et al.* 1992; Hilderson *et al.* 1993*a*; Dorny *et al.* 1997) or both parasites (Nansen *et al.* 1993) had lower faecal egg counts after a homologous artificial challenge infection (Frankena, 1987; Hilderson *et al.* 1993*a*; Dorny *et al.* 1997) or after a natural challenge infection (Christensen *et al.* 1992; Nansen *et al.* 1993) compared to non-immunized animals. In addition, Michel & Sinclair (1969) observed a rise in faecal egg counts following corticosteroid treatment of calves. A reduced faecal egg output in immune animals is the result of a reduced number of female worms and/or reduced egg production per female (Kloosterman *et al.* 1978). Although faecal egg counts do not strictly reflect the fecundity of the parasite population, Albers (1981) and Stear *et al.* (1995*b*) found a very good correlation between faecal egg counts and number of eggs *in utero* of *C. oncophora* and *T. circumcincta* populations, respectively.

Worm counts. A reduced worm burden in immune animals is the result of decreased establishment and/or increased mortality of the worms. Michel, Lancaster & Hong (1973) reported resistance to establishment of *O. ostertagi* in calves, due to previous experience of infection. A lower number of *C. oncophora* (Hilderson *et al.* 1995*b*; Ploeger *et al.* 1995) or *O. ostertagi* (Snider *et al.* 1981; Frankena, 1987; Ploeger *et al.* 1995; Claerebout *et al.* 1997) has been recovered from previously infected calves compared to non-immunized controls. When animals were concurrently immunized against *Ostertagia* and *Cooperia*, the numbers of *Cooperia* that were recovered after a mixed challenge infection were much lower than the numbers of *Ostertagia* (Hilderson *et al.* 1995*b*; Ploeger *et al.* 1995), illustrating that acquired immunity develops earlier and/or more strongly against *Cooperia* than against *Ostertagia* (Armour, 1989).

Acquired immunity can cause arrestment of the development of established *O. ostertagi* larvae in the early L_4 stage (Michel *et al.* 1973, Michel, Lancaster & Hong, 1979; Eysker, 1993). In a number of studies, the proportion of the *Ostertagia* population that was inhibited in the L_4 stage after an artificial challenge infection (Hilderson *et al.* 1993*a*, 1995*b*; Dorny *et al.* 1997) or a natural challenge infection (Eysker, 1993; Claerebout *et al.* 1997) was greater in immunized calves compared to previously uninfected control calves.

Worm length. Stunting of *O. ostertagi* in cattle has been ascribed to the effects of an immune response (Michel, 1963; Michel, Lancaster & Hong, 1972). A reduced length might be due to inhibited growth, a selective expulsion of large worms or shrinkage of worms during the infection (Frankena, 1987). A reduced length of adult male and/or female worms after a homologous challenge infection has frequently been observed in calves that were previously infected with *Cooperia* (Albers, 1981; Kloosterman *et al.* 1984; Frankena, 1987) or *Ostertagia* (Kloosterman *et al.* 1984; Frankena, 1987; Hilderson *et al.* 1991, 1993*a*; Dorny *et al.* 1997).

Number of eggs per female worm. Kloosterman *et al.* (1984) and Frankena (1987) found that the number of eggs per female *Ostertagia* was significantly reduced after previous exposure and could therefore be regarded as a suitable parameter of acquired immunity. The fecundity of *C. oncophora* was not affected by a previous *Cooperia* infection (Kloosterman *et al.* 1984; Frankena, 1987). In contrast, an effect of acquired immune responses on the number of eggs per female *Cooperia* has been observed in other experiments (H. W. Ploeger, personal communication).

Vulval flap development. An example of changes in worm morphology as a consequence of host immunity is the reduced size of the vulvul flap of *O. ostertagi* in previously infected calves (Michel, 1967; Michel *et al.* 1972; Kloosterman *et al.* 1984; Frankena, 1987). The use of the immunosuppressive drug cortisone increased the proportion of worms with well developed flaps (Michel & Sinclair, 1969).

DISCUSSION

In comparison with the wealth of information on immunity against GI nematode infections in rodents and, to a lesser extent, in sheep, our understanding of protective immune responses against GI nematodes in cattle is limited. Although protective immune responses in rodents and ruminants have some features in common, the bovine immune response against *O. ostertagi* appears to differ from the rodent models in several aspects. In contrast to the immune response in mice, which is controlled by T helper 2 (Th_2) lymphocytes and results in mast cell- or goblet cell-mediated expulsion of adult worms, bovine immune responses to *O. ostertagi* do not show a clear Th_2 cytokine profile, and they do not result in rapid expulsion of the parasite. The first manifestation of immunity to *O. ostertagi* in calves is a reduction of worm fecundity, possibly regulated by the local IgA response. Worm numbers are only reduced after a prolonged period of host-parasite contact, and there are indications that *O. ostertagi* actively suppresses the host's immune response. Until the mechanisms of protective immunity against *O. ostertagi* are revealed, the use of immunological parameters to estimate acquired immunity in cattle is primarily based on their correlation with parasitological parameters. Assessing the resistance of calves against a challenge infection by means of parasitological parameters is probably still the most practical way to measure acquired immunity against GI nematodes.

Although the parasitological parameters described above are useful tools for evaluating acquired immunity against GI nematodes in cattle (Klesius, 1988; Vercruysse *et al*. 1994), their interpretation is also not straightforward. Most parasitological parameters are the consequence of different processes in the nematode life cycle. For example, faecal egg counts are the result of the number of worms and the fecundity of the worms, whereas the number of worms depends on the establishment and the survival of the parasites. Furthermore, in natural infections the faecal egg output is composed of eggs from different nematode species. Since acquired immunity against some species (e.g. *C. oncophora*) develops more quickly than against other species (e.g. *O. ostertagi*) (Armour, 1989), the generic composition of the excreted eggs usually changes during the grazing season (Nansen *et al*. 1993; Claerebout *et al*. 1997). Another important drawback is that these parameters are not regulated by acquired immunity alone. Establishment, development, fecundity and mortality of the worms are density-dependent processes, i.e. they are also affected by the size of the present worm population (Quinnell & Keymer, 1990). Inhibition of the development of *O. ostertagi* is influenced by a number of factors, including climate conditioning of the infective larvae on pasture (Armour & Duncan, 1987). In addition, the question can be asked how much a particular parameter (e.g. worm counts) should be reduced after a challenge infection to indicate that an animal or a group of animals is sufficiently 'immune'. This is strongly influenced by the size and the type of challenge infection and by the moment of sampling. As yet, there are no guidelines for evaluating acquired immunity to GI nematodes in cattle. Finally, most parasitological parameters are to be assessed *post mortem* and therefore provide only a static picture of the highly dynamic process of immunity development. As it is more important to evaluate the duration of protective immunity provided by a candidate vaccine than to assess its efficacy at the time of slaughter, this is probably the most important disadvantage of parasitological parameters (except faecal egg counts).

In conclusion, every immunological and parasitological parameter that is currently available can provide only a crude estimation of acquired immunity to GI nematodes in cattle, and different variables can be conflicting (Hilderson *et al*. 1995*b*; Dorny *et al*. 1997). To obtain a reasonable estimation of the acquired immunity against *Ostertagia* and *Cooperia*, different immunological and parasitological parameters should be combined, and parameters measuring changes over time (e.g. faecal

egg counts, serum IgG) should be used together with *post mortem* parameters (e.g. worm counts, mucosal IgA).

REFERENCES

ADAMS, D. B., MERRITT, G. C. & CRIPPS, A. W. (1980). Intestinal lymph and the local antibody and immunoglobulin response to infection by *Trichostrongylus colubriformis* in sheep. *Australian Journal of Experimental Biological and Medical Science* **58**, 167–177.

AHMAD, A., WANG, C. H. & BELL, R. G. (1991). A role for IgE in intestinal immunity. Expression of rapid expulsion of *Trichinella spiralis* in rats transfused with IgE and thoracic duct lymphocytes. *Journal of Immunology* **146**, 3563–3570.

ALBERS, G. A. A. (1981). *Genetic resistance to experimental* Cooperia oncophora *infections in calves*. PhD Thesis, Wageningen, The Netherlands.

ALMERIA, S., CANALS, A., GÓMEZ-MUÑOZ, M. T., ZARLENGA, D. S. & GASBARRE, L. C. (1998). Characterization of protective immune responses in local lymphoid tissues after drug-attenuated infections with *Ostertagia ostertagi* in calves. *Veterinary Parasitology* **80**, 53–64.

ALMERIA, S., CANALS, A., ZARLENGA, D. S. & GASBARRE, L. C. (1997*a*). Isolation and characterization of abomasal lymphocytes in the course of a primary *Ostertagia ostertagi* infection. *Veterinary Immunology and Immunopathology* **57**, 87–98.

ALMERIA, S., CANALS, A., ZARLENGA, D. S. & GASBARRE, L. C. (1997*b*). Quantification of cytokine gene expression in lamina propria lymphocytes of cattle following infection with *Ostertagia ostertagi*. *Journal of Parasitology* **83**, 1051–1055.

ALMOND, N. M. & PARKHOUSE, R. M. E. (1986). Immunoglobulin class-specific responses to biochemically defined antigens of *Trichinella spiralis*. *Parasite Immunology* **8**, 391–406.

ANDERSON, R. M. (1998). Complex dynamic behaviours in the interaction between parasite populations and the host's immune system. *International Journal for Parasitology* **28**, 551–566.

ARMOUR, J. (1989). The influence of host immunity on the epidemiology of trichostrongyle infections in cattle. *Veterinary Parasitology* **32**, 5–19.

ARMOUR, J., BAIRDEN, K., HOLMES, P. H., PARKINS, J. J., PLOEGER, H. & SALMAN, S. K. (1987). Pathophysiological and parasitological studies on *Cooperia oncophora* infections in calves. *Research in Veterinary Science* **42**, 373–381.

ARMOUR, J. & DUNCAN, M. (1987). Arrested larval development in cattle nematodes. *Parasitology Today* **3**, 171–176.

BAIRD, A. W. & O'MALLEY, K. E. (1993). Epithelial ion transport – possible contribution to parasite expulsion. *Parasitology Today* **9**, 141–143.

BAKER, D. G. & GERSHWIN, L. J. (1992). Seasonal patterns of total and *Ostertagia*-specific IgE in grazing cattle. *Veterinary Parasitology* **44**, 211–221.

BAKER, D. G. & GERSHWIN, L. J. (1993). Inverse relationship between IgE and worm burdens in cattle infected with *Ostertagia ostertagi*. *Veterinary Parasitology* **47**, 87–97.

BAKER, D. G., SCOTT, J. L. & GERSHWIN, L. J. (1993). Abomasal lymphatic lymphocyte subpopulations in cattle infected with *Ostertagia ostertagi* and *Cooperia* sp. *Veterinary Immunology and Immunopathology* **39**, 467–473.

BANCROFT, A. J., ELSE, K. J., SYPEK, J. P. & GRENCIS, R. K. (1997). IL-12 promotes a chronic intestinal nematode infection. *European Journal of Immunology* **27**, 2536–2540.

BANCROFT, A. J., McKENZIE, A. N. J. & GRENCIS, R. K. (1998). A critical role for IL-13 in resistance to intestinal nematode infection. *Journal of Immunology* **160**, 3453–3461.

BEHNKE, J. M. (1987). Evasion of immunity by nematode parasites causing chronic infection. *Advances in Parasitology* **26**, 1–70.

BEHNKE, J. M., WAHID, F. N., GRENCIS, R. K., ELSE, K. J., BEN-SMITH, A. & GOYAL, P. K. (1993). Immunological relationships during primary infection with *Heligmosomoides polygyrus* (*Nematospiroides dubius*): down regulation of specific cytokine secretion (IL-9 and IL-10) correlates with poor mastocytosis and chronic survival of adult worms. *Parasite Immunology* **15**, 415–421.

BENDIXSEN, T., EMERY, D. L. & JONES, W. O. (1995). The sensitization of mucosal mast cells during infections with *Trichostrongylus colubriformis* or *Haemonchus contortus* in sheep. *International Journal for Parasitology* **25**, 741–748.

BETTS, C. J. & ELSE, K. J. (1999). Mast cells, eosinophils and antibody-mediated cellular cytotoxicity are not critical in resistance to *Trichuris muris*. *Parasite Immunology* **21**, 45–52.

BIENENSTOCK, J. & BEFUS, A. D. (1980). Mucosal immunology. *Immunology* **41**, 249–270.

BONE, L. W. & KLESIUS, P. H. (1985). Effects of host serum on *in vitro* oviposition by *Trichostrongylus colubriformis* (Nematoda). *International Journal of Invertebrate Reproduction and Development* **10**, 27–32.

BOTTJER, K. P., KLESIUS, P. H. & BONE, L. W. (1985). Effects of host serum on feeding by *Trichostrongylus colubriformis* (Nematoda). *Parasite Immunology* **7**, 1–9.

BRIGANDI, R. A., ROTMAN, H. L., YUTANAWIBOONCHAI, A., LEON, O., NOLAN, T. J., SCHAD, G. A. & ABRAHAM, D. (1996). *Strongyloides stercoralis*: role of antibody and complement in immunity to the third stage larvae in BALB/cByJ mice. *Experimental Parasitology* **82**, 279–289.

BROWN, W. C., DAVIS, W. C., DOBBELAERE, A. E. & RICE-FICHT, A. C. (1994). CD4+ T-cell clones obtained from cattle chronically infected with *Fasciola hepatica* and specific for adult worm antigen express both unrestricted and Th2 cytokine profiles. *Infection and Immunity* **62**, 818–827.

BUDDLE, B. M., JOWETT, G., GREEN, R. S., DOUCH, P. G. C. & RISDON, P. L. (1992). Association of blood eosinophilia with the expression of resistance in Romney lambs to nematodes. *International Journal for Parasitology* **22**, 255–260.

BUTTERWORTH, A. E. (1984). Cell-mediated damage to helminths. *Advances in Parasitology* **23**, 143–235.

CANALS, A. & GASBARRE, L. C. (1990). *Ostertagia ostertagi*: isolation and partial characterization of somatic and metabolic antigens. *International Journal for Parasitology* **20**, 1047–1054.

CANALS, A., ZARLENGA, D. S., ALMERIA, S. & GASBARRE, L. C. (1997). Cytokine profile induced by a primary infection with *Ostertagia ostertagi*. *Veterinary Immunology and Immunopathology* **58**, 63–75.

CHARLEY-POULAIN, J., LUFFAU, G. & PERY, P. (1984). Serum and abomasal antibody response of sheep to infections with *Haemonchus contortus*. *Veterinary Parasitology* **14**, 129–141.

CHEN, Y., KUCHROO, V. K., INOBE, J., HAFLER, D. A. & WEINER, H. L. (1994). Regulatory T cell clones induced by oral tolerance: suppression of autoimmune encephalomyelitis. *Science* **265**, 1237–1240.

CHRISTENSEN, C. M. (1991). *Studies of the immune response in cattle infected with normal or inhibited larvae of* Ostertagia ostertagi (*Trichostrongylidae*), PhD thesis, Royal Veterinary and Agricultural University, Copenhagen, Denmark, pp. 239.

CHRISTENSEN, C. M., NANSEN, P., HENRIKSEN, S. A. A., MONRAD, J. & SATRIJA, F. (1992). Attempts to immunize cattle against *Ostertagia ostertagi* infections employing 'normal' and 'chilled' (hypobiosis-prone) third stage larvae. *Veterinary Parasitology* **44**, 247–261.

CLAEREBOUT, E., AGNEESSENS, J., SHAW, D. J. & VERCRUYSSE, J. (1999*a*). Larval migration inhibition activity in abomasal mucus and serum from calves infected with *Ostertagia ostertagi*. *Research in Veterinary Science* **66**, 253–257.

CLAEREBOUT, E., DORNY, P., VERCRUYSSE, J., AGNEESSENS, J. & DEMEULENAERE, D. (1998*b*). Effects of preventive anthelmintic treatment on acquired resistance to gastrointestinal nematodes in naturally infected cattle. *Veterinary Parasitology* **76**, 287–303.

CLAEREBOUT, E., HILDERSON, H., MEEUS, P., DE MAREZ, T., BEHNKE, J., HUNTLEY, J. & VERCRUYSSE, J. (1996). The effect of truncated infections with *Ostertagia ostertagi* on the development of acquired resistance in calves. *Veterinary Parasitology* **66**, 225–239.

CLAEREBOUT, E., HILDERSON, H., SHAW, D. J. & VERCRUYSSE, J. (1997). The presence of an early L4 population in relation to the acquired resistance of calves naturally infected with *Ostertagia ostertagi*. *Veterinary Parasitology* **68**, 337–346.

CLAEREBOUT, E., RAES, S., GELDHOF, P., AGNEESSENS, J. & VERCRUYSSE, J. (1999*b*). IgA response associated with reduced fecundity of *Ostertagia ostertagi* in cattle. *17th International Conference of the World Association for the Advancement of Veterinary Parasitology*, Copenhagen.

CLAEREBOUT, E., VERCRUYSSE, J., DORNY, P., DEMEULENAERE, D. & DEREU, A. (1998*a*). The effect of different infection levels on acquired resistance to gastrointestinal nematodes in artificially infected cattle. *Veterinary Parasitology* **75**, 153–167.

COFFMAN, R. L., SEYMOUR, B. W. P., HUDAK, S., JACKSON, J. & RENNICK, D. (1989). Antibody to IL5 inhibits helminth induced eosinophilia in mice. *Science* **245**, 308.

CONSTANT, S. L. & BOTTOMLY, K. (1997). Induction of Th1 and Th2 CD4+ T cell responses: the alternative approaches. *Annual Revue of Immunology* **15**, 297–322.

CROSS, D. A., KLESIUS, P. H. & HAYNES, T. B. (1986). Lymphocyte blastogenic responses of calves experimentally infected with *Ostertagia ostertagi*. *Veterinary Parasitology* **22**, 49–55.

CROWLE, P. K. & REED, N. D. (1981). Rejection of the intestinal parasite *Nippostrongylus brasiliensis* by mast cell deficient W/W^v anaemic mice. *Infection and Immunity* **33**, 54–58.

DAWKINS, H. J. S., WINDON, R. G. & EAGLESON, G. K. (1989). Eosinophil responses in sheep selected for high and low responsiveness to *Trichostrongylus colubriformis*. *International Journal for Parasitology* **19**, 199–205.

DE MAREZ, T., COX, E., CLAEREBOUT, E., VERCRUYSSE, J. & GODDEERIS, B. M. (1997). Induction and suppression of lymphocyte proliferation by antigen extracts of *Ostertagia ostertagi*. *Veterinary Immunology and Immunopathology* **57**, 69–77.

DESSEIN, A. J., PARKER, W. L., JAMES, S. L. & DAVID, J. R. (1981). IgE antibody and resistance to infection. I. Selective suppression of the IgE antibody response in rats diminishes the resistance and the eosinophil response to *Trichinella spiralis* infection. *Journal of Experimental Medicine* **153**, 423–436.

DINEEN, J. K., GREGG, P. & LASCELLES, A. K. (1978). The response of lambs to vaccination at weaning with irradiated *Trichostrongylus colubriformis* larvae: segregation into 'responders' and 'non-responders'. *International Journal for Parasitology* **8**, 59–63.

DINEEN, J. K. & WINDON, R. G. (1980). The effect of sire selection on the response of lambs to vaccination with irradiated *Trichostrongylus colubriformis* larvae. *International Journal for Parasitology* **10**, 189–196.

DORNY, P., CLAEREBOUT, E., VERCRUYSSE, J., HILDERSON, H. & HUNTLEY, J. F. (1997). The influence of a *Cooperia oncophora* priming on a concurrent challenge with *Ostertagia ostertagi* and *C. oncophora* in calves. *Veterinary Parasitology* **70**, 143–151.

DOUCH, P. G. C., HARRISON, G. B. L., BUCHANAN, L. L. & BRUNSDON, R. V. (1984). Relationship of histamine in tissues and antiparasitic substances in gastrointestinal mucus to the development of resistance to trichostrongyle infections in young sheep. *International Journal for Parasitology* **16**, 273–288.

DOUCH, P. G. C., HARRISON, G. B. L., BUCHANAN, L. L. & GREER, K. S. (1983). *In vitro* bioassay of sheep gastrointestinal mucus for nematode paralysing activity mediated by substances with some properties characteristic of SRS-A. *International Journal for Parasitology* **13**, 207–212.

DOUCH, P. G. C., HARRISON, G. B. L., ELLIOTT, D. C., BUCHANAN, L. L. & GREER, K. S. (1986). Relationship of gastrointestinal histology and mucus antiparasite activity with the development of resistance to trichostrongyle infections in sheep. *Veterinary Parasitology* **20**, 315–331.

DOUCH, P. G. C., MORUM, P. E. & RABEL, B. (1996). Secretion of antiparasite substances and leukotrienes from ovine gastrointestinal tissues and isolated mucosal mast cells. *International Journal for Parasitology* **26**, 205–211.

DUNCAN, J. L., SMITH, W. D. & DARGIE, J. D. (1978). Possible relationship of levels of mucosal IgA and serum IgG to immune unresponsiveness of lambs to *Haemonchus contortus*. *Veterinary Parasitology* **4**, 21–27.

ELSE, K. J. & FINKELMAN, F. D. (1998). Intestinal nematode parasites, cytokines and effector mechanisms. *International Journal for Parasitology* **28**, 1145–1158.

ELSE, K. J., FINKELMAN, F. D., MALISZEWSKI, C. R. & GRENCIS, R. K. (1994). Cytokine-mediated regulation of chronic intestinal helminth infection. *Journal of Experimental Medicine* **179**, 347–351.

ELSE, K. J. & GRENCIS, R. K. (1991). Cellular immune responses to the murine nematode parasite *Trichuris muris*. I. Differential cytokine production during acute or chronic infection. *Immunology* **72**, 508–513.

ELSE, K. J., HÜLTNER, L. & GRENCIS, R. K. (1992). Cellular immune responses to the murine nematode parasite *Trichuris muris*. II. Differential induction of Th-cell subsets in resistant versus susceptible mice. *Immunology* **75**, 232–237.

EMERY, D. L., McCLURE, S. J. & WAGLAND, B. M. (1993). Production of vaccines against gastrointestinal nematodes of livestock. *Immunology and Cell Biology* **71**, 463–472.

EYSKER, M. (1993). The role of inhibited development in the epidemiology of *Ostertagia* infections. *Veterinary Parasitology* **46**, 259–269.

FAULKNER, H., HUMPHREYS, N., RENAULD, J. C., VAN SNICK, J. & GRENCIS, R. K. (1997). Interleukin-9 is involved in host protective immunity to intestinal nematode infection. *European Journal of Immunology* **27**, 2536–2540.

FINKELMAN, F. D., MADDEN, K. B., CHEEVERS, A. W., KATONA, I. M., MORRIS, S. C., GATELY, M. K., HUBBARD, B. R., GAUSE, W. C. & URBAN, J. F. JR. (1994). Effects of interleukin 12 on immune responses and host protection in mice infected with intestinal nematode parasites. *Journal of Experimental Medicine* **179**, 1563–1572.

FINKELMAN, F. D., PEARCE, E. J., URBAN, J. F. JR. & SHER, A. (1991). Regulation and biological function of helminth-induced cytokine responses. *Immunology Today* **12**, 62–67.

FINKELMAN, F. D., SHEA-DONOHUE, T., GOLDHILL, J., SULLIVAN, C. A., MORRIS, S. C., MADDEN, K. B., GAUSE, W. C. & URBAN, J. F. JR. (1997). Cytokine regulation of host defence against parasitic gastrointestinal nematodes: lessons from studies with rodent models. *Annual Revue of Immunology* **15**, 505–533.

FRANKENA, K. (1987). *The interaction between* COOPERIA *spp. and* OSTERTAGIA *spp. (Nematoda: Trichostrongylidae) in cattle.* PhD thesis, Agricultural University, Wageningen, The Netherlands, pp. 101.

GALLI, S. J., GORDON, J. R. & WERSHIL, B. K. (1991). Cytokine production by mast cells and basophils. *Current Opinion in Immunology* **3**, 865–873.

GAMBLE, H. R. & ZAJAC, A. M. (1992). Resistance of St. Croix lambs to *Haemonchus contortus* in experimentally and naturally acquired infections. *Veterinary Parasitology* **41**, 211–225.

GASBARRE, L. C. (1986). Limiting dilution analyses for the quantification of cellular immune responses in bovine ostertagiasis. *Veterinary Parasitology* **20**, 133–147.

GASBARRE, L. C. (1994). *Ostertagia ostertagi*: changes in lymphoid populations in the local lymphoid tissues after primary or secondary infection. *Veterinary Parasitology* **55**, 105–114.

GASBARRE, L. C. (1997). Effects of gastrointestinal nematode infection on the ruminant immune system. *Veterinary Parasitology* **72**, 327–343.

GASBARRE, L. C. & CANALS, A. (1989). Induction of protective immunity in calves immunized with adult *Oesophagostomum radiatum* somatic antigens. *Veterinary Parasitology* **34**, 223–238.

GASBARRE, L. C., LEIGHTON, E. A. & DAVIES, C. J. (1993b). Influence of host genetics upon antibody responses against gastrointestinal nematode infections in cattle. *Veterinary Parasitology* **46**, 81–91.

GASBARRE, L. C., NANSEN, P., MONRAD, J., GRONVELD, J., STEFFAN, P. & HENRIKSEN, S. A. (1993a). Serum anti-trichostrongyle antibody responses of first and second season grazing calves. *Research in Veterinary Science* **54**, 340–344.

GILL, H. S. (1991). Genetic control of acquired resistance to haemonchosis in Merino lambs. *Parasite Immunology* **13**, 617–628.

GILL, H. S., GRAY, G. D., WATSON, D. L. & HUSBAND, A. J. (1993a). Isotype-specific antibody responses to *Haemonchus contortus* in genetically resistant sheep. *Parasite Immunology* **15**, 61–67.

GILL, H. S., HUSBAND, A. J., WATSON, D. L. & GRAY, G. D. (1994). Antibody-containing cells in the abomasal mucosa of sheep with genetic resistance to *Haemonchus contortus*. *Research in Veterinary Science* **56**, 41–47.

GILL, H. S., WATSON, D. L. & BRANDON, M. R. (1993b). Monoclonal antibody to CD4+ T cells abrogates genetic resistance to *Haemonchus contortus* in sheep. *Immunology* **78**, 43–49.

GRANSMULLER, A., ANTEUNIS, A., VENTURIELLO, S. M., BRUSCHI, F. & BINAGHI, R. A. (1987). Antibody-dependent *in vitro* cytotoxicity of newborn *Trichinella spiralis* larvae: nature of cells involved. *Parasite Immunology* **9**, 281–293.

GRENCIS, R. K. (1996). T cell and cytokine basis of host variability in response to intestinal nematode infections. *Parasitology* **112**, S31–S37.

GRENCIS, R. K., ELSE, K. J., HUNTLEY, J. F. & NISHIKAWA, S. I. (1993). The *in vivo* role of stem cell factor (c-kit) ligand on mastocytosis and host protective immunity to the intestinal nematode *Trichinella spiralis* in mice. *Parasite Immunology* **15**, 55–59.

GRENCIS, R. K. & ENTWISTLE, G. M. (1997). Production of an interferon gamma homologue by an intestinal nematode: functionally significant or interesting artefact? *Parasitology* **115**, S101–S105.

GRENCIS, R. K., HÜLTNER, L. & ELSE, K. J. (1991). Host protective immunity to *Trichinella spiralis* in mice: activation of Th cell subsets and lymphokine secretion in mice expressing different response phenotypes. *Immunology* **74**, 329–332.

GRENCIS, R. K., RIEDLINGER, J. & WAKELIN, D. (1985). L3T4-positive T lymphoblasts are responsible for transfer of immunity to *Trichinella spiralis* in mice. *Immunology* **56**, 213–218.

HERNDON, F. J. & KAYES, S. G. (1992). Depletion of eosinophils by anti-IL5 monoclonal antibody treatment of mice infected with *Trichinella spiralis* does not alter parasite burden or immunological resistance to reinfection. *Journal of Immunology* **149**, 3642–3647.

HILDERSON, H., BERGHEN, P., DE GRAAF, D. C., CLAEREBOUT, E. & VERCRUYSSE, J. (1995a). Immunisation of calves with *Ostertagia ostertagi* fourth stage larval antigens failed to protect calves from infection. *International Journal for Parasitology* **25**, 757–760.

HILDERSON, H., DE GRAAF, D. C., VERCRUYSSE, J. & BERGHEN, P. (1993b). Characterisation of *Ostertagia ostertagi* antigens by the different bovine immunoglobulin isotypes. *Research in Veterinary Science* **55**, 203–208.

HILDERSON, H., DORNY, P., BERGHEN, P., VERCRUYSSE, J., FRANSEN, J. & BRAEM, L. (1991). Gastrin and pepsinogen changes during an *Ostertagia ostertagi* challenge infection in calves. *Journal of Veterinary Medicine B* **38**, 25–32.

HILDERSON, H., VERCRUYSSE, J., CLAEREBOUT, E., DE GRAAF, D. C., FRANSEN, J. & BERGHEN, P. (1995b). Interactions between *Ostertagia ostertagi* and *Cooperia oncophora* in calves. *Veterinary Parasitology* **56**, 107–119.

HILDERSON, H., VERCRUYSSE, J., DE GRAAF, D. C., BASTIAENSEN, P., FRANSEN, J. & BERGHEN, P. (1993a). The presence of an early L4 larvae population in relation to the immune response of calves against *Ostertagia ostertagi*. *Veterinary Parasitology* **47**, 255–266.

HUNTLEY, J. F. (1992). Mast cells and basophils: a review of their heterogeneity and function. *Journal of Comparative Pathology* **107**, 349–372.

HUNTLEY, J. F., GIBSON, S., BROWN, D., SMITH, W. D., JACKSON, F. & MILLER, H. R. P. (1987). Systemic release of a mast cell proteinase following nematode infections in sheep. *Parasite Immunology* **9**, 603–611.

HUNTLEY, J. F., GOODEN, C., NEWLANDS, G. F. J., MacKELLAR, A., LAMMAS, D. A., WAKELIN, D., TUOHY, M., WOODBURY, R. G. & MILLER, H. R. P. (1990). Distribution of intestinal mast cell proteinase in blood and tissue of normal and *Trichinella* infected mice. *Parasite Immunology* **12**, 85–95.

HUNTLEY, J. F., NEWLANDS, G. F. J. & MILLER, H. R. P. (1984). The isolation and characterization of globule leucocytes: their derivation from mucosal mast cells in parasitized sheep. *Parasite Immunology* **6**, 371–390.

HUNTLEY, J. F., PATTERSON, M., MACKELLAR, A., JACKSON, F., STEVENSON, L. M. & COOP, R. L. (1995). A comparison of the mast cell and eosinophil responses of sheep and goats to gastrointestinal nematode infections. *Research in Veterinary Science* **58**, 5–10.

HUNTLEY, J. F., SCHALLIG, H. D. F. H., KOOYMAN, F. N. J., MacKELLAR, A., JACKSON, F. & SMITH, W. D. (1998b). IgE antibody during infection with the ovine abomasal nematode, *Teladorsagia circumcincta*: primary and secondary responses in serum and gastric lymph of sheep. *Parasite Immunology* **20**, 565–571.

HUNTLEY, J. F., SCHALLIG, H. D. F. H., KOOYMAN, F. N. J., MacKELLAR, A., MILLERSHIP, J. & SMITH, W. D. (1998a). IgE responses in the serum and gastric lymph of sheep infected with *Teladorsagia circumcincta*. *Parasite Immunology* **20**, 163–168.

ISHIKAWA, N., HORII, Y. & NAWA, Y. (1993). Immune-mediated alteration of the terminal sugars of goblet cell mucins in the small intestine of *Nippostrongylus brasiliensis*-infected rats. *Immunology* **78**, 303–307.

ISHIKAWA, N., SHI, B.-B., KHAN, A. I. & NAWA, Y. (1995). Reserpine-induced sulphomucin production by goblet cells in the jejunum of rats and its significance in the establishment of intestinal helminths. *Parasite Immunology* **17**, 581–586.

ISHIKAWA, N., WAKELIN, D. & MAHIDA, Y. R. (1997). Role of T helper 2 cells in intestinal goblet cell hyperplasia in mice infected with *Trichinella spiralis*. *Gastroenterology* **113**, 542–549.

JARRETT, E. E. E. & MILLER, H. R. P. (1982). Production and activities of IgE in helminth infection. *Progress in Allergy* **31**, 178–233.

JONAS, W., STANKIEWICZ, M. & RABEL, B. (1995). Influence of sheep eosinophils on third stage larvae of *Trichostrongylus colubriformis*. *Acta Parasitologica* **40**, 103–106.

JONES, W. O., EMERY, D. L., McCLURE, S. J. & WAGLAND, B. M. (1994). Changes in inflammatory mediators and larval inhibitory activity in intestinal contents and mucus during primary and challenge infections of sheep with *Trichostrongylus colubriformis*. *International Journal for Parasitology* **24**, 519–525.

JONES, W. O., HUNTLEY, J. F. & EMERY, D. L. (1992). Isolation and degranulation of mucosal mast cells from the small intestine of parasitised sheep. *International Journal for Parasitology* **22**, 519–521.

KARANU, F. N., McGUIRE, T. C., DAVIS, W. C., BESSER, T. E. & JASMER, D. P. (1997). CD4+ T lymphocytes contribute to protective immunity induced in sheep and goats by *Haemonchus contortus* gut antigens. *Parasite Immunology* **19**, 435–445.

KATONA, I. M., URBAN, J. F. JR. & FINKELMAN, F. D. (1988). The role of L3T4+ and Lyt-2+ T cells in the IgE response and immunity to *Nippostrongylus brasiliensis*. *Journal of Immunology* **140**, 3206–3211.

KEITH, K. A., DUNCAN, M. C., MURRAY, M., BAIRDEN, K. & TAIT, A. (1990). Stage-specific cuticular proteins of *Ostertagia circumcincta* and *Ostertagia ostertagi*. *International Journal for Parasitology* **20**, 1037–1045.

KEUS, A., KLOOSTERMAN, A. & VAN DEN BRINK, R. (1981). Detection of antibodies to *Cooperia* spp. and *Ostertagia* spp. in calves with the enzyme linked immunosorbent assay (ELISA). *Veterinary Parasitology* **8**, 229–236.

KIMAMBO, A. E. & MACRAE, J. C. (1988). Measurement *in vitro* of a larval migration inhibitory factor in gastrointestinal mucus of sheep made resistant to the roundworm *Trichostrongylus colubriformis*. *Veterinary Parasitology* **28**, 213–222.

KING, S. J. & MILLER, H. R. P. (1984). Anaphylactic release of mucosal mast cell protease and its relationship to gut permeability in *Nippostrongylus*-primed rats. *Immunology* **51**, 653–660.

KLESIUS, P. H. (1988). Immunity to *Ostertagia ostertagi*. *Veterinary Parasitology* **27**, 159–167.

KLESIUS, P. H. (1993). Regulation of immunity to *Ostertagia ostertagi*. *Veterinary Parasitology* **46**, 63–79.

KLESIUS, P. H., HAYNES, T. B. & CROSS, D. A. (1985). Chemotactic factors for eosinophils in soluble extracts of L3 stages of *Ostertagia ostertagi*. *International Journal for Parasitology* **15**, 517–522.

KLESIUS, P. H., WASHBURN, S. M., CIORDIA, H., HAYNES, T. B. & SNIDER III, T. G. (1984). Lymphocyte reactivity to *Ostertagia ostertagi* L3 antigen in type I ostertagiasis. *American Journal of Veterinary Research* **45**, 230–233.

KLESIUS, P. H., WASHBURN, S. M. & HAYNES, T. B. (1986). Serum antibody responses to soluble extract of the third-larval stage of *Ostertagia ostertagi* in cattle. *Veterinary Parasitology* **20**, 307–314.

KLOOSTERMAN, A., ALBERS, G. A. A. & VAN DEN BRINK, R. (1978). Genetic variation among calves in resistance to nematode parasites. *Veterinary Parasitology* **4**, 353–368.

KLOOSTERMAN, A., ALBERS, G. A. A. & VAN DEN BRINK, R. (1984). Negative interactions between *Ostertagia ostertagi* and *Cooperia oncophora* in calves. *Veterinary Parasitology* **15**, 135–150.

KLOOSTERMAN, A., BENEDICTUS, J. & AGHINA, H. (1980). Colostral transfer of anti-nematode antibodies and its significance for protection. *Veterinary Parasitology* **7**, 133–142.

KLOOSTERMAN, A., PARMENTIER, H. K. & PLOEGER, H. W. (1992). Breeding cattle and sheep for resistance to gastrointestinal nematodes. *Parasitology Today* **8**, 330–335.

KOOYMAN, F. N. J., VAN KOOTEN, P. J. S., HUNTLEY, J. F., MacKELLAR, A., CORNELISSEN, A. W. C. A. & SCHALLIG, H. D. F. H. (1997). Production of a monoclonal antibody specific for ovine immunoglobulin E and its application to monitor serum IgE responses to *Haemonchus contortus* infection. *Parasitology* **114**, 395–406.

KORENAGA, M., ABE, T. & HASHIGUCHI, Y. (1996). Injection of recombinant interleukin 3 hastens worm expulsion in mice infected with *Trichinella spiralis*. *Parasitology Research* **82**, 108–113.

KORENAGA, M., HITOSHI, Y., YAMAGUCHI, N., SATO, Y., TAKATSU, K. & TADA, I. (1991). The role of interleukin-5 in protective immunity to *Strongyloides venezuelensis* infection in mice. *Immunology* **72**, 502–507.

KORENAGA, M. & TADA, I. (1994). The role of IL-5 in the immune responses to nematodes in rodents. *Parasitology Today* **10**, 234–236.

KOYAMA, K., TAMAUCHI, H. & ITO, Y. (1995). The role of CD4+ and CD8+ T cells in protective immunity to the murine nematode parasite *Trichuris muris*. *Parasite Immunology* **17**, 161–165.

MADDEN, K. B., URBAN, J. F. JR., ZILTNER, H. J., SCHRADER, J. W., FINKELMAN, F. D. & KATONA, I. M. (1991). Antibodies to IL-3 and IL-4 suppress helminth induced intestinal mastocytosis. *Journal of Immunology* **147**, 1387–1394.

MANJILI, M. H., FRANCE, M. P., SANGSTER, N. C. & ROTHWELL, T. L. W. (1998). Quantitative and qualitative changes in intestinal goblet cells during primary infection of *Trichostrongylus colubriformis* high and low responder guinea pigs. *International Journal for Parasitology* **28**, 761–765.

MANSOUR, M. M., DIXON, J. B., CLARKSON, M. J., CARTER, S. D., ROWAN, T. G. & HAMMET, N. C. (1990). Bovine immune recognition of *Ostertagia ostertagi* larval antigens. *Veterinary Immunology and Immunopathology* 24, 361–371.

McCLURE, S. J., DAVEY, R. J., LLOYD, J. B. & EMERY, D. L. (1995). Depletion of IFN-gamma, CD8+ or TCR gamma-delta+ cells *in vivo* during primary infection with an enteric parasite (*Trichostrongylus colubriformis*) enhances protective immunity. *Immunology and Cell Biology* 73, 552–555.

McCLURE, S. J. & EMERY, D. L. (1994). Cell-mediated responses against gastrointestinal nematode parasites of ruminants. In *Cell-mediated Immunity in Ruminants* (eds. Goddeeris, B. M. & Morrison, I. W.), pp. 250. Boca Raton, CRC Press.

McKAY, D. M. & FAIRWEATHER, I. (1997). A role for the enteric nervous system in the response to helminth infections. *Parasitology Today* 13, 63–69.

McKENZIE, G. J., BANCROFT, A., GRENCIS, R. K. & McKENZIE, A. N. J. (1998). A distinct role for interleukin-13 in Th2-cell-mediated immune responses. *Current Biology* 8, 339–342.

MICHEL, J. F. (1963). The phenomena of host resistance and the course of infection of *Ostertagia ostertagi* in calves. *Parasitology* 53, 63–84.

MICHEL, J. F. (1967). Morphological changes in a parasitic nematode due to acquired resistance of the host. *Nature* 215, 520–521.

MICHEL, J. F., LANCASTER, M. B. & HONG, C. (1972). Host induced effects on the vulval flap of *Ostertagia ostertagi*. *International Journal for Parasitology* 2, 305–317.

MICHEL, J. F., LANCASTER, M. B. & HONG, C. (1973). *Ostertagia ostertagi*: Protective immunity in calves. The development in calves of a protective immunity to infection with *Ostertagia ostertagi*. *Experimental Parasitology* 33, 179–186.

MICHEL, J. F., LANCASTER, M. B. & HONG, C. (1979). The effect of age, acquired resistance, pregnancy and lactation on some reactions of cattle to infection with *Ostertagia ostertagi*. *Parasitology* 79, 157–168.

MICHEL, J. F. & SINCLAIR, I. J. (1969). The effect of cortisone on the worm burden of calves infected daily with *Ostertagia ostertagi*. *Parasitology* 59, 691–708.

MILLER, H. R. P. (1984). The protective mucosal response against gastrointestinal nematodes in ruminants and laboratory animals. *Veterinary Immunology and Immunopathology* 6, 167–259.

MILLER, H. R. P. (1987). Gastrointestinal mucus, a medium for survival and for elimination of parasitic nematodes and protozoa. *Parasitology* 94, S77–S100.

MILLER, H. R. P. (1996*a*). Prospects for the immunological control of ruminant gastrointestinal nematodes: natural immunity, can it be harnessed? *International Journal for Parasitology* 26, 801–811.

MILLER, H. R. P. (1996*b*). Mucosal mast cells and the allergic response against nematode parasites. *Veterinary Immunology and Immunopathology* 54, 331–336.

MOSMANN, T. R., CHERWINSKI, H., BOND, M. W., GIEDLIN, M. A. & COFFMAN, R. L. (1986). Two types of murine helper T cell clone. I. Definition according to profiles of lymphokine activities and secreted proteins. *Journal of Immunology* 136, 2348–2357.

MOSMANN, T. R. & COFFMAN, R. L. (1989). Th$_1$ and Th$_2$ cells: different patterns of lymphokine secretion lead to different functional properties. *Annual Revue of Immunology* 7, 145–173.

MURRAY, M., MILLER, H. R. P. & JARRETT, W. F. H. (1968). The globule leucocyte and its derivation from the subepithelial mast cell. *Laboratory Investigation* 19, 222–234.

NANSEN, P., STEFFAN, P. E., CHRISTENSEN, C. M., GASBARRE, L. C., MONRAD, J., GRØNVOLD, J. & HENRIKSEN, S. A. A. (1993). The effect of experimental trichostrongyle infections of housed young calves on the subsequent course of natural infection on pasture. *International Journal for Parasitology* 23, 627–638.

NAWA, Y., ISHIKAWA, N., TSUCHIYA, K., HORII, Y., ABE, T., KHAN, A. I., BING-SHI, ITOH, H., IDE, H. & UCHIYAMA, F. (1994). Selective effector mechanisms for the explusion of intestinal helminths. *Parasite Immunology* 16, 333–338.

NAWA, Y. & MILLER, H. R. P. (1979). Adoptive transfer of intestinal mast cell response in rats infected with *Nippostrongylus brasiliensis*. *Cellular Immunology* 42, 225–239.

NEWLANDS, G. F. J., MILLER, H. R. P. & JACKSON, F. (1990). Immune exclusion of *Haemonchus contortus* larvae in the sheep: effects on gastric mucin of immunization, larval challenge and treatment with dexamethasone. *Journal of Comparative Pathology* 102, 433–442.

OWEN, R. L. (1994). M cells – Entryways of opportunity for enteropathogens. *Journal of Experimental Medicine* 180, 7–9.

PATTERSON, D. M., JACKSON, F., HUNTLEY, J. F., STEVENSON, L. M., JONES, D. G., JACKSON, E. & RUSSEL, A. J. F. (1996). Studies on caprine responsiveness to nematodiasis: segregation of male goats into responders and non-responders. *International Journal for Parasitology* 26, 187–194.

PERNTHANER, A., STANKIEWICZ, M., BISSET, S. A., JONAS, W. E., CABAJ, W. & PULFORD, H. D. (1995). The immune responsiveness of Romney sheep selected for resistance or susceptibility to gastrointestinal nematodes: lymphocyte blastogenic activity, eosinophilia and total white blood cell counts. *International Journal for Parasitology* 25, 523–529.

PLOEGER, H. W., KLOOSTERMAN, A. & RIETVELD, F. W. (1995). Acquired immunity against *Cooperia* spp. and *Ostertagia* spp. in calves: effect of level of exposure and timing of the midsummer increase. *Veterinary Parasitology* 58, 61–74.

PRITCHARD, D. I. (1993). Immunity to helminths: is too much IgE parasite- rather than host-protective? *Parasite Immunology* 15, 5–9.

PRITCHARD, D. I., WILLIAMS, D. J., BEHNKE, J. M. & LEE, T. D. (1983). The role of IgG1 hypergamma-globulinaemia in immunity to the gastrointestinal nematode *Nematospiroides dubius*. The immuno-chemical purification, antigen-specificity and *in vivo* anti-parasite effect of IgG$_1$ from immune serum. *Immunology* 49, 353–365.

QUINNELL, R. J. & KEYMER, A. E. (1990). Acquired immunity and epidemiology. In *Parasites: Immunity and Pathology. The Consequences of Parasitic Infection in Mammals* (ed. Behnke, J. M.), pp. 317–343. London, Taylor and Francis.

RAINBIRD, M. A., MacMILLAN, D. & MEEUSEN, E. N. T. (1998). Eosinophil-mediated killing of *Haemonchus contortus* larvae: effect of eosinophil activation and role of antibody, complement and interleukin-5. *Parasite Immunology* **20**, 93–103.

RAMSAY, A. J., HUSBAND, A. J., RAMSHAW, I. A., BAO, S., MATTHAEI, K. I., KOEHLER, G. & KOPF, M. (1994). The role of interleukin 6 in mucosal IgA antibody responses *in vivo*. *Science* **264**, 561–563.

RITCHIE, J. S. D., ANDERSON, N., ARMOUR, J., JARRETT, W. F. H., JENNINGS, F. W. & URQUHART, G. M. (1966). Experimental *Ostertagia ostertagi* infections in calves: parasitology and pathogenesis of a single infection. *American Journal of Veterinary Research* **27**, 659–667.

ROACH, T. I. A., ELSE, K. J., WAKELIN, D., McLAREN, D. J. & GRENCIS, R. K. (1991). *Trichuris muris*: antigen recognition and transfer of immunity in mice by IgA monoclonal antibodies. *Parasite Immunology* **13**, 1–12.

ROTHWELL, T. L. W. (1989). Immune expulsion of parasitic nematodes from the alimentary tract. *International Journal for Parasitology* **19**, 139–168.

ROTHWELL, T. L. W. & DINEEN, J. K. (1972). Cellular reactions in guinea pigs following primary and challenge infection with *Trichostrongylus colubriformis* with special reference to the roles played by eosinophils and basophils in rejection of the parasite. *Immunology* **22**, 733–745.

ROTHWELL, T. L. W., WINDON, R. G., HORSBURGH, B. A. & ANDERSON, B. H. (1993). Relationship between eosinophilia and responsiveness to infection with *Trichostrongylus colubriformis* in sheep. *International Journal for Parasitology* **23**, 203–211.

ROTMAN, H. L., SCHNYDER-CANDRIAN, S., SCOTT, P., NOLAN, T. J., SCHAD, G. A. & ABRAHAM, D. (1997). IL-12 eliminates the Th-2 dependent protective immune response of mice to larval *Strongyloides stercoralis*. *Parasite Immunology* **19**, 29–39.

SCUDAMORE, C. L., PENNINGTON, A. M., THORNTON, E. M., McMILLAN, L., NEWLANDS, G. F. J. & MILLER, H. R. P. (1995). The basal secretion and anaphylactic release of rat mast cell protease II (RMCP-II) from *ex vivo* perfused rat jejunum; translocation of RMCP-II into the gut lumen and its relationship to mucosal histology. *Gut* **37**, 235–241.

SHAW, R. J., GATEHOUSE, T. K. & McNEILL, M. M. (1998). Serum IgE responses during primary and challenge infections of sheep with *Trichostrongylus colubriformis*. *International Journal for Parasitology* **28**, 293–302.

SINSKI, E., BAIRDEN, K., DUNCAN, J. L., EISLER, M. C., HOLMES, P. H., McKELLAR, Q. A., MURRAY, M. & STEAR, M. J. (1995). Local and plasma antibody responses to the parasitic larval stages of the abomasal nematode *Ostertagia circumcincta*. *Veterinary Parasitology* **59**, 107–118.

SMITH, W. D., JACKSON, F., JACKSON, E., GRAHAM, R. & WILLIAMS, J. (1986). Transfer of immunity to *Ostertagia circumcincta* and IgA memory between identical sheep by lymphocytes collected from gastric lymph. *Research in Veterinary Science* **41**, 300–306.

SMITH, W. D., JACKSON, F., JACKSON, E. & WILLIAMS, J. (1983). Local immunity and *Ostertagia circumcincta*: changes in the gastric lymph of immune sheep after a challenge infection. *Journal of Comparative Pathology* **93**, 479–488.

SMITH, W. D., JACKSON, F., JACKSON, E. & WILLIAMS, J.

(1984a). Resistance to *Haemonchus contortus* transferred between genetically histocompatible sheep by immune lymphocytes. *Research in Veterinary Science* **37**, 199–204.

SMITH, W. D., JACKSON, F., JACKSON, E., WILLIAMS, J. & MILLER, H. R. P. (1984b). Manifestations of resistance to ovine ostertagiasis associated with immunological responses in the gastric lymph. *Journal of Comparative Pathology* **94**, 591–601 .

SMITH, W. D., JACKSON, F., JACKSON, E. & WILLIAMS, J. (1985). Age immunity to *Ostertagia circumcincta*: comparison of the local immune responses of 4½- and 10-month-old lambs. *Journal of Comparative Pathology* **95**, 235–245.

SNIDER, T. G. III, WILLIAMS, J. C., KARNS, P. A., ROMAIRE, T. L., TRAMMEL, H. E. & KEARNEY, M. T. (1986). Immunosuppression of lymphocyte blastogenesis in cattle infected with *Ostertagia ostertagi* and/or *Trichostrongylus axei*. *Veterinary Immunology and Immunopathology* **11**, 251–264.

SNIDER, T. G. III, WILLIAMS, J. C., KNOX, J. W., MARBURY, K. S., CROWDER, C. H. & WILLIS, E. R. (1988). Sequential histopathologic changes of type I, pretype II and type II ostertagiasis in cattle. *Veterinary Parasitology* **27**, 169–179.

SNIDER, T. G. III, WILLIAMS, J. C., SHEEHAN, D. S. & FUSELIER, R. H. (1981). Plasma pepsinogen, inhibited larval development, and abomasal lesions in experimental infections of calves with *Ostertagia ostertagi*. *Veterinary Parasitology* **8**, 176–183.

STANKIEWICZ, M., JONAS, W. E., DOUCH, P. C. G., RABEL, B., BISSET, S. & CABAJ, W. (1993). Globule leucocytes in the lumen of the small intestine and the resistance status of sheep infected with parasitic nematodes. *Journal of Parasitology* **79**, 940–945.

STEAR, M. J., BISHOP, S. C., DOLIGALSKA, M., DUNCAN, J. L., HOLMES, P. H., IRVINE, J., McCRIRIE, L., McKELLAR, Q. A., SINSKI, E. & MURRAY, M. (1995b). Regulation of egg production, worm burden, worm length and worm fecundity by host responses in sheep infected with *Ostertagia circumcincta*. *Parasite Immunology* **17**, 643–652.

STEAR, M. J., BISHOP, S. C., DUNCAN, J. L., McKELLAR, Q. A. & MURRAY, M. (1995a). The repeatability of faecal egg counts, peripheral eosinophil counts, and plasma pepsinogen concentrations during deliberate infection with *Ostertagia circumcincta*. *International Journal for Parasitology* **25**, 375–380.

SVETIC, A., MADDEN, K. B., XIA DI ZHOU, PIN LU, KATONA, I. M., FINKELMAN, F. D., URBAN, J. F. JR. & GAUSE, W. C. (1993). A primary intestinal helmintic infection rapidly induces a gut-associated elevation of Th2-associated cytokines & IL3. *Journal of Immunology* **150**, 3434–3441.

TELFORD, G., WHEELER, D. J., APPLEBY, P., BOWEN, J. G. & PRITCHARD, D. I. (1998). *Heligmosomoides polygyrus* immunomodulatory factor (IMF) targets T-lymphocytes. *Parasite Immunology* **20**, 601–611.

THATCHER, E., GERSHWIN, L. J. & BAKER, N. F. (1989). Levels of serum IgE in response to gastrointestinal nematodes in cattle. *Veterinary Parasitology* **32**, 153–161.

ÜBER, C. L., ROTH, R. L. & LEVY, D. A. (1980). Expulsion of

Nippostrongylus brasiliensis by mice deficient in mast cells. *Nature* **287**, 226–228.

URBAN, J. F. JR., KATONA, I. M. & FINKELMAN, F. D. (1991*a*). *Heligmosomoides polygyrus*: CD4+ but not CD8+ T cells regulate the IgE response and protective immunity in mice. *Experimental Parasitology* **73**, 500–511.

URBAN, J. F. JR., KATONA, I. M., PAUL, W. E. & FINKELMAN, F. D. (1991*b*). Interleukin 4 is important in protective immunity to a gastrointestinal nematode infection in mice. *Proceedings of the National Academy of Sciences, USA* **88**, 5513–5518.

URBAN, J. F. JR., MADDEN, K. B., SVETIC, A., CHEEVER, A., TROTTA, P. P., GAUSE, W. C., KATONA, I. M. & FINKELMAN, F. D. (1992). The importance of Th2 cytokines in protective immunity to nematodes. *Immunological Revue* **127**, 205–220.

URBAN, J. F. JR., MALISZEWSKI, C. R., MADDEN, K. B., KATONA, I. M. & FINKELMAN, F. D. (1995). IL4 treatment can cure established gastrointestinal nematode infection in immunocompetent and immunodeficient mice. *Journal of Immunology* **154**, 4675–4684.

VALLANCE, B. A. & COLLINS, S. M. (1998). The effect of nematode infection upon intestinal smooth muscle function. *Parasite Immunology* **20**, 249–253.

VAN DIEMEN, P. M., PLOEGER, H. W., NIEUWLAND, M. G. B., RIETVELD, F. W., EYSKER, M., KOOYMAN, F. N. J., KLOOSTERMAN, A. & PARMENTIER, H. K. (1997). Low molecular weight *Cooperia oncophora* antigens. Potential to discriminate between susceptible and resistant calves after infection. *International Journal for Parasitology* **27**, 587–593.

VERCRUYSSE, J. & DORNY, P. (1999). Integrated control of nematode infections in cattle: a reality? A need? A future? *International Journal for Parasitology* **29**, 165–175.

VERCRUYSSE, J., HILDERSON, H. & CLAEREBOUT, E. (1994). Effect of chemoprophylaxis on immunity to gastrointestinal nematodes in cattle. *Parasitology Today* **10**, 129–132.

WASHBURN, S. M. (1984). *Cellular and humoral responses to* Ostertagia ostertagi *in type I infection*. PhD thesis, Auburn University, USA, pp. 111.

WASHBURN, S. M. & KLESIUS, P. H. (1984). Leukokinesis in bovine ostertagiasis: stimulation of leukocyte migration by *Ostertagia*. *American Journal of Veterinary Research* **45**, 1095–1098.

WEDRYCHOWICZ, H., MacLEAN, J. M. & HOLMES, P. H. (1984). Secretory IgA responses in rats to antigens of various developmental stages of *Nippostrongylus brasiliensis*. *Parasitology* **89**, 145–157.

WIGGIN, C. J. & GIBBS, H. C. (1987). Pathogenesis of simulated natural infections with *Ostertagia ostertagi* in calves. *American Journal of Veterinary Research* **48**, 274–280.

WIGGIN, C. J. & GIBBS, H. C. (1989). Studies of the immunomodulatory effects of low-level infections with *Ostertagia ostertagi* in calves. *American Journal of Veterinary Research* **50**, 1764–1770.

WIGGIN, C. J. & GIBBS, H. C. (1990). Adverse immune reactions and the pathogenesis of *Ostertagia ostertagi* infections in calves. *American Journal of Veterinary Research* **51**, 825–832.

WILLIAMS, D. J. & BEHNKE, J. M. (1983). Host-protective antibodies and serum immunoglobulin isotypes in mice chronically infected with the nematode parasite *Nematospiroides dubius*. *Immunology* **48**, 37–47.

WINTER, M. D., WRIGHT, C. & LEE, D. L. (1997). The mast cell and eosinophil response of young lambs to a primary infection with *Nematodirus battus*. *Parasitology* **114**, 189–193.

WOODBURY, R. G., MILLER, H. R. P., HUNTLEY, J. F., NEWLANDS, G. F. J., PALLISER, A. C. & WAKELIN, D. (1984). Mucosal mast cells are functionally active during spontaneous expulsion of intestinal nematode infections in rats. *Nature* **312**, 450–452.

YONG, W. K., SAUNDERS, M. J., MORTON, R. E., HERMON, M., McGILLIVERY, D. J., HOOPER, I. J., MORGAN, P., RIFFKIN, G. G., CUMMINS, L. J. & THOMPSON, R. M. (1991). Immunology of *Ostertagia* resistance. In *Breeding for Disease Resistance* (eds. Gray, G. D. & Woolaston, R. R.), p. 87, Melbourne, Australian Wool Corporation.

Development of vaccines against gastrointestinal nematodes

D. P. KNOX

Moredun Research Institute, Bush Loan, Penicuik, Midlothian, UK

SUMMARY

Vaccination against complex metazoan parasites has become a reality with the development and registration of recombinant protein-based vaccines against the cattle tick *Boophilus microplus* and the sheep cestode *Taenia ovis*. Progress towards the development of similar vaccines against gastrointestinal nematodes, primarily of ruminants, is outlined within a framework of defining the practical requirements for successful vaccination, antigen selection, recombinant protein production and antigen delivery, be it mucosal delivery or DNA vaccination. Antigen selection strategies include the fractionation of complex, but protective, parasite extracts, the use of antibody probes, evaluation of excretory-secretory components and gut-expressed hidden antigens as well as antigens targeted on the basis of function such as enzyme activity. The difficulties being encountered in recombinant protein production and their solution are discussed as are the requirements for successful antigen delivery. Recent technological developments such as the use of functional genomics to identify new vaccine candidates and DNA vaccination to present the selected antigen to the host immune system are discussed and are anticipated to have a profound effect on vaccine development in the future.

Key words: Gastrointestinal nematodes, vaccination, antigens, DNA vaccines.

INTRODUCTION

Cattle and sheep are continuously exposed to nematode infection and gradually become resistant to reinfection. This latter fact provides the basis for believing that vaccination is feasible. However, the acquisition of immunity is dependent on numerous different influences. Notwithstanding the complexity of the immune response these influences include breed, the genetic make-up of individuals, age and the differing characteristics of co-infecting nematode species. In addition, management factors such as provision of clean pasture and adequate nutrition can greatly influence attempts to control infection. This brief list could be extended and all of these factors need to be considered when devising vaccination strategies. But why bother with vaccination? Are not currently available anthelmintic drugs quite adequate? Anthelmintic resistance is a stark reality as is increasing consumer awareness of food safety with regard to drug residues in meat products. Problems of resistance and consumer hostility are likely to be encountered by new generations of anthelmintics, if these are developed at all because of the escalating costs which have to be offset by profitability in the market place. However, the current utility of existing anthelmintics, although limited in many regions of the world, provides a window of opportunity for the development of alternative strategies. Recent research shows that it is possible to stimulate practically useful protective immunity against blood-feeding nematode species by vaccination, but not, as yet, with recombinant proteins. Results obtained

with non blood-feeding species are less encouraging for reasons which will be discussed.

It is clear that most progress has been made in developing vaccines against gastrointestinal nematodes of sheep despite the fact that cattle are economically more important. This disparity reflects the relative costs of vaccine trials in sheep and cattle and the relative lack of definition of protective immune responses in cattle.

Vaccine development faces several fundamental challenges not least of which is the isolation of native antigens from non blood-feeders which elicit protective immunity if delivered to the immune system in an appropriate manner. In an appropriate manner? Larval and adult gastrointestinal nematodes stimulate mucosal immune responses and are unlikely to be susceptible to systemic vaccination regimes with the exception of blood-feeding species. This points to three fundamental requirements (1) definition of the mucosal immune mechanisms which result in worm expulsion, (2) identification of worm antigens which stimulate these responses and (3) the development of mucosal antigen delivery systems. Without the latter, how do we know that we have not already isolated the appropriate antigens but have not been able to evaluate efficacy properly? Then there is the problem of recombinant protein production. Not as simple as it may appear but more on that later. Finally, to replace anthelmintics, a subunit vaccine will have to be effective against a number of co-infecting nematodes, be cost-effective and its administration will have to fit in with farm management practices. This review, while sum-

Parasitology (2000), **120**, S43–S61.   Printed in the United Kingdom   © 2000 Cambridge University Press

marizing different rationales and progress made to date in identifying and administering potential nematode antigens as vaccine candidates, will also consider future research directions which are required if multivalent vaccines are to become a practical reality replacing, or at least, reducing anthelmintic usage.

VACCINE ATTRIBUTES AND EFFICACY

When considering the practical development of a vaccine against gastrointestinal nematodes it is necessary to define the required 'performance characteristics' of the vaccine. These issues have been addressed in several recent reviews (e.g. Smith & Munn, 1990; Emery, McLure & Wagland, 1993; Newton, 1995; Emery, 1996; Miller, 1996; Meeusen, 1996; Klei, 1997; Munn, 1997; Smith, 1999). The first question to be addressed is the degree of protection required. Perhaps, the best practical definition of required efficacy is 'reducing parasitism below that which causes a significant production loss' (Klei, 1997). It is unlikely that anti-parasite vaccines will attain the almost 100 % efficacy associated with new anthelmintics and bacterial/viral vaccines (Emery, 1996), but evidence obtained by using computer models of population dynamics of host–parasite interactions (Barnes & Dobson, 1990) indicates that adequate control can be achieved with vaccine efficacies well below 100 % (Barnes, Dobson & Barger, 1995). For example, conventional vaccines, i.e. vaccines based on antigens recognized during the course of natural infection, were predicted to be superior to standard anthelmintic programmes with vaccine efficacies of 60 % in 80 % of the flock. These figures were obtained on the assumption that sheep naturally acquire immunity to re-infection when exposed to parasites on pasture. This was assumed to be more beneficial in the case of vaccines based on conventional antigens compared to those based on hidden antigens and this was upheld by the model which predicted a required vaccine efficacy of 80 % in 80 % of the flock. However, this prediction may be somewhat inaccurate in that lambs which were vaccinated against *Haemonchus contortus* using a defined hidden antigen acquired immunity to re-infection following natural exposure to infection (Smith & Smith, 1993). Notwithstanding, it is clear that sterile immunity is not a prerequisite for a nematode vaccine and, in fact, may prove detrimental in the long term. Next, it is necessary that a commercially viable vaccine must be efficacious against all the principal nematode species co-infecting the gastrointestinal tract. This statement may not apply to Southern Hemisphere countries where anthelmintic resistance in populations of *H. contortus* is particularly prevalent. Here, a vaccine providing protection against this parasite alone is likely to have a significant niche market. However, in general terms, we need to be aiming for vaccines with the same cross-species efficacy of existing anthelmintics to ensure product uptake by the agricultural industry. This may be achieved by identifying protective antigens shared by different nematode species or by developing vaccine formulations containing several different species-specific antigens. This requirement may be facilitated by exploitation of the demonstrable cross-specificity of rapid rejection of incoming larvae from immune animals, an effect promulgated distally but not proximally and consistent with the specific recognition of a parasite antigen followed by a non-specific effector phase (Emery, McClure & Wagland, 1993). This may present future opportunities when the precise mechanisms of worm rejection are further defined. Any vaccine must protect young weaner stock, a difficulty to date only overcome using 'hidden antigen'-based vaccines against the blood feeder, *H. contortus*, progress discussed in more detail below. Finally, an effective vaccine will reduce pasture contamination in successive seasons (Emery *et al.* 1993).

EPIDEMIOLOGICAL AND IMMUNOLOGICAL CONSIDERATIONS

Nematode infections occur on a seasonal basis and this seasonality can be attributed to factors affecting contamination of the environment and those controlling development and survival of the free-living stages of the parasites (Urquhart *et al.* 1987). Differing seasonality between species has potential implications for multivalent vaccine development. For example, in Britain, pasture contamination with *Nematodirus battus* larvae shows an abrupt increase in May and June while contamination with *Teladorsagia circumcincta* peaks between July and October; *Trichostongylus* spp. become prevalent in late summer and early autumn. Therefore, exposure to nematode infection is effectively continuous, but growing lambs will require protection against infection with different nematode species depending on the time of year. An efficacious vaccine would have to provide early protection against *Nematodirus*, and, at the same time, prime immunity to incoming *T. circumcincta* and *Trichostrongylus* spp. later in the grazing season. Pasture contamination will be affected by the ability of larvae on pasture to survive adverse environmental conditions, by arrested larval development in the host, parasite fecundity, stocking density and the immune status of the host. In the case of the latter, the periparturient relaxation of immunity is particularly significant in that it results in an increased level of pasture contamination when the numbers of susceptible young stock are greatest.

Any vaccination regime will have to circumvent the phenomenon of immunological unresponsiveness (generally < 3–6 months of age) of young stock to incoming nematode infections. The reasons for this are unclear and may be influenced by parasite or host species (McClure *et al.* 1995) and the degree and rate of exposure to infection but, given that production losses primarily accrue due to infection in young stock, it provides a major hurdle to vaccine development. Protective immunity takes several months to develop and can result in the rejection of incoming larvae, arrested larval development, reduced fecundity and the expulsion of adult worms.

Host susceptibility to nematode infections is, in part, genetically determined to the extent that selective breeding has been investigated as a means of controlling ovine nematodes (Windon, 1990). Nematode resistance, the ability of the host to reduce establishment and/or delay development, as judged by faecal egg counts, has a heritability of about 0·3 in Romney and Merino sheep. Work has progressed to examine the immune mechanisms underlying within- and between-breed differences in resistance and susceptibility which may help to identify phenotypic and genetic markers for these traits. (Beh & Maddox, 1996; Douch *et al.* 1996). Bissett & Morris (1996) argued that the benefits which may accrue following selection based on resistance would be derived from reduced pasture contamination and that resilience, i.e. the ability to maintain relatively undepressed production in the face of parasite challenge, may be an equally useful criterion for selection. Selective breeding programmes will, undoubtedly, contribute to nematode control in the medium term but they are unlikely to have the broadly based and relatively short-term applicability to counteract developing anthelmintic resistance.

The nutritional status of the host has a profound influence on its ability to mount a successful immune response and this, in turn, is likely to affect individual responses to vaccination. Protein and/or nitrogen supplementation of the diet increases the rate of acquisition of immunity and increases resistance to re-infection (Coop & Holmes, 1996; Knox & Steel, 1996). In addition, the periparturient rise in faecal egg output can be dramatically reduced by dietary protein supplementation. Addition of molybdenum to the diet of lambs given trickle challenge infections of either *H. contortus* or *Trichostrongylus vitrinus* markedly reduced worm burdens at the end of the experiment compared to lambs to given a standard diet (Suttle *et al.* 1992 *a, b*). Mean worm burdens were reduced by 78%, a reduction as good as any obtained by vaccination with a purified antigen(s).

VACCINATION WITH WHOLE WORM MATERIAL AND RADIATION-ATTENUATED PARASITES

Rats have been successfully vaccinated against *Strongyloides ratti* by giving a total of 16 000 heat-killed larvae in 13 subcutaneous injections at intervals of 3 days, a regime which gave resistance to challenge infection equivalent to that induced by a single challenge with 1000 living larvae (Sheldon, 1937). However, the approach proved largely unsuccessful against other infections including *Ancylostoma caninum* in mice, *Ascaris suum* in guinea pigs (Kerr, 1938) and *Trichinella spiralis* in rats (Bachman & Molina, 1933). These early experiments led to the conclusion that the protective antigens were not present in sufficient quantity in extracts of dead parasites but might be actively secreted by the living parasite (Clegg & Smith, 1978).

Parasites attenuated by exposure to X-radiation do not achieve patency, do not induce significant pathology but do stimulate immune responses which can be host protective. This approach has resulted in the development of two commercially available anti-nematode vaccines for the control of lungworm in cattle (*Dictyocaulus viviparus*, Jarrett, *et al.* 1958) and sheep *D. filaria*, Sharma, Bhat & Dhar, 1988). In addition, the same approach was effectively applied to the vaccination of dogs against the intestinal nematode *A. caninum* (Miller, 1971) but the resultant commercial product failed due to respiratory side effects, short shelf life and, in some cases, lack of sterile immunity (Miller, 1978). Irradiated larval vaccines protected mature sheep (> 6 months old) against *H. contortus* (Urquhart *et al.* 1966) or *T. colubriformis* (Gregg *et al.* 1978) but, as in natural infection, immature animals did not develop immunity (Urquhart *et al.* 1966; Gregg *et al.* 1978; Smith & Angus, 1980). Pigs have also been protected against *T. spiralis* infection by vaccination with radiation-attenuated muscle larvae (Cabrera & Gould, 1964). Subsequent work has focused on identifying parasite antigens which stimulate protective immunity.

VACCINATION WITH PARASITE ANTIGENS

Nematodes are multicellular organisms with complex lifecycles. Many parasite proteins are recognized by the host during infection, often in a stage-specific manner, but many of the responses to these antigens will have no functional significance in terms of resistance to re-infection (O'Donnell *et al.* 1989). Four main approaches for identifying parasite antigens which stimulate protective immunity have been defined by Emery & Wagland (1991) and these definitions form the basis for the sub-headings used below. These approaches, though different, are often complementary. Antigens can be selected on an empirical basis where parasite material is fractionated by conventional protein chemistry techniques followed by successive immunization trials. Empirical selection tends to be both time consuming and costly so more rational approaches are being applied, where molecules are selected on the basis of their contribution to parasite survival. Antigens studied to date can be defined as conventional or covert (hidden). Conventional antigens are recognized by the host in the course of natural infection, would augment host natural immunity, but may have limited utility due to selection pressure associated with the host immune response (Emery & Wagland, 1991). Covert antigens are not normally accessible to the host immune system and are less likely to be subjected to selection pressure, reducing the risk of antigenic variation. However, repeated vaccination may be required as specific host immune responses would not be boosted by subsequent natural infections. To date, the antigens tested have been purified in small quantities from parasite extracts or *in vitro* culture fluids (ES). Emery & Wagland (1991) estimated that three donor lambs were required to provide sufficient purified antigen to vaccinate one sheep. ES antigens are even more difficult to harvest, requiring donor animals for the provision of parasites. ES antigens exhibit stage-specificity and are often difficult to purify from complex media while protein yields are poor, adult *T. colubiformis* producing about 1 mg ES protein per 25 000 worms per day (Emery, 1996). Therefore, commercial vaccine production will be dependent on the successful application of recombinant DNA technology, be it recombinant protein production, vaccine vectors or DNA vaccination.

Biochemical fractionation of complex parasite extracts

This approach involves the progressively more refined fractionation of crude, but protective, parasite extracts or ES. Fractionation is achieved using parasite extracts prepared in a variety of solvents combined with relatively standard chromatographic and electrophoretic methodologies. The solvent used to prepare the initial extract, be it an aqueous medium such as PBS or detergents such as Triton or Tween, can have a profound effect on the antigens isolated and the complexity of the resultant extract.

The basic approaches to the isolation of protective antigens in parasite extracts are exemplified by a series of experiments in Australia which led to the identification of tropomyosin as a potentially useful antigen for stimulating protective immunity in sheep against *T. colubiformis* and *H. contortus* infections. Immunization with whole worm homogenates from fourth larval stage *T. colubiformis* accelerated expulsion of the same parasite from guinea pigs (Rothwell & Love, 1974) and homogenate sub-fractions obtained using SDS–PAGE, which still comprised a complex mix of proteins, had a similar effect (O'Donnell *et al.* 1985). A PBS/deoxycholate extract from third larval stage *T. colubiformis* comprised four protein components, one of which, a 41 kDa protein, induced 43–51 % protection against

infection in guinea pigs. Partial amino acid sequence analysis showed this protein was tropomyosin and a cDNA encoding this protein was isolated using oligonucleotide primers based on this sequence (Cobon *et al.* 1989; O'Donnell *et al.* 1989). A 27 kDa subunit was expressed as a β-galactosidase fusion protein in *E. coli* which produced accelerated worm expulsion in guinea pigs following challenge infection. A homologue from *H. contortus*, isolated by DNA hybridization, significantly protected sheep against challenge infection (Cobon *et al.* 1989).

A slightly different approach was used to identify a protective protein fraction from *Oesophagostomum radiatum* which infects the large intestine of calves. Calves acquire strong resistance to re-infection and the protective response primarily targets the late fourth larval stage (Roberts, Elek & Keith 1962). Protective immunity was induced by vaccination with L₄ or adult parasite whole worm homogenates or using excretory gland homogenates from the adult parasite (Keith & Bremner, 1973). Adult parasite extracts were resolved into four fractions using gel filtration, and ELISA analysis showed that antibodies from naturally infected calves predominantly reacted with antigens eluting in the void volume (East, Berrie & Fitzgerald, 1989). Calves immunized with this fraction were significantly protected against challenge infection (East *et al.* 1989). High molecular weight antigen preparations have also been used to vaccinate against *Heligmosomoides polygyrus* (*Nematospiroides dubius*) infection in mice (Monroy & Dobson, 1987), *T. colubiformis* (O'Donnell *et al.* 1985) and *H. contortus* (Neilson & Van de Walle, 1987) in sheep with varying degrees of success.

Isolation of parasite molecules essential for parasite maintenance within the host

This category includes molecules released by the parasite during *in vitro* culture (ES) which are either known to be released or presumed to be released *in vivo*. ES have been ascribed numerous roles including host penetration, parasite feeding and evasion of host anti-parasite immune responses and include enzymes such as proteases, acetylcholinesterases and superoxide dismutases (reviewed by Knox, 1998). In addition, molecules with the capacity to directly alter the local gastrointestinal environment have been identified including a homologue of a porcine intestinal peptide, valosin, secreted by *T. colubiformis* (Savin *et al.* 1990) and a serine proteinase inhibitor with homology to human leukocyte elastase inhibitor from L₄ and adult *Trichostrongylus vitrinus* (MacLennan, K. & Knox, D. P., unpublished).

ES may be derived from the parasite surface, from specialized secretory glands or as by-products of parasite digestion and represent the major antigenic and functional challenge to the host (Lightowlers &

Rickard, 1988). ES are often released in a stage-specific manner. Despite extensive study and significant early successes, surprisingly few ES components have been evaluated as potential as potential protective immunogens.

Antigens purified from L_3 and adult ES from *H. contortus* and *T. colubriformis* have shown considerable potential in trials primarily conducted by Australian workers (Emery, 1996). *T. colubriformis* L_4 and adult ES antigens were fractionated, mainly by SDS–PAGE, and several induced protection in guinea pigs to challenge infection following intra-peritoneal immunization. cDNA analysis showed that the proteins were homologues of a human γ-interferon-induced protein (Dopheide *et al.* 1991), human and insect globins (Frenkel *et al.* 1992) and a porcine intestinal peptide, valosin (Savin *et al.* 1990) and recombinant forms of these proteins gave 30–70% protection against homologous parasite challenge (Emery, 1996). A 94 kDa glycoprotein from *T. colubriformis* L_3 induced variable protection of 30–50% in guinea pigs (O'Donnell *et al.* 1989).

Other ES proteins from ruminant gastrointestinal nematodes which are useful immunogens have been identified using antibody probes from immune animals and are discussed in the next section.

Nippostrongylus brasiliensis and *T. spiralis* have long been used as rodent models for studying gastrointestinal nematode infections. When *N. brasiliensis* was maintained in serum from immune rats, precipitates were formed at the mouth, excretory pore and anus of the worm (Sarles, 1938; Taliaferro & Sarles, 1939) and similar precipitates were detected histologically around the orifices of larval parasites in the skin and lungs of immune rats. These precipitates did not appear to damage the worm directly but may have blocked a secretion, such as an enzyme, required for survival in the host. Rats could be partly protected against *N. brasiliensis* challenge by vaccination with ES released by larval parasites into serum or saline during *in vitro* maintenance (Thorson, 1953) and there was evidence that protection could be enhanced by direct delivery of antigen to the gut. Lipolytic activity in larval ES was inhibited by antiserum from immune rats. (Thorson, 1954) and dogs were partially protected against *A. caninum* challenge by vaccination with oesophageal gland extracts from the same parasite (Thorson, 1956). These extracts contained proteolytic and lipolytic enzymes the activities of which could be inhibited with immune dog serum.

Attention again focused on enzymes in nematode secretions in the 1970s when it was shown that a wide range of gastrointestinal nematodes released acetylcholinesterase (AChE) *in vitro* and that immune animals produce antibodies against the enzyme (Ogilvie *et al.* 1973). As immunity develops, the parasite has the ability to secrete an alternative isoenzyme in greater quantities (Edwards, Burt &

Ogilvie, 1971; Jones & Ogilvie, 1972). Rothwell & Merritt (1975) vaccinated guinea pigs with soluble fractions from *T. colubriformis* greatly enriched for AChE but also containing an undefined low molecular weight allergen. Only the latter was protective. However, it is interesting to note that cattle vaccinated against the bovine lungworm *D. viviparus* produced antibodies which inhibited AChE released by the parasite *in vitro* and a partially purified preparation of this enzyme induced protective immunity in guinea pigs against subsequent challenge (McKeand *et al.* 1995).

Proteases are released during *in vitro* culture of many parasitic helminths which are required for penetration of the host and survival within it (Tort *et al.* 1999). *In vitro*-released proteases from *Fasciola hepatica* induced high levels of protection in sheep (Wijffels *et al.* 1994) and cattle (Dalton *et al.* 1997) but equivalent proteases from gastrointestinal nematodes have not been extensively tested to date. A 35 kDa cysteine protease isolated from glycerol extracts of *H. contortus* conferred significant protection to lambs against challenge infection with the same parasite (Cox *et al.* 1990; Boisvenue *et al.* 1992). The protease had anticoagulant properties and was capable of degrading fibrinogen (Cox *et al.* 1990) but it was not clearly established if the protease was an ES component or a gut-associated enzyme. Recently, Knox, Smith & Smith (1999) reported several trials where the protective efficacy of cysteine protease-enriched extracts from the adult parasite was analysed, enrichment being achieved by thiol-sepharose affinity chromatography. Water-soluble and membrane-associated cysteine proteases were without effect as immunogens and these extracts may have contained cysteine proteases known to be present in adult parasite *in vitro* ES (Karanu *et al.* 1993). This implies that ES cysteine proteases may not be useful immunogens but this would require confirmation by conducting vaccine trials with proteases purified from ES alone.

Glutathione S-transferases (GSTs) induced significant reductions in parasite numbers when used as immunogens in sheep and cattle challenged with *Fasciola hepatica* (Sexton *et al.* 1990; Morrison *et al.* 1996). Antisera raised against GST purified from adult *H. contortus* inhibited enzyme activity *in vitro* but did not affect survival of the parasite *in vitro* (Sharp *et al.* 1991). In parallel experiments, a specific GST inhibitor reduced parasite survival *in vitro* (Sharp *et al.* 1991).

Early reports (Campbell, 1955; Chipman, 1957) indicated that antigenic material secreted by adult and larval *T. spiralis* contained potentially useful immunogens. Despommier & Muller, (1970) showed that a fraction containing mainly β secretory granules derived from the larval stichosome administered once to mice in Freund's complete adjuvant depressed the number of encysted larvae in the muscle of recipients by 95 % following challenge. Subsequent experiments (Despommier, Campbell & Blair, 1977) confirmed these results and indicated that this method of vaccination mainly affected fecundity of the female worm. In further experiments (Despommier & Laccetti, 1981 *a*, *b*), a solubilised extract from the large particle component of the stichocyte (designated S3) and an immunoaffinity purified sub-fraction induced protective immune responses in mice. Despommier (1981) used molecular size chromatography and preparative isoelectric focusing to purify one antigen to homogeneity enabling analysis of amino acid content. Subsequently, monoclonal antibodies were used to isolate 3 antigens, one of which, a 48 kDa protein, induced a high level of protection in mice against challenge infection, protection being comparable to that elicited by exposure to the entire infection (Silberstein & Despommier, 1984).

Swine trichinosis is an important zoonotic infection and it has been established that pigs can expel adult worms from the intestine and do develop resistance to reinfection (Bachman & Molina, 1933; Campbell & Cuckler, 1966). Pigs can be protected by vaccination with radiation-attenuated muscle larvae (Cabrera & Gould, 1964) and with ES antigens from muscle larvae (Vernes, 1976), the latter antigens also being present in S3 stichocyte antigen preparations used by Despommier (1981). Muscle larvae numbers were reduced by about 50 % compared to challenge controls in pigs immunized with S3 antigens, immunization only marginally enhancing gut expulsion (Murrell & Despommier, 1984). The 48 kDa stichocyte antigen described above was ineffective in the natural pig host (Gamble, Murrell & Marti, 1986).

Perhaps amongst the most interesting but least defined group of ES antigens are those released during the moulting process. A very early study (Stoll, 1929) suggested that immune responses capable of reducing egg production and resulting in the expulsion of adult worms were initiated by an early parasite stage in sheep challenged with *H. contortus*. The suggestion was supported later (Stewart, 1953) and expulsion occurred when the larvae in the challenge infection were moulting from the third to the fourth larval stage (Soulsby & Stewart, 1960). This 'self cure' response was also observed in pigs infected with *A. suum* (Taffs, 1968). These early experiments led to several protection trials in different host-parasite systems which give equivocal results. Silverman, Poynter & Podger (1962) protected guinea pigs (> 90%) against *T. colubriformis* challenge infection by immunization with lyophilized whole larvae and ES derived from *in vitro* cultures where the larvae were maintained from the second to the fourth larval stage. However, later experiments (Rothwell & Love, 1974) showed that protective antigens were present in similar

amounts in fourth-stage larval and adult worms and could be readily obtained from whole worm homogenates. A high molecular weight fraction of antigenic material secreted during the *in vitro* moult from third to fourth larval stage *Haemonchus* protected (70–80%) lambs of undefined age against challenge infection (Ozerol & Silverman, 1970) but Neilson (1975) could not reproduce this effect. However, the antigens used in the two studies were not exactly comparable. In addition, antigen in the former study was administered without adjuvant and protective immune responses initiated may have been due to self adjuvanting and been different from those initiated by injection with Freund's adjuvant used in the latter study. Trials of similar material from *Ascaris suum* also gave some conflicting results when used to vaccinate mice prior to challenge (Guerro & Silverman, 1969). Secretions produced during the moult from third to fourth larval stage *A. suum* induced significant protection against challenge in guinea pigs while soluble proteins produced by L_2, L_3, L_4 or adult worms during culture were without effect (Stromberg, Rhoury & Soulsby, 1977).

The surface of parasitic nematodes is a dynamic structure; surface antigens are shed continuously and are often highly antigenic (Maizels, Blaxter & Selkirk, 1993). Cetylmethylammonium bromide (CTAB)-solubilized surface antigens from *T. spiralis* induced significant protection in mice (Grencis *et al.* 1986). However, surface extracts from *H. contortus* L_3 were ineffective (Turnbull *et al.* 1992) and cuticle collagens from third and fourth larval stages of the same parasite did not induce significant protection against homologus challenge (Boisvenue *et al.* 1992). Recently, Jacobs *et al.* (1999) induced protective immunity in sheep against *H. contortus* using a surface antigen purified from the infective L_3 stage and showed that efficacy was dependent on route of administration and adjuvant used.

Antigen selection using antibodies or cells from immune animals

The use of antibodies derived from infected and/or immune animals has, to date, proved the method of choice to identify antigens associated with protective immunity. However, this is fraught with difficulty because of the complexity of the serum response, heterogeneous antigen expression by the parasite within and between stages within the host and the fact that the antibody response reflects both current and previous infections. Despite these difficulties, some progress has been made. As described above, monoclonal antibodies derived from immune mice were used to identify and purify the major immunogens of *T. spiralis* and one of these proteins (M_r 48 kDa), at very low doses, induced significant infection in mice against infection (Silberstein &

Despommier, 1984). Sera from sheep which were defined as resistant or susceptible to *T. circumcincta* were used to identify a 31 kDa protein which was recognised preferentially by resistant animals as early as 3 weeks after experimental infection and was present in and secreted by third larval stage parasites (McGillivery *et al.* 1992). Lambs immunized with the protein were significantly protected against challenge infection (McGillivery *et al.* 1992) but this effect could not be reproduced (Morton *et al.* 1995).

Recently, attention has begun to focus on the development of methods which allow the identification of parasite antigens that trigger local cellular and antibody responses such as T cell Westerns (Haig *et al.* 1989), mast cell stimulation assays (Emery, McClure & Wagland, 1993) and antibody probes derived from lymph nodes draining the site of infection (Meeusen & Brandon, 1994*a*; Meeusen, 1996; Newton & Munn, 1999). These techniques should help to identify the parasite antigens which stimulate immune responses specifically responsible for worm rejection.

Antibody secreting B cells (ASCs) are produced by lymph nodes draining the site of infection in *Taenia hydatigena* infected sheep and form distinct foci around the challenge parasites (Meeusen & Brandon, 1994*a*). These ASCs secreted antibodies with specificity for the susceptible larval stages of the parasite. In rats experimentally infected with liver fluke, ASCs are only produced as long as the parasite is present within the tissue drained by the node and different nodes within the same animal can show contrasting antibody specificities dependent on the parasite stage present in the draining tissues (Meeusen & Brandon, 1994*b*). Techniques have now been developed to harvest ASCs from sheep immune to gastrointestinal nematode infection and maintain them *in vivo* (Meeusen, 1996). Animals immunized by repeated infection are given a single large challenge infection after a period of rest from a priming infection to allow the local cellular response to end. The large challenge infection induces the rapid activation of lymphocytes, including ASCs, in the draining lymph nodes with specificity for antigens presented by the challenge infection. These lymph nodes can be harvested at a time which coincides with parasite rejection and the lymphocytes they contain cultured *in vitro* for at least 5 days without further stimulation. The *in vivo*-induced ASCs continue to secrete antibodies which can then be harvested from the culture fluids and used directly on Western blots, for example, to identify parasite antigens associated with the period of parasite rejection (Meeusen, 1996).

ASC antibody probes have now been used to identify protective antigens from *H. contortus* (Jacobs *et al.* 1999) and *T. circumcincta* (Meeusen, 1995). Antibodies produced by ASCs from abomasal lymph nodes of sheep immune to *H. contortus*

infection recognized L_3 homogenate antigens in the size ranges 44–48 kDa and 70–83 kDa, the latter being purified using immuno-affinity media prepared using ASC antibodies (Bowles, Brandon & Meeusen, 1995). The antigen, designated Hc-sL3, is glycosylated (Ashman *et al.* 1995), developmentally regulated (Raleigh, Brandon & Meeusen, 1996; Raleigh & Meeusen, 1996), is expressed on the surface of exsheathed L_3 (Bowles *et al.* 1995) and can be purified using size-exclusion chromatography (Ashman *et al.* 1995). Merino sheep vaccinated with Hc-sL3 showed 64–69 % and 45–55 % reductions in faecal egg outputs and adult worm burdens respectively after a single challenge with 10000 L_3 (Jacobs *et al.* 1999). The choice of adjuvant is crucial; protection can be induced with aluminium hydroxide and infection exacerbated with Quil A as adjuvant indicating that protection is dependent on the induction of a T helper 2 (Th$_2$)-type response (Newton & Munn, 1999). Studies on the mechanism of rejection (Ashman *et al.* 1995; Rainbird, MacMillan & Meeusen, 1998) indicated that protection was unrelated to IgE-dependent immediate hypersensitivity mechanisms thought to be responsible for rapid expulsion (Newton & Munn, 1999).

The same approach has been used to identify equivalent antigens, in the size range 26–36 kDa from *Teladorsagia circumcincta* (Meeusen, 1995; Raleigh *et al.* 1996) and *T. colubriformis* (Meeusen, 1995). The *T. circumcincta* proteins, purified from gels, induced 40–60 % reductions in faecal egg outputs from lambs given the antigen with Quill A as adjuvant (Meeusen, 1995). These proteins included a prominent doublet at 31–33 kDa and a monoclonal antibody to this doublet was used to isolate cDNAs which encoded tandem-repeat-type β-galactoside binding lectins (galectins; Newton *et al.* 1997). *T. colubriformis* galectin cDNA clones have also been analysed and defined (Greenhalgh, Beckham & Newton, 1999).

By comparing serum antibody responses of sheep partly immune to primary and secondary infections with *H. contortus*, Schallig *et al.* (1994) identified 15 and 24 kDa adult ES antigens specifically recognized by immune animals. These antigens have subsequently been evaluated in protection trials in partly purified form (Schallig & van Leeuwen, 1997) and as essentially pure preparations (Schallig, van Leeuwen & Cornellisen, 1997) giving reductions in faecal egg output and final worm burdens of 77 % and 85 %, respectively. Both antigens are exclusively expressed in the L_4 and adult parasite, and cDNAs encoding them have been isolated and expressed in *E. coli*. The cDNA sequences showed some homology to ES components identified from *T. colubriformis* and *A. caninum*. Early data indicate that denatured, then refolded recombinant versions of these proteins do induce high levels of protection in 8-month-old lambs (Vervelde *et al.* 1999*a*). The same workers reported that protection levels increased with increasing lamb age and that protection was correlated to ES-specific serum IgE levels and increases in abomasal mast cell and eosinophil numbers (Vervelde *et al.* 1999*b*).

Isolation of specific antigens from parasite organs that may serve as targets for antibody or immune effector cells

Activity of host antibodies against the parasite gut wall is well documented. Structural changes in the nematode intestine have been attributed to the host immune response in the rodent *N. brasiliensis* system (Ogilvie & Hockley, 1968) and in *N. battus* harvested from lambs (Lee & Martin, 1980). In addition, Seesee, Wescott & Graham (1976) demonstrated antibodies of various immunoglobulin classes bound to sections of the parasite gut, particularily the microvilli of the intestinal cells, in *N. brasiliensis* retrieved from mice. Although not an obligate blood feeder, this observation might indicate that adult *N. brasiliensis* does, in part, feed on host blood components, possibly as a result of ingesting plasma proteins available at the intestinal mucosal surface as a result of serous exudation resulting from the host inflammatory response to the parasite. Moreover, other economically important genera such as *Ostertagia* and *Teladorsagia*, again not obligate blood feeders, do take in host immunoglobulin (Murray & Smith, 1994). The late larval and adult stages of *H. contortus* actively ingest host blood and artificially-induced systemic antibody responses to these components are clearly detrimental to the parasite (Munn, Graham & Coadwell, 1987, Newton & Munn, 1999).

This approach to the immunological control of blood-feeding endo- and ecto-parasites has received practical confirmation in a number of host–parasite systems and includes the identification of covert (hidden) antigens. It is exemplified by the pioneering work of Munn and co-workers who have isolated several protective antigens from the intestinal luminal surface of blood-feeding stages of *H. contortus* (Munn, 1977; Munn & Greenwood, 1984; Munn, Graham & Coadwell, 1987; Munn & Smith, 1990). It is worthy of note that the initial identification of these proteins was as a result of electronmicroscopy studies of the structure of the intestine of *Haemonchus* and other strongyles (Munn, 1977; Munn & Greenwood, 1984). Targeting gut membrane proteins was pioneered in ticks and this work led to the launch of a commercial vaccine against *Boophilus microplus*, the Australian cattle tick (Willadsen, 1995).

Lambs have been successfully immunized against haemonchosis using several gut antigens fractionated from adult *H. contortus*. A helical polymeric structure, termed contortin (Munn, 1977), loosely associated with the luminal surface of the plasma membrane of the intestinal epithelium from early in the fourth larval stage, was isolated in relatively pure form from saline extracts of the adult parasites by differential centrifugation (Munn *et al.* 1987). Young lambs immunized with this contortin-enriched preparation (CEP) were substantially protected (78·5 % reduction in worm burden) against challenge with the homologous parasite (Munn *et al.* 1987). Lambs which did not mount a significant antibody response to the preparation died of acute haemonchosis, suggesting that antibody was the prime effector mechanism for protection. This work showed, for the first time, that proteins expressed on the surface of the gut could induce protective immune responses and stimulated the search for other antigens associated with the parasite intestine luminal surface.

A major component of the microvilli of adult *H. contortus* is an integral membrane glycoprotein of $M_r = 110$ kDa, designated H11 and this was found to be a major antigenic contaminant of CEP by Western blot analysis. However, only traces of H11 were detectable on gels stained for protein suggesting it could be a 'strong' protective antigen (Smith & Munn, 1990). Injection of sheep with microgram amounts of esssentially pure H11 stimulated substantial (> 90 % reductions in worm burden) protection to challenge infection (Munn & Smith, 1990) which was closely correlated with specific systemic IgG titres to H11.

H11 is an integral membrane glycoprotein expressed exclusively in the intestinal microvilli of the parasitic stages. It can be purified from detergent extracts of the adult parasite using a combination of lectin-affinity and ion-exchange chromatography (Smith & Smith, 1993; Munn *et al.* 1997) and was defined as a microsomal aminopeptidase by cDNA cloning (Smith *et al.* 1997). The native protein exhibits both aminopeptidase A and M-type activities which are attributable to distinct isoforms (Graham *et al.* 1993; Smith *et al.* 1997). On a more historical note, using aminopeptidase purified from the gut of *Ascaris suum*, Ferguson *et al.* (1969) noted a 50 % reduction in larval counts in guinea pigs immunized with the enzyme prior to homologous challenge.

Addressing the practical requirements for a vaccine in the field, H11 is an effective immunogen in very young lambs (Tavernor *et al.* 1992); it is effective in a range of breeds and against anthelmintic-resistant worms (Newton & Munn, 1999). Protection persists for at least 23 weeks after vaccination (Andrews, Rolph & Munn, 1997) and does not interfere with the development of acquired immunity (Smith & Smith, 1993). Pregnant ewes, challenged in their third trimester, were highly (98 %) protected as judged by faecal egg output and a degree of protective immunity was transferred from vaccinated ewes to lambs challenged 5 weeks post-partum (Andrews *et al.* 1995). This latter result indicated that protection was antibody-mediated, a conclusion supported by observations that antisera from vaccinated lambs inhibited the microsomal aminopeptidase activities and that the degree of inhibition was highly correlated to the level of protection obtained (Munn *et al.* 1997). Protection is reduced when H11 is progressively denatured by treatment with SDS alone or SDS + dithiothreitol (Munn *et al.* 1997), results which indicate that conformational epitopes are required for the full expression of protective immunity.

As noted above, cDNAs encoding all three isoforms of H11 have been cloned and have been expressed as full-length enzymically-active recombinant proteins in the baculovirus-S*f*9 insect cell system and defined fragments of the extracellular domains have been expressed in *E. coli* (M. Graham, T. S. Smith and E. A. Munn, unpublished) and the resultant recombinant proteins are currently being evaluated in vaccine trials.

Homologues of H11 have been identified in *T. circumcincta* and *O. ostertagi* and are also being evaluated in protection trials (E. A. Munn and W. D. Smith, unpublished).

Smith, Smith & Murray (1994) specifically labelled glycoproteins on the luminal surface of the gut of adult *H. contortus* with a panel of lectins and also with [^{35}S]methionine. Lectins with affinity for *N*-acetylgalactosamine stained the brush-border membrane of the intestinal cells most intensively and this property was susbsequently used to purify candidate surface proteins by affinity chromatography. The protein complex obtained, designated *Haemonchus* galactose-containing protein complex (H-gal-GP), conferred substantial protective immunity (reductions of 93% in egg output and 72% in worm burdens) on lambs 2–12 months old in a series of protection trials (Smith *et al.* 1994; Smith & Smith, 1996). The complex comprised groups of peptides around 35, 45, 170 and 230 kDa as judged by non-reducing SDS–PAGE. Protection was decreased by reduction of the complex with SDS and DTT prior to immunization and subfractions of the complex prepared by SDS–PAGE yielded partial protection suggesting that more than one component is required for effective vaccination (Smith & Smith, 1996). Protection trials with individual peptides prepared by electroelution form SDS- and native-PAGE and by chromatographic fractionation are continuing. ELISA data indicate that protection is antibody mediated (Smith *et al.* 1999).

Initially, little was known about the functional properties of H-gal-GP. Biochemical studies indicated that it contained proteolytic activity with aspartyl protease activity predominating at acidic pH and metallo-protease activity at neutral pH (Smith *et al.* 1993). These initial findings have been confirmed by a combination of N-terminal sequence analysis of

the individual peptide components and analysis of cDNAs encoding them (Longbottom *et al.* 1997; Redmond *et al.* 1997; Smith *et al.* 1999) and recent work has shown that the complex also contains galectin (Newlands *et al.* 1999), cystatin and thombospondin homologues (Newlands, Skuce, Smith, Smith & Knox, unpublished). Several cDNAs have been expressed in bacteria and yeast and the resultant recombinant proteins are currently being evaluated in protection trials.

Extracts (Knox, Redmond & Jones, 1993) and ES (Karanu *et al.* 1993) from adult *H. contortus* contain several cysteine proteases. As noted above, cysteine protease-enriched preparations from water soluble and membrane-asociated adult parasite extracts were not protective but lambs vaccinated with equivalent preparations from integral membrane protein extracts showed reduced worm burdens and faecal egg outputs of 47 and 77%, respectively in three trials in lambs 3–10 months old at the start of the experiment (Knox *et al.* 1999). The protective proteins, termed thiol sepharose binding proteins (TSBP), are expressed on the microvillar surface and in the cytosol of the intestinal cells and are not recognized by lambs immunized against infection by repeated challenge. The extract comprises a relatively simple group of peptides with a 60 kDa component particularly prominent. To date, full-length cDNAs encoding 3 distinct cathepsin B-like cysteine proteases (Skuce *et al.* 1999*a*) and a glutamate dehydrogenase homologue (Skuce *et al.* 1999*b*) have been isolated and expressed and, again, are being evaluated in protection trials.

O. ostertagi and *T. circumcincta* both contain close homologues of H-gal-GP and TSBP as judged by SDS–PAGE and enzyme analyses, and both induce variable degrees of protection in homologous vaccination/challenge experiments (Knox *et al.* 1995; W. D. Smith & D. P. Knox, unpublished). In some trials, reduced worm burdens and egg outputs of 50% and 85%, respectively have been observed while no significant effects have been noted in other trials. The reasons for this variation are unclear and may reflect antigen stability, dose or may simply indicate that antigens expressed on the surface of worms which are not obligate blood feeders are not appropriate vaccine candidates. Perhaps the worms simply ingest insufficient antibody!

Jasmer and co-workers (Jasmer *et al.* 1993) prepared monoclonal antibodies (mAbs) to target glycoproteins expressed on the gut surface of adult *Haemonchus*. Two of 49 mAbs were analysed further and both bound to the microvillar surface of freshly isolated gut and had specificity for carbohydrate epitopes. Antigen isolated from detergent extracts of adult worms by immunoaffinity chromatography using both mAbs significantly reduced total worm counts by up to 60% and, unusually, to a lesser extent, faecal egg output (34%) in 14-month-old goats given a single challenge infection 5000 L_3. The relevant proteins had M_r values of 100, 52, 46 and 30 kDa. One of the carbohydrate epitopes recognized was also detected in larval *A. caninum* and a mixed population of *C. elegans*. Molecular studies revealed that the 52 and 46 kDa proteins are products of the same gene (designated *GA1*) and are expressed initially as a polyprotein (Jasmer, Perryman & McGuire, 1996). The 52 kDa component shows 47 % sequence identity to the 46 kDa protein and carries a glycerophosphatidyl inositol anchor. The protein can be detected in abomasal mucus of infected lambs and may, in fact, be an antigen recognized by the host in the course of natural infection. These antigens show some similarities to another complex of proteins designated P1 (Smith *et al.* 1993) which comprised peptides of M_r 45, 49 and 53 kDa. Vaccination of lambs with P1 resulted in a 30 % reduction in worm numbers, with the effect being greater in female worms compared to males, and a 69 % reduction in faecal egg output.

RECOMBINANT PROTEIN PRODUCTION

Candidate protective antigens have been isolated and characterized from many gastrointestinal nematodes of veterinary importance and the challenge now is to produce these antigens in an immunologically active form using recombinant DNA technology and to show that the recombinant (sub-unit) vaccine is effective in field conditions. These steps are the key in converting laboratory developments into a commercially available vaccine. This has already been achieved, and commercial vaccines produced, for *Taenia ovis* infections in sheep and *Boophilus microplus* infections of cattle (reviewed by Rickard *et al.* 1995; Willadsen, 1995). There are useful lessons to be learned from these developments. Firstly, strong reactivity in the immunoassay chosen for antigen selection does not always correlate with protective efficacy. A weakly immunopositive cDNA (45W) encoding the host protective 47/52 kDa doublet from *T. ovis* isolated from a λgt11 expression library induced much higher levels of protection when administered as a β-galactosidase fusion protein expressed in *E. coli* compared to a strongly immunopositive cDNA (45S) encoding a smaller portion of the protein (Lightowlers, 1994). The cDNAs encoding the protein of interest should be sub-cloned into a variety of bacterial plasmid vectors to determine efficacy. Neither β-galactosidase fusion proteins stimulated protective immunity against *T. ovis* challenge in recipient sheep while 45W, expressed as a GST fusion protein, was effective (Rickard *et al.* 1995). The GST component was cleaved with thrombin following expression in pGEX-2T and the resultant 45W peptide was equally effective as the GST fusion construct. One problem encountered was that yields of the fusion

protein were low and it was unstable, but this was overcome by modifying the C-terminal end of the protein (Rickard *et al.* 1995). An additional problem encountered was that the GST-fusion was expressed in *E. coli* mostly as insoluble inclusion bodies but this was overcome by solubilisation in urea and its subsequent removal.

The development of the Bm86 antigen-based vaccine against *B. microplus* has involved similar steps (Willadsen, 1995). The antigen is an 89 kDa glycoprotein with an extracellular location on the digest cells of the tick gut. In developing the vaccine, a variety of recombinant proteins was tested, including an *E. coli* expressed β-galactosidase fusion protein that gave significant protection but was not as effective as the native protein (Rand *et al.* 1989). Alternative constructs, baculovirus expression and downstream processing of the bacterially-expressed protein were all tested and yielded products with vaccine efficacies comparable to the native protein (Tellam *et al.* 1992). In addition, Bm86 has been expressed in *Aspergillus* at low levels (Turnbull *et al.* 1990) and in the yeast *Pichia pastoris* (Rodriguez *et al.* 1994), the later being an effective immunogen (Rodriguez *et al.* 1994).

Bacterially-expressed recombinant versions of several gastrointestinal nematode antigens have been produced which induce protective immunity (Emery, 1996). However, difficulties can be anticipated where protective epitopes are conformational in nature, for example the active site region of an enzyme, where glycosylation contributes to protection or, in the case of complexes such as H-gal-GP, where more than one protein may be required to stimulate the full protective response. These difficulties are being addressed by expressing parasite antigens in baculovirus (e.g. H11), in yeast vectors (e.g. cathepsin Ls from *F. hepatica* (Dowd *et al.* 1997)), in the free-living nematode *C. elegans* (D. L. Redmond & D. P. Knox, unpublished) and by testing combinations of recombinant proteins. DNA vaccination may provide the answer (see below).

Another option being investigated is to identify regions of the target molecule which are likely to be accessible to immune effectors and are, in themselves, immunogenic. This can be achieved by, for example, computer modelling of protein structures from primary amino acid sequence data. Regions exposed on the surface of the protein can then be expressed as linear peptides in bacteria and immunogenicity evaluated. This approach mirrors the evaluation of synthetic peptide vaccines which comprise short regions of the protein target. In the case of the *T. ovis* 45W antigen, synthetic peptides were synthesized corresponding to putative host-protective regions at the N- and C-terminal ends of the protein. The N-terminal peptide was clearly more immunogenic and only antibodies to this bound to oncosphere antigens. (Dadleymoore *et al.* 1999).

Finally, the method by which a host-protective recombinant protein is produced must be amenable to commercial scale-up and the protein readily purified, preferably by a one-step procedure. Production costs, their effects on profitability and vaccine cost per dose to the farmer will, ultimately, dictate whether a vaccine ever reaches the market place.

FUNCTIONAL GENOMICS AND ANTIGEN SELECTION

The combination of protein analysis (proteomics), be it by peptide sequencing or analyses of biochemical/immunological function, and gene sequence analyses (genomics) has proven to be a very powerful combination (functional genomics) in the efforts to develop recombinant protein-based vaccines. The precise definition of the peptide components of antigenic mixtures can be exploited to devise more refined fractionation procedures such as substrate/inhibitor-based affinity chromatography for enzymes or lactose solubilization for galectins (Newlands *et al.* 1999). In addition, knowledge of protein function defines the likely requirements for recombinant protein expression. Functional genomics has an even wider application in identifying potential protein targets for vaccination or novel drug targeting. Genome sequencing, expressed sequence tag (EST) analyses and direct protein sequencing have and will continue to provide a range of potential protein targets for further analysis. From the vaccinologists point of view, the first requirement will be that the chosen target molecule is accessible to an immune response stimulated by vaccination. Immunolocalization and *in situ* hybridization analyses can provide an early indication of accessibility exemplified, to an extent, by fractionation trials conducted with *H. contortus* TSBP (Knox *et al.* 1999; Skuce *et al.* 1999 *a*). The prominent component of TSBP is a 60 kDa protein now identified as a GDH by cDNA sequence analysis and subsequent biochemical analyses (Skuce *et al.* 1999 *a*). This protein has been purified, almost to homogeneity, and is not host-protective (Skuce, Smith & Knox, unpublished). Immunolocalization studies demonstrated that GDH was expressed in the cytoplasm of the intestinal cells, not on the microvillar surface, and hence would not be accessible to circulating antibody (Skuce *et al.* 1999 *a*).

MUCOSAL IMMUNE RESPONSES AND ANTIGEN DELIVERY

In many of the host–parasite systems referred to above, the precise mechanisms of worm rejection remain unclear. What is known has been extensively reviewed in recent years (e.g. Miller, 1984; Rothwell, 1989; McClure & Emery, 1994; Miller, 1996). The hidden antigen approach is successful against blood-feeding nematodes but is ineffective against non-blood feeders such as *N. battus* (Smith, 1993) and *T. colubriformis* (Emery, 1996) and, to date, has had only partial efficacy against *Ostertagia* and *Teladorsagia* genera (W. D. Smith & D. P. Knox, unpublished). These parasites contain equivalent antigens but may not ingest sufficient antibody.

Candidate antigens selected from non-blood feeders may need to be presented to the immune system in a manner which stimulates an appropriate local mucosal response. This response is, essentially, a Th_2-like hypersensitivity reaction resulting in rapid worm rejection and featuring IgE production, mast cell proliferation and worm-specific degranulation. The response can be accelerated by depleting γ-interferon (McClure *et al.* 1995), a cytokine which can suppress Th_2 responses during nematode infections in rodents (Else *et al.* 1994). In addition, local IgA responses have been related to reductions in worm length in lambs infected with *T. circumcincta* (Smith, 1988) and are likely to mediate aspects of immune exclusion (Walker, 1994). Therefore, the meaningful evaluation of the ability of any antigen to stimulate protective immune responses is absolutely dependent on it being presented to the immune system in such a way as to stimulate the correct response.

Antigens undergoing evaluation have, until recently, usually been delivered subcutaneously, intramuscularly or intra-peritoneally in association with an adjuvant, the latter, being included to improve immunogenicity. In the past, Freund's adjuvants have been used extensively but a recent study has shown that Freund's adjuvant actually increases the susceptibility of sheep to *T. colubriformis* infection (Wagland *et al.* 1996). Adjuvants can act in five ways namely (1) immunomodulation, (2) antigen presentation, (3) induction of appropriate cellular response, (4) targeting and (5) depot generation (Cox & Coulter, 1997). Immune responses at mucosal sites are linked through the common mucosal immune system so that oral or intra-nasal antigen delivery can result in the induction of secretory immune responses at all mucosal sites. Again, caution is required before advocating any vaccine delivery system because successful oral vaccination of mice against *Trichuris muris* infection was more dependent on host genetics than on the means of antigen delivery (Robinson, Bellaby & Wakelin, 1995). Recent research has identified a variety strategies to optimise mucosal vaccination. This topic has been reviewed recently (Walker, 1994; Husband *et al.* 1996) and discussions included the variety of adjuvants available, their mode of action, antigen delivery in replicating vectors such as *Salmonella* (Chatfield *et al.* 1995) or *Vaccinia*, non-replicating systems such as microspheres or antigen conjugated to an adhesive molecule such as cholera toxin. Antigens can now be delivered in association with cytokines and it can be anticipated that cytokines which support the development of Th_2 responses (IL-4) and inhibit Th_1 responses (IL-4 and IL-10) would be of value in helminth vaccine formulations (Lofthouse *et al.* 1996). Finally, protective immune responses can be stimulated by DNA vaccination (see below) and these responses can be manipulated by co-administration with cytokine fusion constructs (e.g. Maecker *et al.* 1997).

DNA VACCINATION

DNA vaccines have been defined as the 'third generation of vaccines' (Waine & McManus, 1995) and recent developments in this field have been the subject of a thoroughly comprehensive review (Alarcon, Waine & McManus, 1999). DNA vaccines are based on plasmid vectors which express the inserted gene or cDNA of interest under the control of a strong promotor. They eliminate the requirement for the expression and purification of recombinant proteins and are stable, not requiring refrigerated storage. DNA can be taken up by host cells during *in vitro* tissue culture to allow testing confirmation of protein expression before *in vivo* administration. *In vivo*, the DNA can persist for long periods of time, allowing prolonged expression of the encoded protein and the protein is more likely to resemble the normal eukaryotic structure than recombinant proteins expressed in bacterial cells (Alarcon *et al.* 1999). It is not clear precisely how DNA vaccination works but the DNA plasmid is delivered directly to the target cells, is taken up and expressed by the host cells and the protein product is recognized as foreign inducing an immune response.

Attempts to apply DNA vaccination to nematode infections have yet to be reported but encouraging results have been obtained in rodents given DNA vaccines encoding antigens from *S. japonicum* (Yang, Waine & McManus, 1995; Waine *et al.* 1997) and *S. mansoni* (Dupre *et al.* 1997). In both cases, a Th_1-like response was induced and, in the latter, parasite antigen (GST-28) was detectable in the skin of rats at the site of DNA injection. Sera from these rats mediated antibody-dependent cellular cytotoxicity *in vitro* and killing of schistosomula. Parasite challenge induced a rapid increase in the specific IgG antibody response.

The approach has also been applied, with some success, to vaccination of sheep against *T. ovis* with plasmid DNA constructs encoding the 45W antigen (Rothel *et al.* 1997a) and recombinant ovine adenovirus constructs (Rothel *et al.* 1997b), the latter inducing some protective immunity against challenge. Sheep which received the plasmid vaccine and were then boosted with the adenovirus construct mounted IgG_1 responses 65-fold higher than those in sheep which received either vaccine alone.

Humoral responses in the mice following vaccination with DNA constructs encoding glutathione S-transferase from *Fasciola hepatica* have been evaluated (Smooker *et al.* 1999). The DNA constructs directed cytoplasmic or extracellular expression, and the level of response and isotype differed between the groups with intramuscular injection of the cytoplasmic construct generating a Th_1-type response whilst intradermal injection of the extracellular construct gave a Th_2 type response. In addition, the humoral response was highest after injection of the extracellular construct.

Clearly, DNA vaccination has enormous potential for the development of stable and effective vaccines. As our understanding of how DNA vaccination works and how the responses generated can be manipulated and maximized, it is not beyond the bounds of possibility that DNA vaccination will, in the future, replace antigen testing in vaccine trials. Finally, it may not even be necessary to screen for protective antigens given that direct injection of partial expression libraries from *Mycobacterium pulmonis* induced protection against challenge (Barry, Lai & Johnston, 1995) and mice can be partially protected against *Plasmodium chabaudi* infection by a similar approach (Spithill, Setiady & Smooker, 1999).

CONCLUDING COMMENTS

The last decade has seen potentially ground-breaking advances in vaccine development which include antigen selection but, in particular, the refinement of recombinant DNA technologies which are crucial for the production of antigen cheaply and in quantity. Problems are still being encountered in expressing antigens in an immunogenic form in the vectors currently available. Unfortunately, the process still tends to be the subject of trial and error. However, it will be disappointing if a recombinant protein-based vaccine with the desired efficacy against *H. contortus* is not available within the next decade. Whether such a vaccine becomes available to the farmer will depend on commercial prospects and constraints at the time. In parallel with a need for continued antigen isolation and testing, there is a need to develop easy means for mucosal antigen delivery in ruminants. Then new and existing antigens can be tested in the knowledge that the immune response stimulated by vaccination has been appropriate. DNA vaccination has obvious potential but a note of caution is required because there may be consumer resistance to consuming meat products containing foreign DNA. These concerns can be alleviated by defining the precise fate of the DNA innoculum. Finally, multivalent vaccine development will be hastened by collaboration between laboratories working on the same and different nematode species. There is considerable commercial interest in vaccination but their interests are not necessarily best served by secrecy agreements. Our commercial partners need to ask themselves how they are going to develop a multivalent vaccine. It is simple really – collaboration!

REFERENCES

ALARCON, J. B. , WAINE, G. W. & McMANUS, D. P. (1999). DNA Vaccines: Technology and application as anti-parasite and anti-microbial agents. *Advances in Parasitology* **42**, 344–410.

ANDREWS, S. J., HOLE, N. J. K., MUNN, E. A. & ROLPH, T. P. (1995). Vaccination for sheep against haemonchosis with H11 – prevention of the periparturient rise and colostral transfer of protective immunity. *International Journal for Parasitology* **25**, 839–846.

ANDREWS, S. J., ROLPH, T. P. & MUNN, E. A. (1997). Duration of protective immunity against ovine haemonchosis following vaccination with the nematode gut membrane antigen H11. *Research in Veterinary Science* **62**, 223–227.

ASHMAN, K., MATHER, J., WILTSHIRE, C., JACOBS, H. J. & MEEUSEN, E. N. T. (1995). Isolation of a larval surface glycoprotein from *Haemonchus contortus* and its possible role in evading host immunity. *Molecular and Biochemical Parasitology* **70**, 175–179.

BACHMAN, G. W. & MOLINA, R. (1933). Resistance to infestation with *Trichinella spiralis* in hogs. *American Journal of Hygiene* **18**, 76–78.

BARNES, E. H. & DOBSON, R. J. (1990). Population dynamics of *Trichostrongylus colubriformis* in sheep: computer model to simulate grazing systems and the evolution of anthelmintic resistance. *International Journal for Parasitology* **20**, 832–831.

BARNES, E. H., DOBSON, R. J. & BARGER, I. A. (1995). Worm control and anthelmintic resistance: adventures with a model. *Parasitology Today* **11**, 56–63.

BARRY, M. A., LAI, W. C. & JOHNSTON, S. A. (1995). Protection against *Mycoplasma* infection using expression-library immunization. *Nature* **377**, 632–635.

BEH, K. J. & MADDOX, J. F. (1996). Prospects for the development of genetic markers for resistance to gastrointestinal infection in sheep. *International Journal for Parasitology* **26**, 879–898.

BISSETT, S. A. & MORRIS, C. A. (1996). Feasability and implications of breeding sheep for resilience to nematode challenge. *International Journal for Parasitology* **26**, 857–868.

BOISVENUE, R. J., STIFF, M. I., TONKINSON, L. V., COX, G. N. & HAGEMAN, R. (1992). Fibrinogen-degrading proteins from *Haemonchus contortus* used to vaccinate sheep. *American Journal of Veterinary Research* **53**, 1263–1265.

BOWLES, V. M., BRANDON, M. R. & MEEUSEN, E. N. T. (1995). Characterisation of local antibody responses to the gastrointestinal nematode *Haemonchus contortus*. *Immunology* **84**, 669–674.

CABRERA, P. B. & GOULD, S. E. (1964). Resistance to trichinosis in swine induced by administration of irradiated larvae. *Journal of Parasitology* **50**, 681–684.

CAMPBELL, C. H. (1955). The antigenic role of the excretions and secretions of *Trichinella spiralis* in the production of immunity in mice. *Journal of Parasitology* **51**, 185–188.

CAMPBELL, W. C. & CUCKLER (1966). Further studies of the effect of thiabendazole on trichinosis in swine. *Journal of Parasitology* **52**, 260–279

CHATFIELD, S. N., ROBERTS, M., DOUGAN, G., HORMAECHE, C. & KHAN, C. M. A. (1995). The development of oral vaccines against parasitic diseases utilizing live attenuated *Salmonella*. *Parasitology* **110**, S17–S24.

CHIPMAN, P. B. (1957). The antigenic role of the excretions and secretions of *Trichinella spiralis* in the production of immunity in mice. *Journal of Parasitology* **43**, 593–596.

CLEGG, J. A. & SMITH, M. A. (1978). Dead vaccines against helminths. *Advances in Parasitology* **16**, 165–218.

COBON, G. S., KENNEDY, P. K., WAGLAND, B. M., ADAMS, D. B. & O'DONNELL, I. J. (1989). Vaccines against animal parasites. World patent application No. WO 89/00163.

COOP, R. L. & HOLMES, P. H. (1996). Nutrition and parasite interaction. *International Journal for Parasitology* **26**, 951–962.

COX, G. N., PRATT, D., HAGEMAN, R. & BOISVENUE, R. J. (1990). Molecular cloning and sequencing of a cysteine protease expressed by *Haemonchus contortus* adult worms. *Molecular and Biochemical Parasitology* **41**, 25–34.

COX, J. C. & COULTER, A. R. (1997). Adjuvants – a classification and review of their modes of action. *Vaccine* **15**, 248–256.

DADLEYMOORE, D. L., LIGHTOWLERS, M. W., ROTHEL, J. S. & JACKSON, D. C. (1999). Synthetic peptide antigens induce antibodies to *Taenia ovis* oncospheres. *Vaccine* **17**, 1506–1515.

DALTON, J. P., McGONIGLE, S., ROLPH, T. P. & ANDREWS, S. J. (1997). Induction of protective immunity in cattle against infection with *Fasciola hepatica* by vaccination with cathepsin L proteases and haemoglobin. *Infection and Immunity* **64**, 5066–5074.

DESPOMMIER, D. D. (1981). Partial purification and characterisation of protection inducing antigens from the muscle larva of *Trichinella spiralis* by molecular sizing chromatography and preparative flat-bed isoelectric focusing. *Parasite Immunology* **3**, 261–272.

DESPOMMIER, D. D., CAMPBELL, W. C. & BLAIR, L. S. (1977). The *in vivo* and *in vitro* analysis of immunity to *Trichinella spiralis* in mice and rats. *Parasitology* **74**, 109–119.

DESPOMMIER, D. D. & LACCETTI, A. (1981 *a*). *Trichinella spiralis*: partial characterisation of antigens isolated by immunoaffinity chromatography from the large particle fraction of muscle larvae. *Journal of Parasitology* **67**, 332–339.

DESPOMMIER, D. D. & LACCETTI, A. (1981 *b*). *Trichinella spiralis*: proteins and antigens isolated from a large particle fraction derived from the muscle larvae. *Experimental Parasitology* **51**, 279–295.

DESPOMMIER, D. D. & MULLER, M. (1970). The stichosome of *Trichinella spiralis*: Its structure and function. *Journal of Parasitology* **56**, 76–77.

DESPOMMIER, D. D. & MULLER, M. (1976). The stichosome and its secretion granules in the mature larva of *Trichinella spiralis*. *Journal of Parasitology* **62**, 775–783.

DOPHEIDE, T. A. A., TACHEDIJAN, M., PHILLIPS, C., FRENKEL, M. J., WAGLAND, B. M. & WARD, C. (1991). Molecular characterisation of a protective 11 kDa secretory protein from the parasitic stages of *Trichostrongylus colubriformis*. *Molecular and Biochemical Parasitology* **45**, 101–108.

DOUCH, P. G. C., GREEN, R. S., MORRIS, C. A., McEWAN, J. C. & WINDON, R. G. (1996). Phenotypic markers for the selection of nematode-resistant sheep. *International Journal for Parasitology* **26**, 899–914.

DOWD, A. J., TORT, J., ROCHE, L., RYAN, T. & DALTON, J. P. (1997). Isolation of a cDNA encoding *Fasciola hepatica* cathepsin L2 and functional expression in *Saccharomyces cerevisiae*. *European Journal of Biochemistry* **245**, 373–380.

DUPRE, L., POULAIN-GODEGROY, O., BAN, E., IVANOF, N., MEKRANFAR, M., SCHACHT, A., CAPRON, A. & RIVEAU, G. (1997). Intradermal immunisation of rats with plasmid DNA encoding *Schistosoma mansoni* 28 kDa Glutathione S-transferase. *Parasite Immunology* **19**, 505–513.

EAST, I. J., BERRIE, D. A. & FITZGERALD, C. J. (1989). *Oesophagostomum radiatum*: successful vaccination of calves with high molecular weight antigens. *International Journal for Parasitology* **19**, 271–274.

EDWARDS, A. J., BURT, J. S. & OGILVIE, B. M. (1971). The effect of immunity on some enzymes of the parasitic nematode *Nippostrongylus brasiliensis*. *Parasitology* **62**, 339–347.

ELSE, K. J., FINKELMAN, F. D., MALISZEWSKI, C. R. & GRENCIS, R. K. (1994). Cytokine-mediated regulation of chronic helminth infection. *Journal of Experimental Medicine* **179**, 347–351.

EMERY, D. L. (1996). Vaccination against worm parasites of animals. *Veterinary Parasitology* **64**, 31–45.

EMERY, D. L., McCLURE, S. J. & WAGLAND, B. M. (1993). Production of vaccines against gastrointestinal nematodes. *Immunology and Cell Biology* **71**, 463–472.

EMERY, D. L. & WAGLAND, B. M. (1991). Vaccines against gastrointestinal nematode parasites ruminants. *Parasitology Today* **7**, 347–349.

FERGUSON, D. L., RHODES, M. B., MARSH, C. L. & PAYNE, L. C. (1969). Resistance of immunised animals to infection by the larvae of the large roundworm of swine (*Ascaris suum*). *Federation Proceedings* **28**, 497.

FRENKEL, M. J., DOPHEIDE, T. A. A., WAGLAND, B. M. & WARD, C. W. (1992). The isolation, characterisation and cloning of a globin-like, host-protective antigen from the excretory-secretory products of *Trichostrongylus colubriformis*. *Molecular and Biochemical Parasitology* **50**, 27–36.

GAMBLE, H. R., MURRELL, K. D. & MARTI, H. P. (1986). Innoculation of pigs against *Trichinella spiralis* using larval excretory-secretory antigens. *American Journal of Veterinary Research* **47**, 2396–2399.

GRAHAM, M., SMITH, T. S., MUNN, E. A. & NEWTON, S. E. (1993). Recombinant DNA molecules encoding aminopeptidase enzymes and their use in the preparation of vaccines against helminth infections. Patent No. WO 93/23542.

GREENHALGH, C. J., BECKHAM, S. A. & NEWTON, S. E. (1999). Galectins from sheep gastrointestinal nematodes are highly conserved. *Molecular and Biochemical Parasitology* **98**, 285–289.

GREGG, P., DINEEN, J. K., ROTHWELL, T. L. W. & KELLY, J. D. (1978). The effect of age on the response of sheep to vaccination with irradiated *Trichostrongylus colubriformis* larvae. *Veterinary Parasitology* **4**, 35–38.

GRENCIS, R. K., CRAWFORD, C., PRITCHARD, D. I., BEHNKE, J. M. & WAKELIN, D. (1986). Immunisation of mice with surface antigens from muscle larvae of *T. spiralis*. *Parasite Immunology* **8**, 587–596.

GUERRERO, J. & SILVERMAN, P. H. (1969). *Ascaris suum*: immune reactions in mice. II. Metabolic and somatic antigens derived from *in vitro* cultures. *Experimental Parasitology* **29**, 110–115.

HAIG, D. M., WINDON, R. G., BLACKIE, W., BROWN, D. & SMITH, W. D. (1989). Parasite-specific T-cell responses of sheep following live infection with the gastric nematode *Haemonchus contortus*. *Parasite Immunology* **11**, 463–477.

HUSBAND, A. J., BAO, S., McCLURE, S. J., EMERY, D. L. & RAMSAY, A. J. (1996). Antigen delivery strategies for mucosal vaccines. *International Journal for Parasitology* **26**, 825–834.

JACOBS, H. J., ASHMAN, K., BOWLES, V. & MEEUSEN, E. N. T. (1999). Vaccination against the gastrointestinal nematode *Haemonchus contortus* using a purified larval surface antigen. *Vaccine* **17**, 362–368.

JARRETT, W. F. H., JENNINGS, F. W., MARTIN, B., McINTYRE, W. I. M., MULLIGAN, W., SHARP, N. C. C. & URQUHART, G. M. (1958). A field trial of a parasitic bronchitis vaccine. *Veterinary Record* **70**, 451–454.

JASMER, D. P., PERRYMAN, L. P., CONDER, G. A., CROW, S. & McGUIRE, T. C. (1993). Protective immunity to *Haemonchus contortus* induced by immunoaffinity isolated antigens that share a phylogenetically conserved carbohydrate gut surface epitope. *Journal of Immunology* **151**, 5450–5460.

JASMER, D. P., PERRYMAN, L. P. & McGUIRE, T. C. (1996). *Haemonchus contortus* GA1 antigens: related phospholipase C-sensitive, apical gut membrane proteins encoded as a polyprotein and released from the nematode during infection. *Proceedings of the National Academy of Sciences USA* **93**, 8642–8647.

JONES, V. E. & OGILVIE, B. M. (1972). Protective immunity to *Nippostrongylus brasiliensis* in the rat. II. Modulation of worm acetylcholinesterase by antibodies. *Immunology* **22**, 119–129.

KARANU, F. N., RURANGIRWA, F. R., McGUIRE, T. C. & JASMER, D. P. (1993). *Haemonchus contortus*: Identification of proteases with diverse characteristics in adult worm excretory/secretory products. *Experimental Parasitology* **77**, 362–371.

KEITH, R. K. & BREMNER, K. C. (1973). Immunisation of calves against the nodular worm *Oesophagostomum radiatum*. *Research in Veterinary Science* **15**, 23–24.

KERR, K. B. (1938). Attempts to induce an artificial immunity against the dog hookworm *Ancylostoma caninum* and the pig *Ascaris, A. lumbricoides suum*. *American Journal of Hygiene* **27**, 52–59.

KLEI, T. R. (1997). Immunological control of gastrointestinal nematode infections. *Veterinary Parasitology* **72**, 507–523.

KNOX, D. P. (1998). Parasite enzymes and the control of roundworm and fluke infestations in domestic animals. *British Veterinary Journal* **150**, 319–337.

KNOX, D. P., REDMOND, D. L. & JONES, D. G. (1993). Characterisation of proteases in extracts of adult *Haemonchus contortus*, the ovine abomasal nematode. *Parasitology* **106**, 395–404.

KNOX, D. P., SMITH, S. K., SMITH, W. D., REDMOND, D. L. & MURRAY, J. M. (1995). Thiol Binding Proteins. Patent Application No. PCT/GB95/00665.

KNOX, D. P., SMITH, S. K. & SMITH, W. D. (1999). Immunization with an affinity purified protein extract from the adult parasite protects lambs against *Haemonchus contortus*. *Parasite Immunology* **21**, 201–210.

KNOX, M. & STEEL, J. (1996). Nutritional enhancement of parasite control in small ruminant production systems in developing countries of South East Asia. *International Journal for Parasitology* **26**, 963–970.

LEE, D. L. & MARTIN, J. (1980). The structure of the intestine of *Nematodirus battus* and changes during the course of infection in lambs. *Parasitology* **81**, 27–33.

LIGHTOWLERS, M. W. (1994). Vaccination against animal parasites. *Veterinary Parasitology* **54**, 177–204.

LIGHTOWLERS, M. W. & RICKARD, M. D. (1988). Excretory-secretory products of helminth parasites: effects on host immune responses. *Parasitology* **96**, S123–S166.

LOFTHOUSE, S. A., ANDREWS, A. E., ELHAY, M. J., BOWLES, V. M., MEEUSEN, E. N. T. & NASH, A. D. (1996). Cytokines as adjuvants for ruminant vaccines. *International Journal for Parasitology* **26**, 835–842.

LONGBOTTOM, D., REDMOND, D. L., RUSSELL, M., LIDDELL, S., SMITH, W. D. & KNOX, D. P. (1997). Molecular cloning and characterisation of an aspartate protease associated with a highly protective gut membrane protein complex from adult *Haemonchus contortus*. *Molecular and Biochemical Parasitology* **88**, 63–72.

MAECKER, H. T., UMETSU, D. T., DEKRUYFF, R. H & LEVY, S. (1997). DNA vaccination with cytokine fusion constructs biases the immune response to ovalbumin. *Vaccine* **15**, 1687–1696.

MAIZELS, R. M., BLAXTER, M. L. & SELKIRK, M. E. (1993). Forms and functions of nematode surfaces. *Experimental Parasitology* **77**, 380–384.

McCLURE, S. J. & EMERY, D. L. (1994). Cell mediated response against gastrointestinal nematode parasites of ruminants. In *Cell Mediated Immunity in Ruminants* (ed. Morrison, W. I. & Goddeeris, B. M.), pp. 213–227. CRC Press: Boca Raton.

McCLURE, S. J., DAVEY, R. J., LLOYD, J. B. & EMERY, D. L. (1995). Depletion of IFN-γ, CD8$^+$, or Tcr $\gamma\delta$ *in vivo* during primary infection with an enteric parasite (*Trichostrongylus colubriformis*) enhances protective immunity. *Immunology and Cell Biology* **73**, 552–555.

McGHEE, J. R., MESTCKY, J., DERTZBAUGH, M. T., ELDRIDGE, J. H., HIRASAWA, M. & KIYONO, H. (1992). The mucosal immune system: from fundamental concepts to vaccine development. *Vaccine* **10**, 75–88.

McGILLIVERY, D. J., YONG, W. K., ADLER, B. & RIFFKIN, G. G. (1992). A purified stage-specific 31 kDa antigen as a potential protective antigen against *Ostertagia circumcincta* in lambs. *Vaccine* **10**, 607–613.

McKEAND, J. B., KNOX, D. P., DUNCAN, J. L. & KENNEDY, M. W. (1995). Immunisation of guinea pigs against *Dictyocaulus viviparus* using adult ES products enriched for acetylcholinesterases. *International Journal for Parasitology* **25**, 829–837.

MEEUSEN, E. N. T. (1995). Production of antigens. Patent No. WO 95/09182.

MEEUSEN, E. N. T. (1996). Rational design of nematode vaccines: natural antigens. *International Journal for Parasitology* **26**, 813–818.

MEEUSEN, E. N. T. & BRANDON, M R. (1994*a*). Antibody-secreting cells as specific probes for antigen detection. *Journal of Immunological Methods* **172**, 71–76.

MEEUSEN, E. N. T. & BRANDON, M. R. (1994*b*). The use of antibody-secreting cell probes to reveal tissue restricted immune responses during infection. *European Journal for Immunology* **21**, 469–474.

MILLER, H. R. P. (1984). The protective mucosal response against gastrointestinal nematodes in ruminants and laboratory animals. *Veterinary Immunology and Immunopathology* **6**, 167–259.

MILLER, H. R. P. (1996). Prospects for the immunological control of ruminant gastrointestinal nematodes: Natural Immunity, Can it be harnessed? *International Journal for Parasitology* **26**, 801–811.

MILLER, T. A. (1971). Vaccination against the canine hookworm diseases. *Advances in Parasitology* **9**, 153–183.

MILLER, T. A. (1978). Industrial development and field use of the canine hookworm vaccine. *Advances in Parasitology* **16**, 333–342.

MITCHELL, G. F. (1988). Glutathione S-transferases: potential components of antischistosome vaccine. *Parasitology Today* **5**, 34–37.

MONROY, F. G. & DOBSON, C. (1987). Mice vaccinated against *Nematospiroides dubius* with antigens isolated by affinity chromatography from adult worms. *Immunology and Cell Biology* **65**, 223–230.

MORRISON, C. A., COLIN, T., SEXTON, S. & SPITHILL, T. W. (1996). Protection of cattle against *Fasciola hepatica* by vaccination with glutathione S-transferase. *Vaccine* **14**, 1603–1612.

MORTON, R. E., YONG, W. K., RIFFKIN, G. G., BOZAS, S., SPITHILL, T. W., ADLER, B. & PARSONS, J. C. (1995). Inability to reproduce protection against *Teladorsagia circumcincta* in sheep with a purified stage-specific 31 kDa antigen complex. *Vaccine* **13**, 1482–1485.

MUNN, E. A. (1977). A helical polymeric extracellular protein associated with the luminal surface of *Haemonchus contortus* intestinal cells. *Tissue and Cell* **9**, 23–24.

MUNN, E. A. (1997). Rational design of nematode vaccines: hidden antigens. *International Journal for Parasitology* **27**, 359–366.

MUNN, E. A., GRAHAM, M. & COADWELL, W. J. (1987). Vaccination of young lambs by means of a protein fraction extracted from adult *Haemonchus contortus*. *Parasitology* **94**, 385–397.

MUNN, E. A. & GREENWOOD, C. A. (1984). The occurrence of the submicrovillar endotube (modified terminal web) and assoicated cytoskeletal structure in the intestinal epithelia of nematodes. *Philosophical Transactions of the Royal Society* **306**, 1–18.

MUNN, E. A. & SMITH, T. S. (1990). Production and use of anthelmintic agents and protective proteins. Patent No. WO 90/00416.

MUNN, E. A., SMITH, T. S., SMITH, H., SMITH, F. & ANDREWS, S. J. (1997). Vaccination against *Haemonchus contortus* with denatured forms of the protective antigen H11. *Parasite Immunology* **19**, 243–248.

MURRAY, J. M. & SMITH, W. D. (1994). Three important non blood-feeding parasites of ruminants ingest host immunoglobulin. *Research in Veterinary Science* **57**, 387–389.

MURRELL, K. D. & DESPOMMIER, D. D. (1984). Immunisation of swine against *Trichinella spiralis*. *Veterinary Parasitology* **15**, 263–270.

NEILSEN, J. T. M. (1975). Failure to vaccinate lambs against *Haemonchus contortus* with functional metabolic antigens identified by immunoelectro-phoresis. *International Journal for Parasitology* **5**, 427–430.

NEILSEN, J. T. M. & VAN DE WALLE, M. J. (1987). Partial protection of lambs against *Haemonchus contortus* by vaccination with a fractionated preparation from the parasite. *Veterinary Parasitology* **23**, 211–221.

NEWLANDS, G. F. J., SKUCE, P. J., KNOX, D. P., SMITH, S. K. & SMITH, W. D. (1999). Cloning and characterisation of a β-galactoside-binding protein (galectin) from the gastrointestinal nematode *Haemonchus contortus*. *Parasitology*, In Press.

NEWTON, S. E. (1995). Progress on vaccination against *Haemonchus contortus*. *International Journal for Parasitology* **25**, 1281–1289.

NEWTON, S. E., MONTI, J. R., GREENHALGH, C. J., ASHMAN, K. & MEEUSEN, E. N. T. (1997). cDNA cloning of galectins from the third stage larvae of the parasitic nematode *Teladorsagia circumcincta*. *Molecular and Biochemical Parasitology* **86**, 143–153.

NEWTON, S. E. & MUNN, E. A. (1999). The development of vaccines against gastrointestinal nematodes, particularly *Haemonchus contortus*. *Parasitology Today* **15**, 116–122.

O'DONNELL, I. J., DINEEN, J. K., ROTHWELL, T. L. W. & MARSHALL, R. C. (1985). Attempts to probe the antigens and protective immunogens of *Trichostrongylus colubriformis* in immunoblots with sera from infected and hyperimmune sheep and high and low responder guinea pigs. *International Journal for Parasitology* **15**, 129–136.

O'DONNELL, I. J., DINEEN, J. K., WAGLAND, B. M., LETHO, S., WERKMEISTER, J. A. & WARD, C. W. (1989). A novel host protective antigen from *Trichostrongylus colubriformis*. *International Journal for Parasitology* **45**, 101–108.

OGILVIE, B. M. & HOCKLEY, D. J. (1968). Effects of immunity on *Nippostrongylus brasiliensis*: reversible and irreversible changes in infectivity, reproduction and morphology. *Journal of Parasitology* **54**, 1073–1084.

OGILVIE, B. M., ROTHWELL, T. L. W., BREMNER, K. C., SCHNITZERLING, H. J., NOLAN, J. & KEITH, R. K. (1973). Acetylcholinesterase secretion by parasitic nematodes

1. Evidence for the secretion of the enzyme by a number of species. *International Journal for Parasitology* **3**, 589–597.

OZEROL, N. H. & SILVERMAN, P. H. (1970). Further characterisation of active metabolites from histotrophic larvae of *Haemonchus contortus* cultured *in vitro*. *Journal of Parasitology* **56**, 1199–1205.

RAINBIRD, M. A., MacMILLAN, D. & MEEUSEN, E. N. T. (1998). Eosinophil-mediated killing of *Haemonchus contortus* larvae: effects of eosinophil activation and role of antibody, complement and IL-5. *Parasite Immunology* **20**, 93–103.

RALEIGH, J. M., BRANDON, M. R. & MEEUSEN, E. N. T. (1996). Stage-specific expression of surface molecules by the larval stages of *Haemonchus contortus*. *Parasite Immunology* **18**, 125–132.

RALEIGH, J. M. & MEEUSEN, E. N. T. (1996). Developmentally regulated expression of a *Haemonchus contortus* surface antigen. *International Journal for Parasitology* **26**, 673–675.

RAND, K. N., MOORE, T., SRISKANTHA, A., SPRING, K., WILLADSEN, P. & COBON, G. S. (1989). Cloning and expression of a protective antigen from the cattle tick *Boophilus microplus*. *Proceedings of the National Academy of Sciences, USA* **86**, 9657–9661.

REDMOND, D. L., KNOX, D. P., NEWLANDS, G. F. J. & SMITH, W. D. (1997). Molecular cloning of a developmentally regulated putative metallo-peptidase present in a host protective extract of *Haemonchus contortus*. *Molecular and Biochemical Parasitology* **85**, 77–87.

RICKARD, M. D., HARRISON, G. B. L., HEATH, D. D. & LIGHTOWLERS, M. W. (1995). *Taenia ovis* recombinant vaccine – Quo Vadit? *Parasitology* **110**, S5–S9.

ROBERTS, F. H. S., ELEK, P. & KEITH, R. K. (1962). Studies of the resistance in calves with the nodular worm *Oesophagostomum radiatum*. *Australian Journal for Veterinary Research* **13**, 551–573.

ROBINSON, K., BELLABY, T. & WAKELIN, D. (1995). Efficacy of oral vaccination against the murine intestinal nematode parasite *Trichuris muris* is dependent on host genetics. *Infection and Immunity* **63**, 1762–1766.

RODRIQUEZ, M., RUBIERA, R., PENICHET, M., MONTESINOS, R., CREMATA, J., FALCON, V., SANCHEZ, G., BRINGAS, R., CORDOVES, C., VALDES, M., LLEONART, R., HERRARA, L. & DE LA FUENTE, J. (1994). High level expression of the *B. microplus* Bm86 antigen in the yeast *Pichia pastoris* forming highly immunogenic particles for cattle. *Jounal of Biotechnology* **33**, 135–146.

ROTHEL, J. S., BOYLE, D. B., BOTH, G. W., PYE, A. D., WATERKEYN, J. G., WOOD, P. R. & LIGHTOWLERS, M. W. (1997*b*). Sequential nucleic acid and adenovirus vaccination induces host-protective immune responses against *Taenia ovis* infection in sheep. *Parasite Immunology* **19**, 221–227.

ROTHEL, J. S., WATERKEYN, J. G., STRUGNELL, R. A., WOOD, P. R., SEOW, H. F., VADOLAS, J. & LIGHTOWLERS, M. W. (1997*a*). Nucleic acid vaccination of sheep: use in combination with a conventional adjuvanted vaccine against *Taenia ovis*. *Immunology and Cell Biology* **75**, 41–46.

ROTHWELL, T. L. W. (1989). Immune expulsion of parasitic nematodes from the gastrointestinal tract. *International Journal for Parasitology* **19**, 139–168.

ROTHWELL, T. L. W. & LOVE, R. L. (1974). Vaccination

against the nematode *Trichostrongylus colubriformis*. 1. Vaccination of guinea pigs with worm homogenates and soluble products harvested during *in vitro* maintenance. *International Journal for Parasitology* **4**, 293–299.

ROTHWELL, T. L. W. & MERRITT, G. C. (1975). Vaccination against the nematode *Trichostrongylus colubriformis*. II. Attempts to protect guinea pigs with worm acetylcholinesterase. *International Journal for Parasitology* **5**, 453–460.

SARLES, M. P. (1938). The *in vitro* action of immune serum on the nematode *Nippostrongylus muris*. *Journal of Infectious Diseases* **62**, 337–348.

SAVIN, K. W., DOPHEIDE, T. A. A., FRENKEL, M. J., WAGLAND, B. M., GRANT, W. N. & WARD, C. W. (1990). Characterisation, cloning and host-protective activity of a 30 kilodalton glycoprotein secreted by the parasitic stages of *Trichostrongylus colubriformis*. *Molecular and Biochemical Parasitology* **41**, 167–176.

SCHALLIG, H. D. F. H. & VAN LEEUWEN, M. A. W. (1997). Protective immunity to the blood-feeding nematode *Haemonchus contortus* induced by vaccination with parasite low-molecular weight antigens. *Parasitology* **114**, 293–299.

SCHALLIG, H. D. F. H., VAN LEEUWEN, M. A. W. & CORNELLISEN, A. W. C. A. (1997). Protective immunity induced by vaccination with two *Haemonchus contortus* excretory secretory proteins in sheep. *Parasite Immunology* **19**, 447–453.

SCHALLIG, H. D. F. H., VAN LEEUWEN, M. A. W. & HENDRIKX, W. M. L. (1994). Immune responses of sheep to excretory/secretory products of adult *Haemonchus contortus*. *Parasitology* **108**, 351–357.

SCHALLIG, H. D. F. H., VAN LEEUWEN, M. A. W., VERSTREPEN, B. E. & CORNELLISEN, A. W. C. A. (1997). Molecular characterisation and expression of two putative protective excretory secretory proteins of *Haemonchus contortus*. *Molecular and Biochemical Parasitology* **88**, 203–213.

SEESEE, F. M., WESCOTT, R. B. & GRAHAM, J. R. (1976). *Nippostrongylus brasiliensis*: indirect fluorescent antibody studies of immunity in mice. *Experimental Parasitology* **39**, 214–221.

SEXTON, J. L., MILNER, A. R., PANACCIO, M., WADDINGTON, J., WIJFELS, G., CHANDLER, D., THOMPSON, C., WILSON, L., SPITHILL, T. W., MITCHELL, G. F. & CAMPBELL, N. J. (1990). Glutathione S-transferase. Novel vaccine against *Fasciola hepatica* infection in sheep. *Journal of Immunology* **145**, 3905–3910.

SHARMA, R. L., BHAT, T. K. & DHAR, D. N. (1988). Control of sheep lungworm in India. *Parasitology Today* **4**, 33–36.

SHARP, P. J., SMITH, D. R. J., BACH, W., WAGLAND, B. M. & COBON, G. S. (1991). Purified glutathione S-transferases from parasites as candidate protective antigens. *International Journal for Parasitology* **21**, 839–846.

SHELDON, A. J. (1937). Studies on active acquired resistance, natural and artificial, in the rat to infection with *Strongyloides ratti*. *American Journal of Hygiene* **25**, 53–65.

SILBERSTEIN, D. S. & DESPOMMIER, D. D. (1984). Antigens from *Trichinella spiralis* that induce a protective response in the mouse. *Journal of Immunology* **132**, 898–904.

SILVERMAN, P. H., POYNTER, D. & PODGER, K. R. (1962). Studies on larval antigens derived by cultivation of some parasitic nematodes in simple media: protection tests in laboratory animals. *Journal of Parasitology* **48**, 562–570.

SKUCE, P. J., REDMOND, D. L., LIDDELL, S., STEWART, E. M., NEWLANDS, G. F. J., SMITH, W. D. & KNOX, D. P. (1999*b*). Molecular cloning and characterisation of gut-derived cysteine proteases associated with a host-protective extract from *Haemonchus contortus*. *Parasitology* **119**, 405–412.

SKUCE, P. J., STEWART, E. M., SMITH, W. D. & KNOX, D. P. (1999*a*). Cloning and characterisation of glutamate dehydrogenase (GDH) from the gut of *Haemonchus contortus*. *Parasitology* **118**, 297–304.

SMITH, S. K. & SMITH, W. D. (1996). Immunisation of sheep with an integral membrane glycoprotein complex of *Haemonchus contortus* and its major polypeptide components. *Research in Veterinary Science* **60**, 1–6.

SMITH, S. K., PETTIT, D., NEWLANDS, G. F. J., REDMOND, D. L., SKUCE, P. J., KNOX, D. P. & SMITH, W. D. (1999). Further immunisation and biochemical studies with a protective antigen complex from the microvillar membrane of the intestine of *Haemonchus contortus*. *Parasite Immunology* **21**, 187–199.

SMITH, T. S., GRAHAM, M., MUNN, E. A., NEWTON, S. E., KNOX, D. P., COADWELL, W. J., McMICHAEL-PHILLIPS, D., SMITH, H., SMITH, W. D. & OLIVER, J. J. (1997). Cloning and characterisation of a microsomal aminopeptidase from the intestine of the nematode *Haemonchus contortus*. *Biochimica et Biophysica Acta* **1338**, 295–306.

SMITH, T. S. & MUNN, E. A. (1990). Strategies for vaccination against gastrointestinal nematodes. *Revues of the Scientific and Technical Office for International Epizooitology* **9**, 577–595.

SMITH, T. S., MUNN, E. A., GRAHAM, M., TRAVERNOR, A. & GREENWOOD, C. A. (1993). Purification and evaluation of the integral membrane protein H11 as a protective antigen against *Haemonchus contortus*. *International Journal for Parasitology* **23**, 271–277.

SMITH, W. D. (1988). Mechanisms of immunity to gastrointestinal nematodes of sheep. In *Increasing Small Ruminant Productivity in Semi-arid Areas* (ed. Tompson, E. F. & Thompson, F.), pp. 275–286. ICARDA: Netherlands.

SMITH, W. D. (1993). Protection in lambs immunised with *Haemonchus contortus* gut membrane proteins. *Research in Veterinary Science* **54**, 94–101.

SMITH, W. D. (1999). Prospects for vaccines of helminth parasites of grazing ruminants. *International Journal for Parasitology* **29**, 17–24.

SMITH, W. D. & ANGUS, K. W. (1980). *Haemonchus contortus*: Attempts to immunise lambs with irradiated larvae. *Research in Veterinary Science* **29**, 45–50.

SMITH, W. D. & SMITH, S. K. (1993). Evaluation of aspects of the protection afforded to sheep immunised with a gut membrane protein from *Haemonchus contortus*. *Research in Veterinary Science* **55**, 1–9.

SMITH, W. D., SMITH, S. K. & MURRAY, J. M. (1994). Protection studies with integral membrane fractions of *Haemonchus contortus*. *Parasite Immunology* **16**, 231–241.

SMITH, W. D., SMITH, S. K., MURRAY, J. M., LIDDELL, S. & KNOX, D. P. (1993) Vaccines against metazoan parasites. Patent Application No. PCT/GB/93/01:521.

SMOOKER, P. M., STEEPER, K. R., DREW, D. R., STRUGNELL, R. A. & SPITHILL, T. W. (1999). Humoral responses in mice following vaccination with DNA encoding glutathione S-transferase of *Fasciola hepatica*: effects of mode of vaccination and the cellular compartment of antigen expression. *Parasite Immunology* **21**, 357–364.

SOULSBY, E. J. L. & STEWART, D. F. (1960). Serological studies of the self-cure reaction in sheep infected with *Haemonchus contortus*. *Australian Journal for Agricultural Research* **11**, 595–603.

SPITHILL, T. W., SETIADY, Y. & SMOOKER, P. M. (1999). Expression library immunisation protects mice against lethal *Plasmodium chabaudi* infection. *Proceedings of the British Society for Parasitology*, Spring Meeting No. 67.

STEWART, D. F. (1953). Studies on resistance of sheep to infestation with *Haemonchus contortus* and *Trichostrongylus* species and on the immunological reactions of sheep exposed to infestation. V. The nature of the self-cure phenomenon. *Australian Journal for Agricultural Research* **4**, 100–117.

STOLL, N. R. (1929). Studies with the strongylid nematode *Haemonchus contortus*. I. Acquired resistance of hosts under natural reinfection conditions out-of-doors. *American Journal of Hygiene* **10**, 384–418.

STROMBERG, B. E., RHOURY, A. & SOULSBY, E. J. L. (1977). *Ascaris suum*: Immunisation with soluble antigens in the guinea pig. *International Journal for Parasitology* **7**, 287–291.

SUTTLE, N. F., KNOX, D. P., ANGUS, K. W., JACKSON, F. & COOP, R. L. (1992*a*). Effects of dietary molybdenum on nematode and host during *Haemonchus contortus* infection in lambs. *Research in Veterinary Science* **52**, 225–229.

SUTTLE, N. F., KNOX, D. P., ANGUS, K. W., JACKSON, F. & COOP, R. L. (1992*b*). Effects of dietary molybdenum on nematode and host during *Trichostrongylus vitrinus* infection in lambs. *Research in Veterinary Science* **52**, 230–235.

TAFFS, L. F. (1968). Immunological studies on experimental infection of pigs with *Ascaris suum*. VI. The histopathology of the liver and lung. *Journal of Helminthology* **42**, 157–172.

TALIAFERRO, W. H. & SARLES, M. P. (1939). The cellular reactions in the skin, lungs and intestine of normal and immune rats after infection with *Nippostrongylus muris*. *Journal of Infectious Diseases* **64**, 157–192.

TAVERNOR, A. S., SMITH, T. S., LANGFORD, F., MUNN, E. A. & GRAHAM, M. (1992). Vaccination of young Dorset lambs against haemonchosis. *Parasite Immunology* **14**, 645–656.

TELLAM, R. L., SMITH, D., KEMP, D. H. & WILLADSEN, P. (1992). Vaccination against ticks. In *Animal Parasite Control Utilising Biotechnology* (ed. Yong, W. K.), pp. 303–331. CRC Press: Boca Raton.

THORSON, R. E. (1953). Studies on the mechanisms of immunity in rats to the nematode *Nippostrongylus muris*. *American Journal of Hygiene* **58**, 1–15.

THORSON, R. E. (1954). Absorption of protective antibodies from serum of rats immune to *Nippostrongylus muris*. *Journal of Parasitology* **42**, 21–25.

THORSON, R. E. (1956). The stimulation of acquired immunity in dogs by the injection of extracts of the oesophagus of adult hookworms. *Journal of Parasitology* **42**, 501–504.

TORT, J., BRINDLEY, P. J., KNOX, D. P., WOLFE, K. H. & DALTON, J. P. (1999). Proteases and associated genes of parasitic helminths. *Advances in Parasitology* **43**, 162–266.

TURNBULL, I. F., BOWLES, V. M., WILTSHIRE, C. J., BRANDON, M. R. & MEEUSEN, E. N. T. (1992). Immunisation of sheep with surface antigens from *Haemonchus contortus* larvae. *International Journal for Parasitology* **22**, 537–542.

TURNBULL, I. F., SMITH, D. R., SHARP, P. J., COBON, G. S. & HYNES, M. J. (1990). Expression and secretion in *Aspergillus nidulans* and *Aspergillus niger* of a cell surface glycoprotein from the cattle tick *Boophilus microplus* by using the fungal *amdS* promotor system. *Applied and Environmental Microbiology* **56**, 2847–2852.

URQUHART, G. M., ARMOUR, J., DUNCAN, J. L., DUNN, A. M. & JENNINGS, F. W. (1987). *Veterinary Parasitology*. Longman Scientific and Technical, Essex: UK.

URQUHART, G. M., JARRETT, W. F. H., JENNINGS, F. W., McINTYRE, W. I. M. & MULLIGAN, W. (1966). Immunity to *Haemonchus contortus* infection: Relationship between age and successful vaccination with irradiated larvae. *American Journal for Veterinary Research* **27**, 1645–1648.

VERNES, A. (1976). Immunisation of the mouse and mini-pig against *Trichinella spiralis*. In *Biochemistry of Parasites and Host–Parasite Relationships* (ed. Van den Bossche, H.), pp. 319–324. Elsevier/North Holland Press: Amsterdam.

VERVELDE, L., KOOYMAN, F., SCHALLIG, H. D. F. H., VAN LEEUWEN, M. & CORNELLISEN, A. W. C. A. (1999*a*). Young sheep vaccinated but not protected against *Haemonchus contortus* infections show IgG1 but no IgE responses. Conference abstract No. 114 from British Society of Parasitology, Spring Meeting.

VERVELDE, L., VAN LEEUWEN, M., KOOYMAN, F. & CORNELLISEN, A. W. C. A. (1999*b*) Vaccination against *Haemonchus contortus* with native and recombinant excretory/secretory proteins. Conference abstract NoP30 from British Society for Parasitology, Spring Meeting.

WAGLAND, B. M., McCLURE, S. J., COSSEY, S. G., EMERY, D. L. & ROTHWELL, T. L. W. (1996). Effects of Freund's adjuvants on guinea pigs infected with, or vaccinated against, *Trichostrongylus colubriformis*. *International Journal for Parasitology* **26**, 85–90.

WAINE, G. J. & MCMANUS, D. P. (1995). Nucleic acids: vaccines of the future. *Parasitology Today* **11**, 113–116.

WAINE, G. J., YANG, W., SCOTT, J. C., MCMANUS, D. P. & KALINNA, B. H. (1997). DNA-based vaccination using *Schistosoma japonicum* genes. *Vaccine* **15**, 846–848.

WALKER, R. I. (1994). New strategies for using mucosal vaccination to achieve more effective immunisation. *Vaccine* **12**, 387–400.

WIJFFELS, G. L., SALVATORE, L., DOSEN, M., PANACCIO, M. & SPITHILL, T. W. (1994). Vaccination of sheep with purified cysteine portliness of *Fasciola hepatica* decreases worm fecundity. *Experimental Parasitology* **78**, 132–148.

WILLADSEN, P. (1995). Commercialisation of a recombinant vaccine against *Boophilus microplus*. *Parasitology* **110**, S43–S50.

WINDON, R. G. (1990). Selective breeding for the control of nematodiasis in sheep. *Revues of the Science and Technical Office for International Epizooitology* **9**, 555–576.

YANG, W., WAINE, G. J. & McMANUS, D. P. (1995). Antibodies to *Schistosoma japonicum* paramyosin induced by nucleic acid vaccination. *Biochemical and Biophysical Research Communications* **212**, 1029–1039.

Immunological responses of sheep to *Haemonchus contortus*

H. D. F. H. SCHALLIG*

Willem de Zwijgerlaan 261, 1055 PW Amsterdam, The Netherlands

SUMMARY

Infections with *Haemonchus contortus* are a major constraint on ruminant health world-wide. Young lambs are very sensitive to *Haemonchus* infection. Older lambs and sheep acquire immunity after a continuous or seasonal exposure to the parasite. The mechanisms underlying immunity are still not completely understood. Antibodies, in particular local IgA and IgE, certainly play a role. The role of IgG is less clear. Lymphocyte proliferation responses seem to correlate to immunity. Sheep that have high antigen-induced lymphocyte responses have a low susceptibility to infection. Furthermore, several studies have demonstrated that immunity against *H. contortus* is associated with mastocytosis and hypersensitivity reactions. More recently, increasing attention is being paid to the role of cytokines (interleukins and γ-interferon) in the activation of specific defence mechanisms. Reverse transcriptase–polymerase chain reaction (RT–PCR) assays to study cytokine mRNA expression have become available. The inability of young lambs to mount a significant Th_2 response, which is normally characterized by high IgE levels, mastocytosis and eosinophilia, may account for the phenomenon of unresponsiveness in these animals.

Key words: *Haemonchus contortus*, sheep, immunity, antibody, cellular mechanisms.

INTRODUCTION

Infections with the blood-feeding nematode *Haemonchus contortus* are a major constraint on sheep and goat health and production in many parts of the world. The parasite mainly affects the abomasal mucosa of its host. Adult worms feed on blood and can cause severe anaemia, resulting in poor growth rate and weight loss, and heavy infections can result in death. The control of gastrointestinal nematodes in general is at present dependent on the repeated use of anthelmintics and, where possible, pasture management. However, clean pastures are not readily available under intensive grazing conditions and, perhaps more importantly, there is an increasing occurrence of parasites resistant to the action of anthelmintics (Jackson, 1993; Waller, 1994; Borgsteede *et al*. 1997; van Wyk, Malan & Randles 1997). Furthermore, there are concerns regarding drug residues in meat and the environment (Madsen *et al*. 1990; Lumaret *et al*. 1993).

The above-mentioned problems could be overcome by the development of immunological methods to control gastrointestinal helminths. A good knowledge of the mechanisms underlying protective immunity in sheep is a prerequisite for the development of such methods. The immune response against intestinal nematodes has been studied extensively in humans and rodent models (Miller, 1984, 1996; Cox & Liew, 1992; Sher & Coffman, 1992). Traditionally, it has been accepted that immunity against these parasites comprises the production of specific IgE antibodies, eosinophilia and mucosal mastocytosis

* Tel: (++) 31 20 6865808.
E-mail: rita.henk@consunet.nl

(Miller, 1984, 1996; Urban, Madden & Svetic, 1992) and that it is dependent on the activation of T helper (Th) 2 cells (Finkelman *et al*. 1991).

As far as the ruminants are concerned, most studies have been mainly restricted to studying systemic and peripheral antibody responses. Recently, important advances have been made in obtaining tools to dissect the ruminant immune responses to nematode infections and our understanding of this response has increased considerably in the last years. For example, monoclonal antibodies specific for ovine IgE have become available (Shaw *et al*. 1996; Kooyman *et al*. 1997), several cytokines have been cloned, sequenced and expressed (Wood & Seow, 1996) and potential protective antigens have been identified (Newton & Munn, 1999).

The purpose of this brief review is to focus on some of the recent insights in the immune responses of sheep to gastrointestinal nematode infections in ruminants in general and of *Haemonchus* infections in sheep in particular. Furthermore, some recent observations regarding the possible cause of unresponsiveness in young lambs will be presented. Finally, the immune responses induced by vaccination of sheep against gastrointestinal nematodes will be discussed.

ANTIBODY

Ovine immunoglobulins

The ovine immune system comprises IgG_1, IgG_2, IgM, IgA and IgE isotype antibodies. During the last decades many research groups have studied the possible role of these antibodies in immunity against gastrointestinal nematodes. Research has mainly focused on both systemic and locally produced IgA and IgG (reviewed by Miller, 1996) . The role of IgM is often considered of minor importance (Schallig, van Leeuwen & Hendrikx, 1995). Recently, monoclonal antibodies against ovine IgE have also become available (Shaw *et al.* 1996; Kooyman *et al.* 1997) and evidence for the important role of this particular antibody in immunity against gastrointestinal nematodes is rapidly growing.

IgA and IgG

A possible relationship between IgA and IgG antiparasite antibodies and resistance to *Haemonchus* has been described in many studies. In general, an increase in serum antibodies against larval and adult antigens after primary or secondary infection is observed (Smith, 1977; Duncan, Smith & Dargie, 1978; Smith & Christie, 1978; Charley-Poulain, Luffau & Perry, 1984; Gill, 1991; Schallig *et al.* 1994*a*, 1995; Gomez-Munoz *et al.* 1998). However, a direct relationship between the serum antibody levels and the immune status of sheep is questioned (Bowels, Brandon & Meeusen, 1995; Gomez-Munoz *et al.* 1999). This is probably due to the fact that *H. contortus* is confined to the surface of the abomasum and its mucosa and that the peripheral immune response is probably a poor reflection of the local mucosal response. For example, Schallig *et al.* (1994*a*, 1995) found only low levels of anti-*H. contortus* serum IgA in immune sheep. In contrast, several other studies demonstrated that IgA plays an important role in the mucosal response of immune sheep (Smith *et al.* 1983, 1984, 1986; Charley-Poulain *et al.* 1984; Gill, Husband & Watson, 1992*a*).

A problem with studying the local immune response is the *in vivo* accessibility of the abomasum for experimental sampling. The development of the technique to cannulate the gastric lymph duct which contains efferent lymph from the ovine stomachs made it possible to monitor local immune responses to abomasal nematodes (Smith, 1988). Gill *et al.* (1992*a*) used the technique of abomasal cannulation followed by serial biopsy to study the kinetics of local immune responses to *H. contortus* in sheep. The number of IgA-, IgG$_1$-, IgG$_2$- or IgM-containing cells in the abomasum of sheep before infection was low. Following infection, increased numbers of immunoglobulin-containing cells were observed in the abomasum and peak values were found 21 and 28 days after infection. IgA-containing cells were the most frequently observed cell types, followed by IgG$_1$, suggesting an important role for IgA, and to a lesser extent for IgG$_1$, in the immune response against haemonchosis.

The mechanisms by which IgG and IgA antibodies attribute to immunity against gastrointestinal nematodes is not completely clear. The antibodies could have a direct effect on the parasite. For instance, Gill *et al.* (1993*a*) suggested that antiparasite IgA and IgG antibodies contribute to resistance by neutralising or inactivating vital metabolic enzymes of *H. contortus*. Smith *et al.* (1985) observed a negative correlation between the magnitude of the gastric lymph IgA response of *Teladorsagia circumcincta*-infected sheep and worm length suggesting that IgA antibodies could interfere with the worms' ability to feed (Smith, 1988). Furthermore, it has also been shown that antiparasite IgG suppresses *Trichostrongylus colubriformis* feeding *in vitro* (Bottjer, Klesius & Bone, 1985). A more general role for IgG and IgA antibodies by participating in hypersensitivity reactions has also been suggested (Gill *et al.* 1993*a*; Miller, 1996). Examples are the ability of secretory IgA to induce eosinophil degranulation (Abu-Ghazaleh *et al.* 1989), the binding of IgA/antigen immune complexes to inflammatory cells in the mucosa and the subsequent release of cytokines and inflammatory mediators (Dubucquoi *et al.* 1994) and the cytophilic properties of IgG$_1$ for mast cells (Askenase, 1977).

IgE

An increase in total and parasite-specific IgE is generally regarded as an important factor in the host response to helminth infection (Jarrett & Miller, 1982; Hagan, 1993; Pritchard, 1993). The majority of studies on IgE responses to helminth infections have been in humans and rodents. In humans, high levels of serum IgE are thought to be associated with protection against gastrointestinal nematodes (Pritchard, Quinell & Walsh, 1995). Until recently, the role of IgE in protection against helminths in ruminants has been studied less extensively. However, specific antibodies against bovine (Thatcher & Gershwin, 1988) and ovine (Shaw *et al.* 1996; Kooyman *et al.* 1997) IgE have now become available.

Studies on the role of IgE in helminth-infected sheep were aided by the development of a specific monoclonal antibody against ovine IgE by Kooyman *et al.* (1997). Specific oligonucleotide primers were designed on the basis of the nucleotide and amino acid sequences of an ovine ε-chain cDNA (Engwerda *et al.* 1992) which allowed the amplification of a part of the C region (Cε3–Cε4, nucleotides 1111–1575) of ovine IgE. The amplification product was cloned into a suitable expression vector and the recombinant protein was purified by affinity chromatography. Specific polyclonal and monoclonal antibodies were produced. The monoclonal antibody, designated IE7 which is specific for sheep and goat IgE, was used to study IgE responses in *H. contortus* or *Teladorsagia circumcincta* infected sheep (Kooyman *et al.* 1997; Huntley *et al.* 1998*a, b*).

Kooyman *et al.* (1997) developed an IgE ELISA to monitor total and antigen-specific IgE serum levels during a controlled infection experiment. Infection of sheep with *H. contortus* resulted in increased total serum IgE levels at 2–4 weeks after infection. A negative correlation between worm counts and total IgE serum levels at necropsy was found in repeatedly infected sheep. This is in line with observations in *Ostertagia ostertagi* infected calves (Thatcher, Gershwin & Baker, 1989; Baker & Gershwin, 1993). In addition, significantly increased levels of excretory-secretory (ES) adult antigens specific IgE titres were found after infection. In contrast, no significant changes in third stage larvae (L_3) antigen-specific IgE levels in sera could be detected after infection. This indicates that L_3 antigens of *H. contortus* are probably less allergenic than ES antigens.

Huntley *et al.* (1998*a*) studied the IgE response of naive or primary infected sheep to a challenge with 50000 L_3 of *T. circumcincta* (for details of the design of this experiment see Stevenson *et al.* 1994) in serum and gastric lymph using a dot blot assay based on the monoclonal antibody IE7. In naive sheep, lymph and serum IgE concentrations increased from days 8 and 14 after infection, respectively. The IgE response of previously infected sheep to the challenge was more rapid, but not necessarily greater, than that following a primary infection. IgE concentrations were usually approximately four-fold higher in gastric lymph than in serum irrespective of whether the sheep were responding to a primary or a challenge infection. This indicates local production of IgE in the regional lymph nodes.

Serum and gastric lymph samples obtained in the experiment described above were also analysed using an ELISA to investigate the IgE antibody responses (Huntley *et al.* 1998*b*). During a primary response, IgE antibody to antigens obtained from L_3 and adult *T. circumcincta* were negligible, with low levels of IgE antibody detected in the gastric lymph and serum samples. There was, however, a marked IgE antibody response in 50% of the sheep to L_3 antigens during 2–8 days after challenge of primary infected sheep. In contrast, low levels of IgE antibody to adult antigens were found (Huntley *et al.* 1998*b*). This observation is in contrast to the findings of Kooyman *et al.* (1997), who demonstrated a clear response against adult ES antigens of *H. contortus*. In addition, Shaw *et al.* (1998) described the systemic IgE response in sheep to primary and challenge infections with *Trichostrongylus colubriformis* and demonstrated peak IgE responses to adult antigens between 20–27 days post infection and a secondary IgE response to both adult and L_3 ES antigens. Apparently there are differences in the ovine IgE responses to these species. Whether these relate to differences in the biology of the parasites is unclear. *T. colubriformis* dwells in the small intestine and,

although *Haemonchus* and *Teladorsagia* both live in the abomasum, they have different habits. *H. contortus* feeds on blood whereas *T. circumcincta* browses the epithelium and mucosa, and may therefore induce different immune mechanisms. Alternatively, these differences may simply reflect differences in breed of sheep or infection regimes.

LYMPHOCYTE PROLIFERATION RESPONSES

It is well known that the lymphocyte plays an important role in the generation of immune responses against helminth parasites. Sheep that are repeatedly infected with or immunised against *H. contortus* generally have lymphocytes that respond by proliferation *in vitro* to antigens from this parasite (Haig *et al.* 1989; Schallig, van Leeuwen & Hendrikx, 1994*b*; Schallig & van Leeuwen, 1997). Furthermore, Torgerson & Lloyd (1992, 1993) showed that lymphocytes from lambs, even animals totally naive to *H. contortus*, proliferate in response to soluble L_3 antigens. Riffkin & Dobson (1979) suggested that such lymphocytes might be important in the innate resistance of sheep to this parasite. Thus, naive sheep that had the highest antigen-induced lymphocyte responses had a low susceptibility to experimental infection. In addition, it has been found that vaccinated sheep not responding with a protective immune response to *H. contortus* challenge infection had lower proliferation responses against the vaccine antigens compared to those that were protected (Schallig & van Leeuwen, 1997).

EOSINOPHILS AND MAST CELLS

Mast cell responses

One of the most marked features of a gastrointestinal nematode infection is the recruitment and hyperplasia of mucosal mast cells (MMCs). Mucosal mastocytosis, including the presence of intra-epithelial globule leucocytes, is invariably associated with gastrointestinal helminthiasis (Huntley *et al.* 1992), suggesting that type I immediate hypersensitivity reactions are important in worm expulsion (Miller, 1984).

Earlier studies on the kinetics of ovine mast cell responses were based on histochemical detection, toluidine blue staining, and counting of these cells. Recently, antibodies against sheep mast cell protease (SMCP) have become available. This enzyme is exclusively located in the MMCs and globule leucocytes (Huntley *et al.* 1986). The development of an ELISA for SMCP has made it possible to measure its concentration in gastrointestinal tissues. The concentration of SMCP is correlated with the number of MMCs and globule leucocytes present in the tissues (Huntley *et al.* 1992).

Huntley *et al.* (1995) studied the mucosal mast cell responses of sheep previously maintained on pasture and treated with anthelmintic when housed and of worm-free lambs to a mixed challenge infection with *Trichostrongylus vitrinus* and *Teladorsagia circumcincta*. Eleven days after challenge, the ewes had significantly lower worm burdens than the naive lambs, but significantly higher tissue concentrations of SMCP. Toluidine blue staining revealed significant increased numbers of MMCs in sections of the abomasum and jejunum from ewes when compared with the lambs.

Numbers of MMCs in the abomasal tissue of non-, primary- or secondary-*H. contortus* infected sheep were also studied (Schallig, van Leeuwen & Cornelissen, 1997*a*). The numbers of MMCs in non-infected sheep were low, approximately 17·5/0·2 mm^{-2}. Primary infection resulted in a significant increase to around 40/mm^{-2}. The number of toluidine blue-stained cells were significantly increased after secondary infection, 52·1/mm^{-2}, compared to the counts after primary infection, suggesting a correlation between protection and the numbers of MMCs found in the abomasum after infection.

Eosinophils

Circulating and tissue eosinophilia is a common feature of helminthiasis. Eosinophils have been shown to be involved *in vivo* with helminth rejection by treating mice or guinea pigs with anti-eosinophil serum or anti-interleukin 5 (IL-5) monoclonal antibodies during infections with several helminths (Rainbird, MacMillan & Meeusen, 1998). Eosinophils accumulate around the tissue of invasive L$_3$ of sheep gastrointestinal parasites, including *H. contortus*, *in vivo* (Rainbird *et al.* 1998). In addition, eosinophils obtained from mammary washes of sheep, were shown to immobilize and kill *H. contortus* larvae *in vitro* in the presence of antibody specific against a L$_3$ surface antigen (Rainbird *et al.* 1998). The level of larval immobilization in the presence of antibody was increased when complement and IL-5 were added. Ultrastructural analysis of the eosinophil–larvae interaction at 6 h of incubation showed degranulation of adhering eosinophils onto the surface of the larvae. By 24 h of incubation, many larvae showed signs of damage and most eosinophils had degenerated. This suggests that eosinophil-mediated killing may be an effector mechanism for the elimination of L$_3$ in immune sheep (Rainbird *et al.* 1998).

In contrast, little or no difference was observed in the number of circulating or tissue eosinophils in non-infected, primary or secondary infected sheep in the studies by Huntley *et al.* (1995) and Schallig *et al.* (1997*a*). This indicated that eosinophils *per se* do not have a direct effector function against gastrointestinal parasites, although an indirect role cannot be discounted as shown above.

CYTOKINES

Th$_1$/Th$_2$ responses

The immune response in sheep to gastrointestinal nematodes is thought to be mediated by CD4+ T cells generated in the mesenteric lymph nodes (Gill, Watson & Brandon, 1992*b*). Many experimental, mainly murine, models demonstrated that CD4+ T cells can often be classified into two subsets, T helper type 1 (Th$_1$) and Th$_2$, based on the cytokines that they secrete (Mosmann *et al.* 1986; Mosmann & Coffman, 1989). Th$_1$ cells produce a number of cytokines most notably interleukin 2 (IL-2), interferon-γ (IFN-γ) and tumour necrosis factor β (TNF-β) resulting in a cell mediated immune response. Th$_2$ cells are defined by the production of IL-4, IL-5, IL-6 and IL-10 among others. A typical Th$_2$ response is characterized by increased immunoglobulin secretion by B cells, in particular IgG$_1$ and IgE, and proliferation of eosinophils and mast cells.

In the last 5 years significant progress has been made in the cloning and characterization of ovine cytokines (Wood & Seow, 1996), but their role in the ruminant immune response is still largely unknown. However, reverse transcriptase–polymerase chain reaction (RT–PCR) assays to study the cytokine mRNA expression have now been developed by several research groups and information on cytokine responses in the ruminant system is now becoming available.

Cytokines in nematode-infected ruminants

Canals *et al.* (1997) studied the cytokine profile induced by a primary non-protective infection with *Ostertagia ostertagi* in cattle. This infection resulted in decreased levels of IL-2 mRNA expression and increases in IL-4 and IL-10 transcription. Furthermore, a reduction in the percentages of T cells and an increase in B cells was observed. These observations are consistent with a Th$_2$ response, but did not protect the calves against the *O. ostertagi* infection.

The cytokine mRNA expression in nematode-resistant and -susceptible line lambs artificially infected with *Trichostrongylus colubriformis* was studied by Pernthaner *et al.* (1997). Four weeks after infection mesenteric lymph node cells (MLNC) from both lines expressed high levels of mRNA coding for IL-2, IL-4 and IFN-γ. MLNC from resistant lambs, when stimulated for 1 day with excretory/secretory antigen from adult *T. colubriformis*, had higher mRNA expression of IL-2 and IFN-γ and after 3 days of culture had higher levels of IL-4 mRNA than MLNC from susceptible-line lambs. This suggests, according to Pernthaner *et al.* (1997), that after an initial enhanced IFN-γ-mediated inflammatory response, regulatory Th$_2$-like

Table 1. Oligonucleotide sequence of forward (F) and reverse (R) primers used for RT–PCR

Product	Primer	Expected fragment size
IL-2	F: CGGGATCC-GCTGCTGGATTTACAGTTGC	370
	R: GGGGTACC-GTCATTGTTGAGTAGATGCTTTGAC	
IL-4	F: GCAAGGATCC-GCCCCAAAGAACACAACTGAGAAG	194
	R: GATCGGTACC-CTTTCCAAGAGGTCTTTCAGAGTA	
IL-5	F: CCGGATCC-CTCATCGAACTCTGCTGATAG	262
	R: GCGGTACC-CTTGCAGGTAGTCTAGGAATTG	
IFN-γ	F: CGGGATCC-TGGCCAGGGCCCATTTTTTAAAG	413
	R: CGGGTACC-TTACATTGATGCTCTCCGGCCTCG	
GAPDH	F: CCGGATCC-ATCACTGCCACCCAGAAGACT	168
	R: GGGGTACC-ATGCCAGTGAGCTTCCCGT	

IL, interleukin; IFN-γ, interferon-gamma; GAPDH, glyceraldehyde-3-phosphate dehydrogenase; oligonucleotide sequence before '-', restriction site (*Bam*HI or *Kpn*I in 'F' or 'R' respectively).

Table 2. Optimal [MgCl$_2$] and annealing temperatures used for amplification of different cytokines

Primer set	[MgCl$_2$] in mM	Annealing temperature (°C)
IL-2	1·8	55
IL-4	1·2	55
IL-5	2·1	60
IFN-γ	1·5	55
GAPDH	2·1	60

cytokines such as IL-4 become predominant leading to a typical Th$_2$-like response.

The cytokine expression profiles in sheep harbouring *H. contortus* infection are still mainly unknown. Réchards, van Leeuwen and Schallig (unpublished results obtained at the laboratory for Parasitology and Tropical Veterinary Medicine, Utrecht University, The Netherlands) have set up RT–PCR to study cytokine expression profiles in *Haemonchus*-infected sheep. cDNA-specific oligo-nucleotide primers for IL-2, IL-4, IL-5 and IFN-γ glyceraldehyde-3-phosphate dehydrogenase (GAPDH), which served as a control in all the RT–PCR assays, were designed using published sequences (Table 1). A *Bam*HI or a *Kpn*I restriction site was added to the forward or the reversed primer, respectively, to facilitate the cloning of the mRNA fragments after the RT–PCR. The cloned fragments were sequenced to confirm the identity and sequence of the amplification products. PCR conditions were optimized using genomic DNA extracted from peripheral blood mononuclear cells of non-infected sheep. Table 2 summarizes the optimal PCR conditions that were used in the preliminary study presented below.

The RT–PCR assays were used to analyse the cytokine expression of MLNC after a primary or secondary infection with *H. contortus*. The results of the RT–PCR are summarized in Table 3. The results obtained suggested that the primary infected

sheep responded with a non-protective Th$_1$ response, characterized by high levels of IL-2 and IFN-γ expression. In contrast, 3 out of 4 of the secondary infected sheep responded with a protective Th$_2$ response, with no expression of IL-2 and IFN-γ. The one animal in this group that did show expression of IL-2 and IFN-γ showed signs of another infection at necropsy. The fact that all animals had high levels of IL-4 and low to moderate IL-5 levels is at present difficult to explain.

The preliminary data obtained in the study described above suggest that not the quantity of each individual cytokine but the ratio of the different cytokines is of importance to the final outcome of the Th response. Further studies of cytokine profiles induced by infection and/or vaccination either in the form of quantitative RT–PCR assays or cytokine ELISAs are required.

THE POSSIBLE CAUSE OF UNRESPONSIVENESS

Unresponsiveness of young lambs against infectious diseases

Young lambs under 6 months are in general more susceptible to infectious diseases than mature sheep. This phenomenon, called unresponsiveness or hyporesponsiveness has been described for some viral diseases and various bacterial infections (Blood & Henderson, 1974; Weis, Chanana & Joel, 1986), but it is most evident for infections with gastrointestinal nematodes (Manton *et al.* 1962; Urquhart *et al.* 1966; Neilson, 1975; Dineen, Gregg & Lascelles, 1978). The lower resistance of lambs against infectious diseases in general appears to be due largely to immunological hyporesponsiveness; it is not simply a consequence of underexposure to pathogens and antigens to develop active immunity or immunosuppressive activities of the micro-organisms (Watson *et al.* 1994). The most important component of the immunological unresponsiveness of lambs seems to be constitutive. The immune system appears to progress through a maturation process

Table 3. Cytokine profiles expressed by mesenteric lymph nodes cells from non-infected (N.I.), primary infected (P.I.) or secondary infected (S.I.) sheep

	IL-2	IL-4	IL-5	IFN-γ
N.I. sheep 1	+ +	+ +	+	±
N.I. sheep 2	−	+ +	+	−
P.I. sheep 1	+ +	+ +	+	+ +
P.I. sheep 2	+ +	+ +	+	+ +
S.I. sheep 1	− −	+ +	±	−
S.I. sheep 2	+ +	+ +	+ +	+ +
S.I. sheep 3	−	+ +	±	− −
S.I. sheep 4	− −	+ +	±	− −

Detected cytokine levels: − −, not detectable; −, very low; ±, low; +, moderate; + +, high.

which starts during foetal life and continues during the first 12 months of life (Watson & Gill, 1991). Lambs have significantly lower proportions of CD4+ and CD8+ lymphocytes and greater numbers of $\gamma\delta$ T cells in blood and lymph (Hein & Mackay, 1991; Watson *et al.* 1994). In addition, blood lymphocytes from lambs produce less γ-interferon *in vitro* and young sheep mount, in general, smaller antibody responses than do mature animals (Colditz *et al.* 1996). These findings may explain, in part, why lambs are in general more susceptible to infectious diseases than adult sheep.

Unresponsiveness against gastrointestinal nematodes

In the case of gastrointestinal nematode infections unresponsiveness is thought to be associated also with lower numbers of CD4+ and CD8+ cells in blood and skin and lower levels of specific antibodies (Watson *et al.* 1994). However, in a pilot study Schallig and co-workers (unpublished results obtained at the laboratory for Parasitology and Tropical Veterinary Medicine, Utrecht University, The Netherlands) could not find significant lower levels of IgG, IgA or IgM antibodies in serum of infected lambs compared to adult sheep. In addition, Kooyman and Schallig (preliminary unpublished results obtained at the laboratory for Parasitology and Tropical Veterinary Medicine, Utrecht University, The Netherlands) compared the immune responses of lambs (3 or 6 months of age) and adult sheep (9 months of age) following vaccination with an experimental adult excretory/secretory (ES) vaccine (Schallig *et al.* 1997*a*) and subsequent challenge with *H. contortus*. The 9 or 6 month-old sheep were protected after vaccination against the challenge infection, whereas the young lambs became heavily infected. Serum antibody levels of the animals were measured and the number of eosinophils (both local and circulating) and mast cells in the mucosa of the abomasum were determined. No significant differences in antibody responses were observed between the three age groups. In contrast, peripheral blood eosinophils and mast cell counts in the abomasum were significant higher in the 9 month-old sheep compared to the young lambs. These data suggest that young lambs lack a Th$_2$ response which is characterised by eosinophilia and mastocytosis (Finkelman *et al.* 1991; Urban *et al.* 1992; Miller, 1996). This may be due to the relatively low numbers of CD4+ in the abomasum of young lambs (Hein & Mackay, 1991; Watson *et al.* 1994), resulting in a low or insufficient production of IL-4, the cytokine which is probably crucial for the Th$_2$ response. Other studies have also indicated that a reduction in CD4+ cells results in a reduction in immunity against *H. contortus* (Gill, Watson & Brandon, 1993*b*; Karanu *et al.* 1997). In addition, the abomasum of young lambs contains relatively high numbers of $\gamma\delta$ T cells (Hein & Mackay, 1991). However, the full cytokine repertoire of these cells has not been determined. Recent studies using flow cytometry have shown that $\gamma\delta$ T cells from mice infected with *Listeria monocytogenes* produce IFN-γ. In contrast, $\gamma\delta$ T cells of mice infected with *Nippostrongylus brasiliensis* produce IL-4 (Ferrick *et al.* 1995). Furthermore, it has been reported that bovine $\gamma\delta$ T cells stimulated with concanavalin A can express IL-2, IFN-γ and TNF-α (Wood & Seow, 1996). If these same cytokines are expressed by the $\gamma\delta$ T cells in young lambs, a typical Th$_1$ response would be induced which may not be sufficient to protect against gastrointestinal nematodes. This, however, remains to be supported with experimental data.

IMMUNITY INDUCED BY VACCINATION

The increasing occurrence of anthelmintic resistance (Jackson, 1993; Waller, 1994; Borgsteede *et al.* 1997; van Wyk *et al.* 1997) has prompted the need for the development of alternative methods, vaccines, to control gastrointestinal nematodes. At present, there are several research groups trying to develop a vaccine against *H. contortus*. Such a vaccine must meet several criteria before it will be commercially viable. First, it must be safe and efficacious, especially in young lambs. Second, it must be easy and cheap to produce. The current academic research at this moment is mainly concerned with the first item.

There are predominantly mainly two types of antigen preparations currently being evaluated: (1) natural antigens and (2) hidden antigens. Both types of antigen preparations afford varying degrees of immunity to lambs and sheep, but they also have their limitations.

Natural antigens

Numerous attempts have been made to induce immunity to *H. contortus* with irradiated larvae, somatic extracts of different life stages of the parasite and E/S preparations of larvae or adults all with variable results. It is out of the scope of this review to extensively discuss all these efforts in detail. Two recent examples with promising outcome are the following. Firstly, vaccination with a 70–83 kDa surface antigen of exsheated L_3 generated protection in five-month-old sheep (Jacobs *et al.* 1999). The protection induced with this natural antigen is dependent on the induction of a Th_2 response (Jacobs *et al.* 1999). Secondly, Schallig *et al.* (1997*a*) demonstrated that immunisation with purified 15 and 24 kDa E/S products afforded significant protection against a challenge infection in older lambs and sheep. The immune mechanisms induced by vaccination with the 15/24 kDa E/S products is Th_2 related and characterized by mastocytosis (Schallig *et al.* 1997*a*). Unfortunately, the 15/24 kDa E/S products did not provide protection to young lambs.

Both antigen preparations described above are at present obtained from living parasites. Harvesting these antigens is labour intensive and thus commercially not attractive. Recombinant expression of the putative protective antigens may facilitate their large-scale production. However, both antigen preparations contain glycosylated proteins (Ashman *et al.* 1995; Schallig *et al.* 1997*b*) which may hamper their production by conventional recombinant DNA technology.

Hidden antigens

Considerable effort has been applied to developing vaccines based on so-called hidden antigens, especially gut molecules. These antigens are normally hidden from the immune system and an immune response is not induced against these molecules during an infection. Vaccination with hidden antigen preparations has resulted in good protection against *H. contortus* infections in young lambs, older lambs, sheep and pregnant ewes (extensively reviewed by Newton & Munn, 1999). The protection induced by this type of vaccination is based on the induction of antibodies directed against these hidden antigens

(Newton & Munn, 1999). The antibodies are taken up during a blood meal and probably affect the gut of the parasite (Smith, 1993). However, the level of protection obtained by this approach is dependent on the maintenance of high antibody levels over a rather long period. This must be achieved by vaccination alone, since natural boosting does not occur.

SOME CONCLUDING REMARKS

The ovine immune response against *Haemonchus contortus* can be in general characterized as a typical Th_2 response with high levels of serum IgE and mastocytosis. In addition, local IgA and, to a lesser extent, IgG are probably also important in protection. Furthermore, *in vitro* lymphocyte proliferation responses seem to correlate to immunity. Young lambs cannot mount a protective Th_2 response, as reflected by the reduced number of mast cells and eosinophils in the abomasal mucosa. This may be due to the fact that young lambs have relatively low numbers of CD4+ cells and relatively high numbers of $\gamma\delta$ T cells in their abomasum, resulting in a cytokine profile that is typical for a Th_1 response, which is probably not sufficient to protect against gastrointestinal nematodes. In this light, studies on cytokine expression profiles in ruminants have become increasingly relevant because they may provide more insight in the possible application of (recombinant) cytokines as adjuvanting agents in vaccination experiments.

ACKNOWLEDGEMENTS

I would like to thank my former colleagues, in particular Marianne van Leeuwen, Frans Kooyman, Wim van der Aar and Maarten Eysker, at Utrecht University (Utrecht, The Netherlands), Dr John Huntley, Dr David Smith, Dr Dave Knox and their colleagues at Moredun Research Institute (Edinburgh, Scotland) and Dr Dante Zarlenga and his colleagues at USDA ARS (Beltsville, U.S.A.) for their valuable contributions to parts of the research described in this paper.

REFERENCES

ABU-GHAZALEH, R. I., FUJISAVA, T., MESTEKY, J., KYLE, R. A. & GLEICH, G. J. (1989). IgA-induced eosinophil degranulation. *Journal of Immunology* **142**, 2393–2397.

ASHMAN, K., MATHER, J., WILTSHIRE, C., JACOBS, H. J. & MEEUSEN, E. (1995). Isolation of a larval surface glycoprotein from *Haemonchus contortus* and its possible role in evading host immunity. *Molecular and Biochemical Parasitology* **70**, 175–179.

ASKENASE, P. W. (1977). Role of basophils, mast cells, vasomines in hypersensitivity reactions with delayed time course. *Progress in Allergy* **23**, 199–216.

BAKER, D. G. & GERSHWIN, L. J. (1993). Inverse relationship between IgE and worm burdens in cattle infected with *Ostertagia ostertagi*. *Veterinary Parasitology* **47**, 87–97.

BLOOD, D. C. & HENDERSON, J. A. (1974). *Veterinary Medicine*. 4th edition. London, Baillière, Tindall, pp. 504–548.

BORGSTEEDE, F. H. M., PEKELDER, J. J., DERCKSEN, D. P., SOL, J., VELLEMA, P., GAASENBEEK, C. P. & VAN DER LINDEN, J. N. (1997). A survey of anthelmintic resistance in nematodes of sheep in The Netherlands. *Veterinary Quarterly* **19**, 167–172.

BOTTJER, K. P., KLESIUS, P. H. & BONE, L. W. (1985). Effects of host serum on feeding by *Trichostrongylus colubriformis* (Nematoda). *Parasite Immunology* **7**, 1–7.

BOWLES, V. M., BRANDON, M. R. & MEEUSEN, E. (1995). Characterization of local antibody responses to the gastrointestinal parasite *Haemonchus contortus*. *Immunology* **84**, 69–674.

CANALS, A., ZARLENGA, D. S., ALMERIA, S. & GASBARRE, L. C. (1997). Cytokine profile induced by a primary infection with *Ostertagia ostertagi* in cattle. *Veterinary Immunology and Immunopathology* **58**, 63–75.

CHARLEY-POULAIN, J., LUFFAU, G. & PERRY, P. (1984). Serum and abomasal antibody response of sheep to infections with *Haemonchus contortus*. *Veterinary Parasitology* **14**, 129–141.

COLDITZ, I. G., WATSON, D. L., GRAY, G. D. & EADY, S. J. (1996). Some relationships between age, immune responsiveness and resistance to parasites in ruminants. *International Journal for Parasitology* **26**, 869–877.

COX, F. E. G. & LIEW, E. Y. (1992). T-cell subsets and cytokines in parasitic infections. *Parasitology Today* **8**, 371–374.

DINEEN, J. K., GREGG, P. & LASCELLES, A. K. (1978). The response of lambs to vaccination at weaning with irradiated *Trichostrongylus colubriformis* larvae: segregation into responders and non-responders. *International Journal for Parasitology* **8**, 59–63.

DUBUCQUOI, S., DESREUMAUX, P., JANIN, A., KLEIN, O., GOLDMAN, M., TAVERNIER, J., CAPRON, A. & CAPRON, M. (1994). Interleukin 5 synthesis by eosinophils: association with granules and immunoglobulin-dependent secretion. *Journal of Experimental Medicine* **179**, 703–708.

DUNCAN, J. L., SMITH, W. D. & DARGIE, J. D. (1978). Possible relationship of levels of mucosal IgA and serum IgG to immune unresponsiveness of lambs to *Haemonchus contortus*. *Veterinary Parasitology* **4**, 21–27.

ENGWERDA, C. R., SANDEMAN, R. A., STUART, S. J. & SANDEMAN, R. M. (1992). Isolation and sequence of sheep immunoglobulin E heavy-chain complementary DNA. *Veterinary Immunology and Immunopathology* **37**, 115–126.

FERRICK, D. A., SCHRENZEL, M. D., MULVANIA, T., HSIEH, B., FERLIN, W. G. & LEPPER, W. (1995). Differential production of interferon-gamma and interleukin-4 in response to Th1- and Th2-stimulating pathogens by gamma delta T cells *in vivo*. *Nature* **373**, 255–257.

FINKELMAN, F. D., PEARCE, E. J., URBAN, J. F. & SHER, A. (1991). Regulation and biological function of helminth induced cytokine responses. In *Immuno-Parasitology Today* (ed. Ash, C. & Gallagher, R.), pp. A62–A66 Elsevier Trends Journal: Cambridge.

GILL, H. S. (1991). Genetic control of acquired resistance to haemonchosis in Merino lambs. *Parasite Immunology* **13**, 617–628.

GILL, H. S., HUSBAND, A. J. & WATSON, D. L. (1992a). Localization of immunoglobulin-containing cells in the abomasum of sheep following infection with *Haemonchus contortus*. *Veterinary Immunology and Immunopathology* **31**, 179–187.

GILL, H. S., GRAY, G. D., WATSON, D. L. & HUSBAND, A. J. (1993a). Isotype-specific antibody responses to *Haemonchus contortus* in genetically resistant sheep. *Parasite Immunology* **15**, 61–67.

GILL, H. S., WATSON, D. L. & BRANDON, M. R. (1992b). *In vivo* inhibition by a monoclonal antibody to CD4+ T cells of humoral and cellular immunity in sheep. *Immunology* **77**, 38–42.

GILL, H. S., WATSON, D. L. & BRANDON, M. R. (1993b). Monoclonal antibody to CD4+ T cells abrogates genetic resistance to *Haemonchus contortus* in sheep. *Immunology* **78**, 43–49.

GOMEZ-MUNOZ, M. T., CUQUERELLA, M., DE LA FUENTE, C., GOMEZ-IGLESIAS, L. A. & ALUNDA, J. M. (1998). Infection-induced protection against *Haemonchus contortus* in merino and manchego sheep. Relationship to serum antibody response. *Zentralblat Veterinarmedicin [B]* **45**, 449–459.

GOMEZ-MUNOZ, M. T., CUQUERELLA, M., GOMEZ-IGLESIAS, L. A., MENDEZ, S., FERNANDEZ-PEREZ, F. J., DE LA FUENTE, C. & ALUNDA, J. M. (1999). Serum antibody response of Castellana sheep to *Haemonchus contortus* infection and challenge: relationship to abomasal worm burdens. *Veterinary Parasitology* **81**, 281–293.

HAGAN, P. (1993). IgE and protective immunity to helminth infections. *Parasite Immunology* **15**, 1–4.

HAIG, D. M., WINDOW, R., BLACKIE, W., BROWN, D. & SMITH, W. D. (1989). Parasite-specific T cell responses of sheep following live infection with the gastric nematode *Haemonchus contortus*. *Parasite Immunology* **11**, 463–477.

HEIN, W. R. & MACKAY, C. R. (1991). Prominence of γδ T cells in the ruminant immune system. *Immunology Today* **12**, 30–34.

HUNTLEY, J. F., GIBSON, S., KNOX, D. & MILLER, H. R. P. (1986). The isolation and purification of a proteinase with chymotrypsin-like properties from ovine mucosal mast cells. *International Journal for Biochemistry* **18**, 673–682.

HUNTLEY, J. F., NEWLANDS, G. F. J., JACKSON, F. & MILLER, H. R. P. (1992). The influence of challenge dose, duration of immunity, or steroid treatment on mucosal mast cells and on the distribution of sheep mast cell proteinase in *Haemonchus contortus* infected sheep. *Parasite Immunology* **14**, 429–440.

HUNTLEY, J. F., PATTERSON, M., MacKELLAR, A., JACKSON, F., STEVENSON, L. M. & COOP, R. L. (1995). A comparison of the mast cell and eosinophil responses of sheep and goats to gastrointestinal nematode infections. *Research in Veterinary Science* **58**, 5–10.

HUNTLEY, J. F., SCHALLIG, H. D. F. H., KOOYMAN, F. N. J., MacKELLAR, A., MILLERSHIP, J. & SMITH, W. D. (1998a). IgE responses in the serum and gastric lymph of sheep infected with *Teladorsagia circumcincta*. *Parasite Immunology* **20**, 163–168.

HUNTLEY, J. F., SCHALLIG, H. D. F. H., KOOYMAN, F. N. J.,

MACKELLAR, A., JACKSON, F. & SMITH, W. D. (1998b). IgE antibody during infection with the ovine abomasal nematode, *Teladorsagia circumcincta*: primary and secondary responses in serum gastric lymph of sheep. *Parasite Immunology* **20**, 565–571.

JACKSON, F. (1993). Anthelmintic resistance – the state of play. *British Veterinary Journal* **149**, 123–138.

JACOBS, H. J., WILTSHIRE, C., ASHMAN, K. & MEEUSEN, E. (1999). Vaccination against the gastrointestinal parasite *Haemonchus contortus* using a purified larval surface antigen. *Vaccine* **17**, 362–368.

JARRETT, E. E. E. & MILLER, H. R. P. (1982). Production and activities of IgE in helminth infection. *Progress in Allergy* **31**, 178–233.

KARANU, F. N., McGUIRE, T. C., DAVIS, W. C., BESSER, T. E. & JASMER, D. P. (1997). CD4+ T lymphocytes contribute to protective immunity induced in sheep and goats by *Haemonchus contortus* gut antigens. *Parasite Immunology* **19**, 435–445.

KOOYMAN, F. N. J., VAN KOOTEN, P. J. S., HUNTLEY, J. F., MACKELLAR, A., CORNELISSEN, A. W. C. A. & SCHALLIG, H. D. F. H. (1997). Production of a monoclonal antibody specific for ovine immunoglobulin E and its application to monitor serum IgE responses to *Haemonchus contortus* infection. *Parasitology* **114**, 395–406.

LUMARET, J. P., GALANTE, E., LUMBRERAS, C., MENA, J., BERTRAND, M., BERNAL, J. L., COOPER, J. F., KADIRI, N. & CROWE, D. (1993). Field effects of ivermectin residues on dung beetles. *Journal of Applied Ecology* **30**, 428–436.

MADSEN, M. B., NIELSEN, B. O., HOLTER, P., PEDERSEN, O. C., JESPERSEN, J. B., VAGN JENSEN, K. M., NANSEN, P & GRONVOLD, J. (1990). Treating cattle with ivermectin: the effects on the fauna and decomposition of dung pats. *Journal for Applied Ecology* **27**, 1–15.

MANTON, V. J. A., PEACOCK, B., POYNTER, D., SILVERMANN, P. H. & TERRY, R. J. (1962). The influence of age on naturally acquired resistance to *Haemonchus contortus* in lambs. *Research in Veterinary Science* **3**, 308–314.

MILLER, H. R. P. (1984). The protective mucosal response against gastrointestinal nematodes in ruminants and laboratory animals. *Veterinary Immunology and Immunopathology* **6**, 167–259.

MILLER, H. R. P. (1996). Prospects for the immunological control of ruminant gastrointestinal nematodes: natural immunity, can it be harnessed? *International Journal for Parasitology* **26**, 801–811.

MOSMANN, T. R., CHERWINSKI, H., BOND, M. W., GIEDLIN, M. A. & COFFMAN, R. L. (1986). Two types of murine helper T cell clone. 1. Definition according to profiles of lymphokine activities and secreted proteins. *Journal of Immunology* **136**, 2348–2357.

MOSMANN, T. R. & COFFMAN, R. L. (1989). Th1 and Th2 cells: Different patterns of lymphokine secretion lead to different functional properties. *Annual Review of Immunology* **7**, 145–173.

NEILSON, J. T. M. (1975). Failure to vaccinate lambs against *Haemonchus contortus* with functional metabolic antigens identified by immunoelectrophoresis. *International Journal for Parasitology* **5**, 427–430.

NEWTON, S. E. & MUNN, E. A. (1999). The development of vaccines against gastrointestinal nematode parasites, particularly *Haemonchus contortus*. *Parasitology Today* **15**, 116–122.

PERNTHANER, A., VLASSOFF, A., DOUCH, P. G. C. & MAASS, D. R. (1997). Cytokine mRNA expression and IFN-γ production in nematode resistant and susceptible line lambs artificially infected with gastro-intestinal nematodes. *Acta Parasitologica* **42**, 55–61.

PRITCHARD, D. I. (1993). Immunity to helminths: is too much IgE parasite- rather than host-protective? *Parasite Immunology* **15**, 5–9.

PRITCHARD, D. I., QUINELL, R. J. & WALSH, E. A. (1995). Immunity in humans to *Necator americanus*: IgE, parasite weight and fecundity. *Parasite Immunology* **17**, 71–75.

RAINBIRD, M. A., MacMILLAN, D. & MEEUSEN, E. N. Y. (1998). Eosinophil-mediated killing of *Haemonchus contortus* larvae: effect of eosinophil activation and role of antibody, complement and interleukin-5. *Parasite Immunology* **20**, 93–103.

RIFFKIN, G. D. & DOBSON, C. (1979). Predicting resistance of sheep to *Haemonchus contortus* infections. *Veterinary Parasitology* **5**, 365–378.

SCHALLIG, H. D. F. H. & VAN LEEUWEN, M. A. W. (1997). Protective immunity to the blood-feeding nematode *Haemonchus contortus* induced by vaccination with parasite low molecular weight antigens. *Parasitology* **114**, 293–299.

SCHALLIG, H. D. F. H., VAN LEEUWEN, M. A. W., BERNADINA, W. E. & HENDRIKX, W. M. L. (1994a). Serum antibody responses of Texel sheep experimentally infected with *Haemonchus contortus*. *Parasitology* **108**, 351–357.

SCHALLIG, H. D. F. H., VAN LEEUWEN, M. A. W. & CORNELISEN, A. W. C. A. (1997a). Protective immunity induced by vaccination with two *Haemonchus contortus* excretory secretory proteins in sheep. *Parasite Immunology* **19**, 447–453.

SCHALLIG, H. D. F. H., VAN LEEUWEN, M. A. W. & HENDRIKX, W. M. L. (1994b). Immune responses of Texel sheep to excretory/secretory products of adult *Haemonchus* contortus. *Parasitology* **108**, 351–357.

SCHALLIG, H. D. F. H., VAN LEEUWEN, M. A. W. & HENDRIKX, W. M. L. (1995). Isotype specific serum antibody responses of sheep to *Haemonchus contortus* antigens. *Veterinary Parasitology* **56**, 149–162.

SCHALLIG, H. D. F. H., VAN LEEUWEN, M. A. W., VERSTREPEN, B. E. & CORNELISEN, A. W. C. A. (1997b). Molecular characterization and expression of two putative protective excretory secretory proteins of *Haemonchus contortus*. *Molecular and Biochemical Parasitology* **88**, 203–213.

SHAW, R. J., GRIMMITT, D. J., DONAGHY, M. J. & DOUCH, P. G. C. (1996). Production and characterisation of monoclonal antibodies recognising ovine IgE. *Veterinary Immunology and Immunopathology* **51**, 235–251.

SHAW, R. J., GATEHOUSE, T. K. & McNEILL, M. M. (1998). Serum IgE responses during primary and challenge infections of sheep with *Trichostrongylus colubriformis*. *International Journal for Parasitology* **28**, 293–302.

SHER, A. & COFFMAN, R. L. (1992). Regulation of immunity to parasites by T cells and T cell-derived cytokines. *Annual Review of Immunology* **10**, 385–409.

SMITH, W. D. (1977). Anti-larval antibodies in the serum and abomasal mucus of sheep hyperinfected with *Haemonchus contortus*. *Research in Veterinary Science* **22**, 334–338.

SMITH, W. D. (1988). Mechanisms of immunity to gastrointestinal nematodes of sheep. In *Increasing Small Ruminant Productivity in Semi-arid Areas* (ed. Thomson, E. F. & Thomson, F. S.), pp. 275–286. ICARDA: The Netherlands.

SMITH, W. D. (1993). Protection in lambs immunised with *Haemonchus contortus* gut membrane proteins. *Research in Veterinary Science* **52**, 94–101.

SMITH, W. D. & CHRISTIE, M. G. (1978). *Haemonchus contortus*: local and serum antibodies in sheep immunised with irradiated larvae. *International Journal for Parasitology* **8**, 219–223.

SMITH, W. D., JACKSON, F., JACKSON, E. & WILLIAMS, J. (1983). Local immunity and *Ostertagia circumcincta*: changes in the gastric lymph of sheep after primary infection. *Journal of Comparative Pathology* **93**, 471–478.

SMITH, W. D., JACKSON, F., JACKSON, E., WILLIAMS, J. & MILLER, H. R. P. (1984). Manifestations of resistance to ovine ostertagiasis associated with immunological response in the gastric lymph. *Journal of Comparative Pathology* **94**, 591–601.

SMITH, W. D., JACKSON, F., JACKSON, E. & WILLIAMS, J. (1985). Age immunity to *Ostertagia circumcincta*: comparison of the local immune response of 4 1/2 and 10 month-old lambs. *Journal of Comparative Pathology* **95**, 235–245.

SMITH, W. D., JACKSON, F., JACKSON, E., WILLIAMS, J., WILLADSEN, S. M. & FEHILLY, C. B. (1986). Transfer of immunity to *Ostertagia circumcincta* and IgA memory between identical sheep by lymphocytes collected from the gastric lymph. *Research in Veterinary Science* **41**, 300–306.

STEVENSON, L. M., HUNTLEY, J. F., SMITH, W. D. & JONES, D. G. (1994). Local eosinophil- and mast cell-related responses in abomasal nematode infections of lambs. *FEMS Immunology and Medical Microbiology* **8**, 167–174.

THATCHER, E. F. & GERSHWIN, L. J. (1988). Generation and characterization of murine monoclonal antibodies specific for bovine immunoglobulin E. *Veterinary Immunology and Immunopathology* **18**, 53–66.

THATCHER, E. F., GERSHWIN, L. J. & BAKER, N. F. (1989). Levels of serum IgE in response to gastro-intestinal nematodes in cattle. *Veterinary Parasitology* **32**, 153–161.

TORGERSON, P. R. & LLOYD, S. (1992). The B-cell dependence of *Haemonchus contortus* antigen-induced lymphocyte proliferation. *International Journal for Parasitology* **23**, 925–930.

TORGERSON, P. R. & LLOYD, S. (1993). The same fractions of *Haemonchus contortus* soluble antigen induce lymphocyte responses in naive lambs and immune sheep. *Research in Veterinary Science* **54**, 244–246.

URBAN, J. F., MADDEN, K. B. & SVETIC, A. (1992). The importance of Th2 cytokines in protective immunity to nematodes. *Immunological Reviews* **127**, 204–220.

URQUHART, G. M., JARRETT, W. F. H., JENNINGS, F. W., MacINTYRE, W. I. M. & MULLIGAN, W. (1966). Immunity to *Haemonchus contortus* infection. Relationship between age and successful vaccination with irradiated larvae. *American Journal of Veterinary Research* **27**, 1645–1648.

VAN WYK, J. A., MALAN, F. S. & RANDLES, J. L. (1997). How long before resistance makes it impossible to control some field strains of *Haemonchus contortus* in South Africa with any of the modern anthelmintics? *Veterinary Parasitology* **70**, 111–122.

WALLER, P. J. (1994). The development of anthelminthic resistance in ruminant livestock. *Acta Tropica* **56**, 233–243.

WATSON, D. L. & GILL, H. S. (1991). Post natal ontogeny of immunological responsiveness in Merino sheep. *Research in Veterinary Science* **51**, 88–93.

WATSON, D. L., COLDITZ, I. G., ANDREW, M., GILL, H. S. & ALTMANN, K. G. (1994). Age-dependent immune response in Merino sheep. *Research in Veterinary Science* **57**, 152–158.

WEISS, R. A., CHANANA, A. D. & JOEL, D. D. (1986). Postnatal maturation of pulmonary antimicrobial defence mechanisms in conventional and germ-free lambs. *Pediatric Research* **20**, 496–504.

WOOD, P. R. & SEOW, H.-F. (1996). T cell cytokines and disease prevention. *Veterinary Immunology and Immunopathology* **54**, 33–44.

Impact of nutrition on the pathophysiology of bovine trypanosomiasis

P. H. HOLMES*, E. KATUNGUKA-RWAKISHAYA[1], J. J. BENNISON[2], G. J. WASSINK[3] and J. J. PARKINS

University of Glasgow Veterinary School, Bearsden Road, Glasgow G61 1QH, UK

SUMMARY

Trypanosomiasis is a major veterinary problem over much of sub-Saharan Africa and is frequently associated with under-nutrition. There is growing evidence that nutrition can have a profound effect on the pathophysiological features of animal trypanosomiasis. These features include anaemia, pyrexia, body weight changes, reduced feed intake and diminished productivity including reduced draught work output, milk yield and reproductive capacity. Anaemia is a principal characteristic of trypanosomiasis and the rate at which it develops is influenced by both protein and energy intakes. Pyrexia is associated with increased energy demands for maintenance which is ultimately manifested by reductions in voluntary activity levels and productivity. Weight changes in trypanosomiasis are markedly influenced by the levels of protein intake. High intakes allow infected animals to grow at the same rate as uninfected controls providing energy intake is adequate whilst low energy levels can exacerbate the adverse effects of trypanosomiasis on body weight. Reductions in feed intake are less apparent in animals which are provided with high protein diets and where intake is limited by the disease animals will often exhibit preferential selection of higher quality browse. Further studies are required to evaluate the minimum levels of protein and energy supplementation required to ameliorate the adverse effect of trypanosomiasis, the nature and quality of protein supplement to achieve these benefits and the influence these have on digestive physiology.

Key words: Bovine trypanosomiasis, nutrition/parasite interactions, pathophysiology of trypanosomiasis.

INTRODUCTION

Trypanosomiasis, a protozoan disease transmitted by tsetse flies, is probably the most serious veterinary and animal production problem in sub-Saharan Africa and prevents, or seriously curtails, the keeping of ruminants and equines over millions of square kilometres of potentially productive land. The most pathogenic trypanosome species affecting cattle in Africa are the vascular trypanosomes *Trypanosoma congolense* and *T. vivax*. Through the absence of cattle to provide draught power, milk, manure and meat, rural development is severely impaired. Much research and extension application effort has been devoted to the control of this disease through the use of drugs and control of the fly vector yet only 5 % of the affected area has been cleared of tsetse flies over the past century.

BACKGROUND

Attempts to control the disease are compromised by the problem of geographical scale and the often unusual biology of the trypanosome itself. Drug resistance is apparently increasing and there seems

to be little hope of producing a conventional vaccine within the foreseeable future. The difficulty here is that trypanosomes possess multiple mechanisms for immune evasion since, in the infective state, they are in constant contact with the immune systems of the host which they are able to evade by antigenic variation by switching their major variant surface glycoproteins (VSG). Added to this problem, each trypanosome species comprises a number of different strains or serodemes, all capable of eliciting a different repertoire of VSG variation (Murray & Black, 1985; Barry & Turner, 1991).

These factors and the sheer enormity of the tsetse fly control problem largely restricts the keeping of cattle and other ruminants to areas where tsetse flies are less abundant, such as the edges of semi-arid areas where the supplies of water and feed are frequently compromised. The only exception to this general rule is in West Africa where a number of breeds of cattle, e.g. N'Dama and West African Shorthorn and some breeds of small ruminants have developed over many centuries a degree of innate resistance or trypanotolerance to the pathogenic effects of trypanosomes, and as a result those breeds can at least survive in areas with high tsetse populations. However, even in these areas, feed supplies through the prolonged dry season can be severely restricted and trypanotolerance can be compromised. Therefore across Africa, cattle infected with trypanosomes frequently also suffer from undernutrition.

* To whom all correspondence should be addressed.
Present address: [1] Dean, Faculty of Veterinary Medicine, Makere University, PO Box 7062, Kampala, Uganda. [2] Agrimin Limited, Elsham Wold Estate, Brigg, Lincolnshire DN20 0SP, UK. [3] Groom's Cottage, Knowlegate, Ludlow, Shropshire S78 5JA, UK.

Parasitology (2000), **120**, S73–S85.   Printed in the United Kingdom   © 2000 Cambridge University Press

THE DISEASE

Bovine trypanosomiasis is characterized by the development of a moderate to severe anaemia, loss of condition and pyrexia, which is especially prominent during the early phases of the disease when the waves of parasitaemia are particularly high. Although the disease can occur in an acute form, it is normally associated with infections which last weeks or months and a slow and insidious loss of condition resulting in eventual death.

ROLE OF NUTRITION

There can be no doubt, in the face of increasing drug resistance, concerns about pesticide residues entering the food chain and the difficulties with the ongoing search for effective practical selection markers for resistance for use in cattle breeding programmes, that host nutrition remains an important factor influencing the host–parasite relationship and the ability to withstand the impact of parasitic infection. The purpose of this paper is to review the impact that nutrition has on the pathophysiology of bovine trypanosomiasis.

TRYPANOSOMIASIS AND PARASITIC GASTROINTESTINAL INFECTIONS

Until relatively recently studies on the interactions between nutrition and the pathophysiology of parasitic infections have largely been confined to helminth infections and there is now a large body of literature on this subject which has been reviewed periodically in more recent years (e.g. Parkins & Holmes, 1989; Coop & Holmes, 1996; Van Houtert & Sykes, 1996). These have clearly demonstrated that poor nutrition, especially low protein intake, can have a profound effect on the pathophysiology of gastrointestinal nematode infections. The earlier experimental studies evolved from original field observations in Australia which were later confirmed in controlled laboratory studies in Scotland and elsewhere. Similarly designed studies have only recently been conducted in ruminants infected with pathogenic trypanosomes and they are still far from complete.

As with the helminth infections, the studies with trypanosomes have been stimulated by field observations which showed that in areas of endemic trypanosomiasis the disease is exacerbated when there is obvious undernutrition of the livestock. One series of important observations was made in Ethiopia in the 1970s in oxen at a settlement scheme in a lowland valley heavily infested with tsetse flies (Bourn & Scott, 1978). These oxen could only be maintained by frequent and judicious use of trypanocidal drugs and good husbandry. As a result the cattle acquired a degree of 'resistance' and could continue working despite the presence of trypano-

somes in their blood. However, during a period of drought when the level of feed provision fell, the oxen developed severe signs of trypanosomiasis (despite a decrease in the numbers of tsetse flies) and many succumbed to the disease despite the drug treatments. Similarly, in West Africa it has been reported that the clinical signs of trypanosomiasis in trypanotolerant cattle are more severe at times of the year when they are under nutritional stress, for example during the dry season (Agyemang *et al.* 1992). Interestingly, in the same area of West Africa when peasant farmers were asked whether they thought trypanosomiasis in their cattle was associated with tsetse flies or poor nutrition, nearly half of them replied that it was associated with poor nutrition rather than tsetse flies (Snow, personal communication).

However, in contrast to these observations, other studies have suggested that the onset of parasitaemia in cattle infected with trypanosomes may be earlier in animals which are receiving a supplemented diet (Little *et al.* 1990), possibly because of the raised levels of trypanosome nutrients in the blood. It is known from laboratory studies that the onset and degree of parasitaemia can influence the blood levels of various constituents used metabolically by trypanosomes such as lipoproteins and cholesterol (Black & Vanderweed, 1989). However, it is also recognized that host nutrition can have profound effects on many aspects of the immune response e.g. antibody production which in turn is known to influence the level of parasitaemia.

Feed and water intakes

A complication of most parasitic infections is varying degrees of anorexia. Another important feature of parasite infection is the change in preference or rather the specific selection of particular dietary components by infected animals. A number of observations have noted that decreased intake of the 'poorer' components (i.e. those with most fibre and least protein contents) occurs with a counter-balancing preference for the most 'nutritious' feeds in an otherwise overall depressed dry matter intake in both helminth (Kyriazakis *et al.* 1994) and trypanosome infections (Romney *et al.* 1997). There can also be marked changes in water intake and loss, and, as a result, changes in water retention and body composition. Again, observations on these changes have been largely restricted to studies conducted with helminth infections (Holmes & Bremner, 1971; Parkins, Bairden & Armour, 1982) and it is clearly important to see if similar effects are observed with trypanosomiasis.

Although gastrointestinal helminth and trypanosome infections are fundamentally quite different in their respective pathophysiological characteristics, it is worth briefly reconsidering here the known effects of gastrointestinal parasitism on host metabolism and digestive function so that these observations may act as a comparison template for the evaluation of nutrition-host effects with trypanosomiasis.

Reduction of voluntary feed intake is a major feature of the pathogenesis of both infections but for apparently different reasons. It is now accepted that there is no convincing evidence, in either infection type, of a single mechanism controlling anorexia. In gastrointestinal infection, local pain, changes in abomasal pH, alterations in gut motility, digesta flow rates and elevation of the hormone gastrin, have all been implicated in reduced feed intakes in such parasitized animals. Water intake and retention are also commonly increased and may confuse observations on body weight as an indicator of tissue loss in gastrointestinal infections (Parkins & Holmes, 1989). However, in trypanosomiasis, anorexia is a noted feature of the acute phase of infection and is possibly partly due to a release of interleukin-1 (McCarthy, Kluger & Vander, 1985; van Miert, van Duin & Koot, 1990) acting on the hypothalamus. Gut motility may also be affected by interleukins.

Effects on feed utilization

It is entirely reasonable to predict that gastrointestinal parasitisms of cattle and sheep produce serious disruption, or at least some evident impairment of digestive and physiological efficiency when abomasal and small intestinal function is clearly challenged by these disease processes and leads to the devastations to the nitrogen economy of the host caused by losses of blood plasma proteins (Parkins & Holmes, 1989). This is not so predictable in the case of trypanosomiasis.

Also, reduced nitrogen retention in gastrointestinal infections of both cattle and sheep has been clearly attributed to increased urinary nitrogen losses and logically gave rise to the interpretation that somehow a reduced efficiency of utilization of absorbed amino acids and possibly also tissue protein catabolism in the host were ultimately the main causes of the decreased production performance of infected livestock.

The limited work performed with trypanosome infections to date has implied that the effects of the disease on digestive function (i.e. apparent digestibility which is strictly defined as that part of the feed intake fraction not excreted in the faeces) are minimal and not allied to the degree of trypanotolerance exhibited by the breed type under investigation in each particular trial. For example, trypanotolerant West African Dwarf goats showed no apparent differences in the digestibility of the organic matter and nitrogen fraction of the feed despite reduced feed intake as a result of infection (Akinbamijo *et al.* 1990, 1992, 1994*a, b*; van Dam *et al.* 1996). With Gobra and N'Dama cattle,

Akinbamijo *et al.* (1997) concluded that infection had no effect on the digestive physiology due to similar *in vivo* dry matter digestibilities and rate of passage.

RECENT STUDIES WITH TRYPANOTOLERANT N'DAMA CATTLE

Most work on the interactions between nutrition and trypanosomiasis in cattle have been conducted at the International Trypanotolerance Centre in The Gambia using local N'Dama cattle where early studies (Agyemang *et al.* 1990*a, b*; Little *et al.* 1990) provided clear evidence of an interaction between nutrition and trypanosomiasis. However, the feeding regimes were not always representative of traditional livestock systems so, in later studies emphasis was placed on assessing the minimum quantities of supplementary feed required, the timing of supplementation in relation to stage of infection and the effects of trypanosomiasis on animals subjected to diets that were close to or below those required for maintenance purposes (Bennison *et al.* 1998*a*, 1999).

Anaemia and parasitaemia

A significant decline in PCV in response to trypanosomiasis is a characteristic feature of the disease and a primary criterion for assessing its severity (Murray & Dexter, 1988). Trypanotolerance is characterized by an ability to regulate the parasite population and a capacity to control anaemia (Dwinger *et al.* 1992). This was clearly demonstrated in the study by Akinbamijo *et al.* (1997) where Gobra (susceptible *Bos indicus*) and N'Dama bulls were infected with *T. congolense*. The degree of anaemia was more severe in the Gobra animals and parasitaemias were correspondingly higher.

However, Bennison (1997) demonstrated in N'Dama cows that the degree of anaemia responds to short-term changes in the plane of nutrition after the onset of infection. Cows, infected with *T. congolense* in the late dry season were offered either 0, 1 or 2 kg/d fresh groundnut hay as a supplement to native pastures. Results are shown in Fig. 1. The supplementation with groundnut hay did not commence until peak parasitaemia, 2 weeks post-infection. During the 8 week period of infection cows were withdrawn from the trial if their PCV fell below 15% (a critical level) but only 1/8 animals was withdrawn from the group given 2 kg/d supplement compared to 5/8 and 4/8 in the nil and 1 kg/d groups, respectively. The necessity of withdrawing 41% of the infected animals was unexpected. All were mature, multiparous females originally from areas of low to medium tsetse fly challenge. Of the previous experiments with grazing N'Damas in The Gambia, none had reported the loss or the need to treat animals.

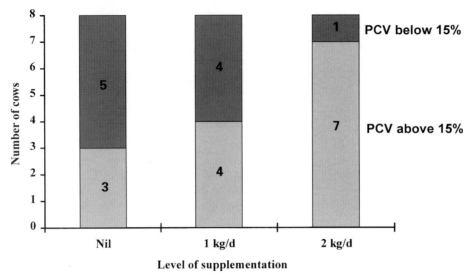

Fig. 1. Effect of feed supplementation of N'Dama cows infected with *T. congolense* on anaemia status.

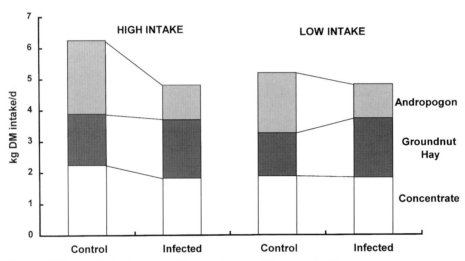

Fig. 2. Effect of infection with *T. congolense* and level of feeding on the voluntary intake of different dietary components in N'Dama cows.

Fever

Fluctuating fever, particularly in susceptible animals, is a typical sign of trypanosomiasis and is the result of an increase in metabolic rate. There is some evidence that trypanotolerance is linked to the suppression of fever during infection. Murray *et al.* (1981) found that the N'Dama, in contrast to the Zebu, did not become febrile even during waves of parasitaemia. The study by Akinbamijo *et al.* (1997) adopted a pair-feeding approach in which the food intake of uninfected control animals was matched to the intake of their infected partners. This ensured that the effects of anorexia on liveweight gain could be isolated from those associated with a change in the host's metabolism during infection. The trial showed that the decrease in liveweight gain in trypano-tolerant animals is principally due to changes in intake, whereas in susceptible animals, the cause was a combination of decreased intake and increased

energy requirements for maintenance which con-firmed a previous finding by van Dam (1996).

Voluntary feed intake

Romney *et al.* (1997) observed that N'Dama heifers maintained intakes of groundnut hay and concentrate while apparently selectively decreasing intake of the poorest quality component, *Andropogon* hay, which was high in fibre and low in crude protein. It was suggested that animals may have selected against the feed expected to have the greatest effect on gut fill. Evidence that the intake response to infection might be influenced by factors other than those affecting gut fill is provided by Bennison *et al.* (1999) who offered lactating N'Dama cows two different total amounts of a diet consisting of fixed ratios of groundnut hay, concentrate and *Andropogon* hay. Results are shown in Fig. 2. When animals were

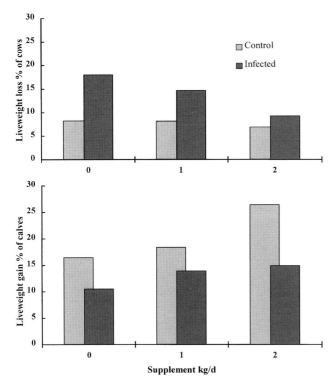

Fig. 3. Liveweight changes (%) of N'Dama cows and their calves following infection with *T. congolense* and supplemented with 0, 1 or 2 kg/d groundnut hay.

offered the high ration allowance, infection reduced intakes of all three dietary components by at least 20%, however, when the low ration was offered only the intake of *Andropogon* hay was reduced (by 26%). Furthermore, a draught animal study demonstrated significant interactions between infection, diet and work on food intake. Animals responding to the demands of work, appeared to be able to maintain feed intake during infection if the diet included a higher quality component equivalent to groundnut hay (Bennison *et al.* 1998*a*).

However, it now appears unlikely that trypanotolerance is linked to a superior ability to maintain intake during infection. In a comparison of pair-fed infected and control Zebu and N'Dama bulls, the relative decline in food intake was similar for both breeds (Akinbamijo *et al.* 1997).

Liveweight

The majority of trials provide evidence of the negative effect of trypanosomiasis on liveweight. Nevertheless, there are marked variations in the liveweight response to infection in N'Dama cattle. Paling *et al.* (1991) reported an impressive (for the N'Dama) average daily gain of 360 g/d over 694 days using 1 year old bulls and heifers with an initial weight of 150 kg. During this period the bulls were subjected to a sequential challenge of four different serodemes of *T. congolense*. The rate of liveweight

gain was not significantly different from the uninfected controls. Unfortunately, no details of the diet were provided but it is thought that the animals received *ad libitum* hay and 1 kg/d of concentrate with a crude protein (CP) content of 140–160 g/kg DM (Leak, S.G.A, personal communication). One of the *T. congolense* clones used in the study by Paling *et al.* (1991) was IL1180. Romney *et al.* (1997) used the same strain in a study of the effects of infection and diet on 12 month old N'Dama heifers weighing 89–146 kg. The rations were devised such that animals in the basal group received less than the estimated ME requirements with a CP content of approximately 80 g/kg DM. The second group received in addition to groundnut hay and *Andropogon* hay, supplementary groundnut cake at 3·9 g/kg liveweight. This ensured ME intake was above the requirements for maintenance and the CP content was between 140–150 g/kg DM. Infected animals on the basal diet lost significantly more weight than their non-infected counterparts (71 g compared with 14 g/d). While infected animals supplemented with groundnut cake still gained weight, it was at a lower rate than the controls (52 g compared with 168 g/d). Both Romney *et al.* (1997) and Akinbamijo *et al.* (1997) suggested that in N'Dama cattle, weight change was primarily a function of a change in feed intake as a consequence of infection and not a change in metabolism. This might suggest that the quality of diet in the study by Paling *et al.* (1991) was sufficient for the infected animals to maintain liveweight gain despite a change in intake.

Feeding behaviour

It might be expected that these effects of infection would also be associated with detectable differences in animal behaviour, activity levels and diet selection by free-grazing animals. In The Gambia, Wacher (1993) provided circumstantial evidence that infected animals appeared to move more slowly, rest more and select higher proportions of browse and fruit. Although, these effects were often not statistically significant, the trends were most noticeable in the harshest conditions i.e. the dry season. These responses however, were not observed in a later study by Bennison *et al.* (1998*b*). Trypanosomiasis infection had no significant effect on diet selection, although there was a significant interaction between supplementation and infection on behaviour. Infected cows supplemented with groundnut hay spent a greater proportion of the day resting. In experimental conditions, infected goats masked part of the energy costs of infection by reducing their standing time (van Dam, 1996). This suggests that animals adopt energy-saving behaviour patterns in response to infection.

Interactions between infection, nutrition and milk yield

Although trypanosomiasis has a direct effect on milk yield of N'Dama cows, the relative decline in milk output and liveweight response to infection may be related to the plane of nutrition at the time of infection. Work from The Gambia reported by Bennison (1997), shown in Fig. 3, depicts the liveweight changes in N'Dama cows and their calves (expressed as a % of the original liveweights) following infection of the dam with *T. congolense* and dietary treatments which consisted of supplementation with 0, 1 or 2 kg groundnut hay per day. The figure shows the *relatively* reduced loss of liveweight in infected cows with increasing levels of groundnut hay supplementation and the mirrored *relative* increases of liveweight gain in their suckling calves. Thus the liveweight gain increases observed in the calves reflect the milk output benefits of supplementation in both control and infected cows. However, Bennison (1997) also showed that supplementation with groundnut hay of diets well below maintenance alleviated liveweight loss of infected cows but had little effect on milk yield as represented by calf liveweight change but when breeding cows were given rations close to, or marginally above requirements, there were significant interactions between diet and infection on the productivity response. Animals given a higher plane of nutrition attempted to maintain milk yield at the expense of liveweight, whereas the infected cows on the basal diet had insufficient liveweight reserves with detriment to the milk yield and consequently the calf (Bennison *et al.* 1999). *Pre-partum* nutrition had no effect on the relative change in milk yield and liveweight during a trypanosomiasis infection *post-partum*. However, despite differential *pre-partum* feeding, the cows were still in a lean but moderate condition at parturition and so the trial was unable to determine whether a marked improvement in body condition would influence the response. The study by Akinbamijo *et al.* (1994*b*) on West African Dwarf sheep suggests that this is possible.

STUDIES IN SMALL RUMINANTS

Trypanosome infection does sometimes appear, from the available literature, to exhibit remarkably similar nutritional outcomes to gastrointestinal infections in that negative nitrogen balance and retention in the host are adversely affected by both parasitic diseases (Parkins & Holmes, 1989; Akinbamijo *et al.* 1990; Verstegen *et al.* 1991; van Dam, 1996) but interpretation of some of these results is difficult where reduced feed intakes are not matched by a system of pair-feeding the control animals.

Most work investigating the influence of nutrition on the pathogenesis of trypanosome infections has used two contrasting planes of nutrition. Under such conditions, the respective roles of protein and energy are difficult to evaluate. A novel series of experiments were conducted in Glasgow, (Katunguka-Rwakishaya *et al.* 1993, 1995) which investigated the direct influences of protein and energy on various aspects of the pathophysiology of trypanosome infection using Scottish Blackface sheep, a breed which displays a significant degree of trypanotolerance. The sheep, aged 4–6 months, were given a diet which consisted of a mixture of sugar beet pulp, barley siftings (these are the 'awns' of the thrashed seed only equal in nutritive value to the barley straw itself), soyabean meal and vitamin/minerals in differing proportions. For the protein studies, animals on a 'high protein' diet received 1 kg of fresh matter per day which provided 116 g digestible crude protein (DCP) and 9·8 MJ of Metabolisable Energy (MJME) per day. Animals on 'low' protein intake received 51·5 g DCP and 10·0 MJME/d.

For the energy studies, the diets and feeding regimes were formulated so that animals on a 'high' energy ration received 9·9 MJME with 109 g crude protein (CP) per day and the animals given the 'low' energy ration received 6·1 MJME and 109 g CP/d. The animals in each study were divided into two groups and placed on either high or low protein or energy intake. Each group was further subdivided into an infected and control group. Trypanosome infection was achieved by inoculation with about 10^5 *T. congolense 1180* parasites, a cloned derivative of an isolate made in the Serengeti, Tanzania (Nantulya *et al.* 1984). Parasitaemia, haematological and blood biochemical changes were observed for periods ranging from 10 to 14 weeks.

Parasitaemia

It was observed that neither dietary energy intake had a significant effect on the prepatent periods or the intensity of parasitaemia that followed trypanosome establishment. These observations are in agreement with those of Little *et al.* (1990) and Reynolds & Ekwuruke (1988).

Bodyweight

Where energy intakes offered were adequate for normal growth (*c.* 10 MJME/d), the low protein intake significantly reduced growth in infected animals but not in their pair-fed controls, whereas the high protein intake apparently overcame any effect of infection and animals grew at the same rate as their controls. However, when a reduced intake of energy was offered, growth rates were greatly reduced in both infected and their control partners despite an adequate protein intake. The findings in the protein study are in agreement with those of

Hecker *et al.* (1991) and Agyemang *et al.* (1990*b*) in sheep and cattle infected with trypanosomes respectively. The poorer growth in the low protein infected group could not be attributed to any observed decrease in feed intake which contrasts with the reports of reduced feed intakes in goats infected with *T. brucei* (van Miert *et al.* 1990) or *T. vivax* (Zwart *et al.* 1991). Experiments in goats infected with *T. vivax* (Verstegen *et al.* 1991) have demonstrated that development of fever during a course of trypanosome infection is associated with increased heat production and increased metabolizable energy for maintenance. The consequence of this is that a proportion of nutrients that could be used for growth is reduced, as it is metabolized to provide the extra energy required. This effect appears to be greater for animals receiving a low protein diet, which was just above maintenance, compared to those on a high protein intake. It is interesting that this effect was not observed in the dietary energy study where infected animals on low energy intake grew at the same rate as their controls, which was significantly lower than the rates of infected and control animals on high energy intake. It can therefore be concluded that increased dietary protein ameliorates the effect of trypanosome infection on growth rate where energy intakes are not limiting.

Packed red cell volume (PCV)

In all dietary treatments infection caused a significant drop in PCV, the decline beginning with the appearance of trypanosomes in the circulation. The mean PCV values of the infected animals given either the low or the high protein diets were not significantly different but infected animals given the low energy diet had a significantly lower PCV than the high energy diet. The observation of similar degrees of anaemia in the high and low protein groups is consistent with the observations of Agyemang *et al.* (1990*a*). It would suggest that trypanosome establishment and the rate of development of anaemia are not influenced by dietary protein. In contrast, the observation from the energy study supports those of Little *et al.* (1990) in N'Dama cattle experimentally infected with *T. congolense* and given either a low or a high plane of nutrition and those of Fagbemi *et al.* (1990), and Makinde, Otesile & Fagbemi (1991) in pigs infected with *T. brucei* and given either a low or a high energy intake. In addition, Makinde *et al.* (1991) observed that pigs given a low protein diet developed significant increases in plasma volume, while the increase in plasma volume was not significant in pigs that were given a high energy ration. An increase in plasma volume as a factor in the development of anaemia in trypanosome infected animals has been

documented (Katunguka-Rwakishaya, Murray & Holmes, 1992). It would appear that high energy intake may not prevent the rapid establishment of trypanosomes following inoculation but the ensuing anaemia is less severe than that observed in animals receiving a low energy ration.

Recovery from anaemia and erythropoietic responses

Studies conducted in cattle suggested that the rate of recovery from anaemia was faster in animals supplemented with groundnut cake (Agyemang *et al.* 1990*a*). In sheep infected with *T. congolense* and given either a low or a high protein diet, it was observed that following treatment with a trypanocidal drug (isometamidium chloride) at 70 days after infection, animals given a high protein diet recovered from anaemia much faster than those receiving a low protein diet (Katunguka-Rwakishaya *et al.* 1993). This was associated with macrocytosis in the high protein infected group as opposed to normocytosis in the low protein infected group. A further experiment was conducted using similar protein levels to the previous studies, to investigate the influence of dietary protein on erythropoietic responses in sheep by using the Evans Blue dilution technique to measure plasma volume and [^{59}Fe] ferric citrate to assess red cell synthesis based on plasma iron turnover rates (PITR) and red cell iron utilization (Katunguka-Rwakishaya, 1992). Infected animals given the low protein diet had a significantly lower circulating red cell volume than uninfected control animals but there were no differences between the high protein groups. Similarly, infection and dietary protein apparently had no influence on total blood volumes.

Ferrokinetic measurements indicated that the plasma [^{59}Fe] half lives and estimated red cell lifespan were lower while the PITR, [^{59}Fe] utilization and RBC [^{59}Fe] incorporation rates were higher in infected animals than in control animals. This is consistent with enhanced erythropoietic activity in infected animals particularly during the early stages of infection. Another interesting observation was that PITR and RBC [^{59}Fe] incorporation rates were higher in the infected animals given high protein than those given low protein although the differences were not statistically significant. This supports previous observations and suggests that improved protein nutrition enhances erythropoietic activity even as early as 4 weeks after infection. This could explain the faster rate of recovery from anaemia recorded in previous studies (Katunguka-Rwakishaya *et al.* 1993). It can therefore be concluded that dietary protein has a major influence on erythropoietic activity which accounts for improvement in PCV values especially in long-standing infections.

Serum lipid fractions

It was observed that *T. congolense* infection in sheep is associated with profound biochemical changes some of which are modulated by dietary protein or energy. Infected animals showed significant decreases in plasma cholesterol and serum phospholipid concentrations with resultant decrease in serum total lipids. The decline in these fractions was greater in animals receiving low protein or low energy rations compared to those receiving high protein or energy rations.

A reduction in serum cholesterol and phospholipids confirms previous observations in sheep infected with *T. congolense* by Roberts *et al.* (1977) and Traore-Leroux, Fumoux & Pinder (1987). There is evidence that lipids constitute 15–20 % of trypanosomal dry weight (Venkatessan & Ormerod, 1976) and that trypanosomes obtain cholesterol from the host by uptake and degradation of low density (Coppens *et al.* 1987; Gillet & Owen, 1987) or high density lipoproteins (Traore-Leroux *et al.* 1987). It has also been demonstrated that trypanosomes require cholesterol for growth and multiplication (Black & Vanderweed, 1989) and that it is the main sterol in trypanosomes (Carrol & McCroire, 1986). In addition, the trypanosomes take up free fatty acids which may be circulating freely (Tizard *et al.* 1978) or bound to albumin (Vickerman & Tetley, 1979). The decline in serum cholesterol commences with the appearance of trypanosomes in the circulation. While uptake by trypanosomes might seem to largely account for the decline in serum cholesterol, it may not be entirely so because even at the peak of parasitaemia, total cholesterol content of the trypanosomes may not be sufficiently high to have any impact on the serum concentration.

It was observed that control animals receiving a high protein diet had higher concentrations of serum total lipids, cholesterol and non-esterified fatty acids than those given the low protein diet. It has also been observed that high cholesterol concentrations are associated with higher parasite numbers (Traore-Leroux *et al.* 1987). It therefore appears that provision of higher protein has a sparing effect on lipid metabolism of the host, making these nutrients available for trypanosome growth while at the same time, the host is able to continue gaining weight.

Albumin metabolism

Infected animals on all dietary treatments developed hypoalbuminaemia and hypoproteinaemia which were more severe in animals receiving low protein and energy ration compared to those on high protein and energy rations. In addition, control animals on high protein diet had higher concentrations of plasma albumin and serum urea. These observations

support those of Otesile, Fagbemi & Adeyemo (1991) in boars placed on different energy levels and infected with *T. brucei*. It has been suggested that the degree of hypoalbuminaemia may be related to the severity of trypanosome infection (Holmes, 1976) and it is possible that trypanosomal uptake of albumin-bound fatty acids and lipoproteins for their metabolism and growth (Vickerman & Tetley, 1979) and haemodilution (Holmes, 1976) may account for the decrease in plasma albumin. The observation that the decrease in plasma albumin concentration was mild in animals receiving high protein and energy rations indicates that adequate protein and energy nutrition enhances the ability of trypanosome infected animals to withstand the adverse effects of infection.

INFLUENCE OF TRYPANOSOME INFECTIONS ON THE DIGESTIVE PHYSIOLOGY OF SHEEP

Similarly controlled studies were conducted in Scottish Blackface wethers given diets of different compositions in order to investigate the effects of trypanosome infections on feed intake, digestive function and gross nitrogen balance. Studies were performed on the apparent digestibility of the diets and also the rate of passage of the roughage component of the diet using chromium as a marker (Uden *et al.* 1980, 1982).

In order to distinguish the effects of a reduced feed intake from the other effects of infection, the animals were matched in pairs and the control animal was offered the average amount of food consumed by its infected counterpart during the previous 2 days (pair-feeding regime). In one such study (A), eight sheep arranged in four pairs, were each offered a diet of chopped barley straw *ad libitum* plus a pelleted barley/soyabean mixture supplying 70 g metabolizable protein (MP) and 8·3 MJME/d. Another eight sheep similarly arranged in four pairs, were offered a diet of chopped lucerne hay *ad libitum* plus the same quantity (366 g DM) of a pelleted barley concentrate supplying 140 g MP and 13·3 MJME/d (Wassink *et al.* 1997). One animal of each pair was infected with *T. congolense* 1180 (Nantulya *et al.* 1984).

In a second study (B) also reported by Wassink *et al.* (1997), each sheep of four pairs received 200 g DM grass hay and 425 g DM crushed barley grain (plus minerals) in the morning and barley straw *ad libitum* in the afternoon (low roughage/high concentrate diet; 60 g MP, 10 MJME/d). The sheep in the other four pairs were offered 400 g DM grass hay and 315 g DM crushed barley grain (plus minerals) in the morning and barley straw *ad libitum* in the afternoon (high roughage/low concentrate diet; 60 g MP, 10 MJME/d). Again, one animal of each pair was infected with *T. congolense 1180* (Nantulya *et al.* 1984).

The third study (C) described by Wassink (1997) was designed so that each sheep of four pairs received 150 g DM grass hay and 319 g DM crushed barley grain (plus mineral mix) in the morning and barley straw *ad libitum* in the afternoon (low roughage/high concentrate diet; 45 g MP, 7 MJME/d). The sheep in the other four pairs were offered 300 g DM grass hay and 236 g DM crushed barley grain (plus mineral mix) in the morning and barley straw *ad libitum* in the afternoon (high roughage/low concentrate diet; 45 g MP, 7 MJME/d). One animal of each pair was infected with *T. vivax* strain Y486, isolated by Leeflang, Ige & Olatunde (1976). In contrast to study A, in which there was a large difference in the level of nutrition between the two dietary groups, in studies (B) and (C) the levels of energy and protein were similar between the two dietary groups. However, as the source of protein was different between the dietary groups, the proportion of effective rumen degradable protein (ERDP) and digestible undegraded protein (DUP) intake was different.

The general course of the *T. congolense* infections in the first two experiments followed the same pattern as previously found by Katunguka-Rwakishaya (1992) using the same strain of trypanosomes in Scottish Blackface sheep. Parasitaemia levels were relatively low and packed cell volume (PCV) decreased only moderately, but significantly ($P < 0.01$), between day 10 and 20 post infection. PCV levels stabilized after 20 days post infection.

The effects of the disease were much more pronounced in the sheep infected with *T. vivax* (C). These sheep had higher levels of parasitaemia throughout the study period compared to those infected with *T. congolense*. PCV levels declined significantly ($P < 0.01$) post infection and, unlike the *T. congolense* infection, continued to decline towards the end of the experiment. The effects of the disease on plasma cholesterol and albumin levels were also greater in the *T. vivax*-infected sheep compared to those infected with *T. congolense*. However, no significant interaction effects were found between the diets and the blood parameters measured.

Feed intake

In all three studies, the depression in feed intake due to the trypanosome infection was relatively small compared to previous experiments in both sheep (Akinbamijo *et al.* 1994 *a, b*) and goats (Verstegen *et al.* 1991; Akinbamijo *et al.* 1992). The apparently more virulent *T. vivax* infection (C) did not result in a greater decrease in feed intake compared with the milder *T. congolense* infection (A and B). The slight depression in total feed intake that was observed in the last two experiments was mostly due to the reduction in the intake of the *ad libitum* barley straw.

Body weight gain

The relatively small effect of the infection on the feed intake may be partly reflected in the data on body weight gain, which in A was no different between infected animals and their pair-fed controls. Although significant differences were found in body weight gains in (B) ($P < 0.05$) and (C) ($P < 0.01$) between infected sheep and their pair-fed controls, those differences were small and not thought to be of biological significance.

Apparent digestibility

Organic matter digestibility, OMD, was significantly lower in the *T. congolense*-infected sheep in study (A) ($P < 0.01$) and (B) ($P < 0.05$) with no interaction effect being observed between the diet and infection. In study (A), OMD of the pair-fed control animals increased and this may have been due to the below potential feed intake in these animals as a result of being paired to the infected ones. The OMD of the *T. vivax*-infected sheep was not found to be significantly different to the OMD of their pair-fed controls. These results indicate no apparent effect of infection on the OMD of the diets. Van Dam (1996) also found no evidence of a decrease in organic matter digestibility in *T. vivax* infected West African Dwarf goats fed lucerne pellets or grass straw.

Crude protein digestibility was also significantly lower in the infected sheep of study (A). However, unlike OMD, the digestibility of crude protein of the pair-fed control animals did not change during the experiment, but the crude protein digestibility values of the infected animals decreased. Again, no interaction between the diet and infection was observed. In studies (B) and (C), the nitrogen digestibility values also appeared to be lower in the infected sheep compared to their pair-fed controls, but most differences were not statistically significant. As the infected animals and their controls were pair-fed, the difference could be attributed mostly to a slightly higher faecal nitrogen excretion in the infected animals of studies (B) and (C) compared to their pair-fed controls. Whether the source of this extra faecal nitrogen excretion was dietary or endogenous was not clear and requires further investigation.

Mean retention time of the roughage in the digestive tract

The mean retention time of the roughage in the digestive tract was found to be significantly longer in the *T. congolense*-infected sheep in study (A). No significant interaction effect was found between the type of diet and infection. A similar effect was observed in study (B) with the *T. congolense*-infected animals having a significantly longer mean retention time than their pair-fed controls and this effect was more pronounced in the animals on the low roughage/high concentrate diet.

The mean retention time of the *T. vivax*-infected sheep (study C) was only slightly longer than that of their pair-fed controls ($P < 0.05$) and in this experiment no interaction between diet and infection was observed. These results indicate an effect of trypanosome infections on the mean retention time of roughage in the digestive tract irrespective of the pathogenicity of the disease. There was some evidence to suggest that the type of diet influenced the effects of a *T. congolense* infection on the mean retention time. No such evidence was found in the *T. vivax*-infected animals of study (C).

Nitrogen retention

Although there was a tendency towards a lower nitrogen retention in *T. congolense*-infected sheep compared to their respective pair-fed controls none of the differences were statistically significant. In the *T. vivax* study (C), however, differences were found to be statistically significant, with the infected sheep having a lower nitrogen retention than their pair-fed controls. Significantly higher urinary nitrogen excretion and to a lesser extent a higher faecal nitrogen excretion appear to be responsible for this lower nitrogen retention in the infected animals. Van Dam (1996) and Akinbamijo *et al.* (1994*a*, *b*) also reported a lower nitrogen retention in *T. vivax*-infected West African Dwarf goats and West African Dwarf sheep, respectively, compared with the controls. However, no change in the efficiency of nitrogen utilization was found and the difference could be attributed to a decreased feed intake in the infected animals.

A significant interaction effect was also found between infection and diet with the infected animals on the low roughage/high concentrate diet being affected more by the infection in terms of nitrogen retention than the ones on the high roughage/low concentrate diet. Close examination of the data revealed that the lower nitrogen retention in the *T. vivax*-infected sheep offered the low roughage/high concentrate diet was mainly caused by a higher urinary nitrogen excretion in these animals. There was, however, a high variation in response between animals. It was calculated that the digestible undegraded protein (DUP) intake of the sheep was lower in the sheep fed low roughage/high concentrate diet compared to those on the high roughage/low concentrate diet (1 and 1.5 g/kg $M^{0.75}$/day respectively). Whether this factor was involved in the difference between the dietary groups needs further investigation.

It was concluded in these carefully controlled nutrient intake studies that trypanosome infections did indeed affect the digestive function of sheep, an outcome that previously had been partly demon-

strated but with some doubts being raised over the experimental designs employed. Here clearly nitrogen digestibility was lower in the infected animals, but organic matter digestibility was unaffected. The mean retention time of the roughage in the digestive tract was significantly longer in infected animals, which was found to be affected by the type of diet in the case of *T. congolense*. However, the pathogenicity of the trypanosome infection did not appear to affect the level of changes in digestive function.

The nitrogen retention of the *T. congolense*-infected sheep was similar to that of their pair-fed controls. In contrast, the nitrogen retention of the *T. vivax*-infected sheep was significantly lower than that of their pair-fed controls. This lower nitrogen retention was mainly due to a higher urinary nitrogen excretion and was significantly affected by the type of diet.

CONCLUSIONS

Across Africa, cattle infected with trypanosomes frequently also suffer from undernutrition and the impact that nutrition has on the pathophysiology of bovine trypanosomiasis is recognized as a key area for investigation where the outcomes may afford practical husbandry strategies to counter the effects of the disease. This review has attempted to collate relevant recent work specifically designed to determine the direct and compounded effects of nutrition on the disease, digestive efficiency and productivity.

It is apparent from the studies described in this review that nutrition can have a profound effect on the pathophysiology of bovine trypanosomiasis. The severe problems of impaired growth or weight loss, reduced productivity and increased mortality associated with trypanosomiasis are considerably exacerbated by undernutrition. It is noteworthy that almost all the studies carried out to date have been conducted in breeds of ruminants which show significant resistance or trypanotolerance to trypanosomiasis such as N'Dama cattle, West African sheep and goats and Scottish Blackface sheep. Nevertheless, even in such breeds the impact of undernutrition is clearly evident. It can be anticipated that the effects in trypanosusceptible breeds are likely to be even more pronounced.

Although there are differences between the various studies described in terms of the species, age and history of the hosts, species and strains of trypanosomes, and the level and composition of the feed, the results of the numerous studies do allow a number of tentative conclusions to be drawn. These are best described by examining the effects that nutrition has on the principal clinical and pathophysiological features of trypanosomiasis. These features include anaemia, pyrexia, body weight changes, reduced feed intake and diminished productivity.

Anaemia is characteristic of trypanosomiasis and begins with the first waves of parasitaemia and is progressive. It is clear that nutrition does not influence the onset of the anaemia but the rate of its subsequent development and severity may be affected by the levels of energy intake with lower levels being frequently associated with more rapid development of the anaemia. Following treatment higher protein levels are associated with a more rapid recovery of the packed cell volume.

Pyrexia also follows the onset of parasitaemia and is cyclical with the waves of parasitaemia. It is also associated with increased energy demands for maintenance which is ultimately manifested by reduction in voluntary activity levels and reduced productivity.

Body weight changes (rate of gain or loss) in trypanosomiasis are particularly influenced by the levels of protein intake. High intakes allow infected animals to grow at the same rate as uninfected control animals and indicate that as in the case of gastrointestinal helminthiasis high protein intakes can largely ameliorate the adverse affects of parasitism providing energy intake is adequate whilst low energy levels can exacerbate the adverse effects of trypanosomiasis on body weight.

Reductions in feed intake are common in parasitized animals and this is a feature of trypanosomiasis. However, this is less apparent in animals which are provided with high protein diets, since there are benefits from both the quality and quantity of the dietary intake resulting in significantly reduced impairment from the disease. Where intake is limited by the disease, animals will often exhibit preferential selection of higher quality browse. Although feed selection effects are still difficult to evaluate and the apparent preference for the 'better food' does not always occur.

A common feature of trypanosomiasis is impaired productivity including reduced draught work output, lower milk yields and reduced reproductive capacity. All of these adverse features can be ameliorated to a considerable extent by the provision of higher quality feeds which helps maintain appetite, reduces the severity of the anaemia and supplies energy for work in addition to that required for increased maintenance.

It can therefore be concluded that increased dietary protein and energy intakes ameliorate the effect of trypanosome infection on growth rate and that it would appear that high energy intake may not prevent the rapid establishment of trypanosomes following inoculation but the ensuing anaemia is less severe than that observed in animals receiving a low energy ration. The work reviewed here reinforces the finding that dietary protein has a major influence on erythropoietic activity which accounts for improvement in PCV values especially in long-standing infections.

Despite these tentative conclusions many aspects of the interactions between nutrition and trypanosomiasis remain unclear. The areas which require further study include detailed evaluations of the minimum levels of protein and energy supplementation required to ameliorate the adverse effects of trypanosomiasis, the nature and quality of the protein supplement to achieve these benefits and the influence these have on digestive physiology. Through such knowledge, practical husbandry strategies may be developed, which will allow animals to survive and be productive in many of the large areas of Africa infested with tsetse flies.

REFERENCES

AGYEMANG, K., DWINGER, R. H., JEANNIN, P., LAPERRE, P., GRIEVE, A. S., BAH, M. L. & LITTLE, D. A. (1990a). Biological and economic impact of trypanosomes infections on milk production in N'Dama cattle managed under village conditions in The Gambia. *Animal Production* **50**, 383–389.

AGYEMANG, K., DWINGER, R. H., LITTLE, D. A., LAPERRE, P. & GRIEVE, A. S. (1992). Interactions between physiological status in N'Dama cows and trypanosome infections and its effect on health and productivity of cattle in The Gambia. *Acta Tropica* **50**, 91–99.

AGYEMANG, K., DWINGER, R. H., TOURAY, B. N., JEANNIN, P., FOFANA, D. & GRIEVE, A. S. (1990b). Effects of nutrition on the degree of anaemia and liveweight changes in N'Dama cattle infected with trypanosomes. *Livestock Production Science* **26**, 39–51.

AKINBAMIJO, O. O., ADEMOSUM, A. A., ZWART, D., TOLKAMP, B. J. & BROUWER, B. O. (1990). Effect of *Trypanosoma brucei* infection on liveweight, organic matter intake, digestibility and N-balance in West African Dwarf goats. *Tropical Veterinarian* **8**, 140–148.

AKINBAMIJO, O. O., BENNISON, J. J., ROMNEY, D. L., WASSINK, G. J., JAITNER, J., CLIFFORD, D. J. & DEMPFLE, L. (1997). An evaluation of food intake, digestive physiology and live-weight changes of N'Dama and Gobra zebu bulls following experimental *Trypanosoma congolense* infection. *Animal Science* **65**, 151–158.

AKINBAMIJO, O. O., HAMMINGA, B. J., WENSING, T., BROUWER, B. O., TOLKAMP, B. J. & ZWART, D. (1992). The effect of *Trypanosoma vivax* infection in West African Dwarf goats on energy and nitrogen metabolism. *The Veterinary Quarterly* **14**, 95–100.

AKINBAMIJO, O. O., REYNOLDS, L. & GORT, G. (1994a). Effects of *Trypanosoma vivax* infection during pregnancy on food intake, nitrogen retention and liveweight changes in West African Dwarf ewes. *Journal of Agricultural Science* **123**, 379–385.

AKINBAMIJO, O. O., REYNOLDS, L., SHERINGTON, J. & NSAHLAI, I. V. (1994b). Effects of post-partum *Trypanosoma vivax* infection on feed intake, liveweight changes, milk yield and composition in West African Dwarf ewes and associated lamb growth rates. *Journal of Agricultural Science* **123**, 387–392.

BARRY, J. D. & TURNER, C. M. R. (1991). The dynamics of antigenic variation and growth of African trypanosomes. *Parasitology Today* **7**, 207–211.

BENNISON, J. J. (1997). *The Effects of Nutrition and Trypanosomiasis on the Productivity of Trypanotolerant N'Dama Cattle.* Ph.D. thesis, Wye College, University of London.

BENNISON, J. J., AKINBAMIJO, O. O., JAITNER, J., DEMPFLE, L., HENDY, C. R. & LEAVER, J. D. (1999). Effects of nutrition *pre-partum* and *post-partum* on subsequent productivity and health of N'Dama cows infected with *Trypanosoma congolense. Animal Science* **68**, 819–829.

BENNISON, J. J., CLEMENCE, R. G., ARCHIBALD, R. F., HENDY, C. R. C. & DEMPFLE, L. (1998*a*). The effects of work and two planes of nutrition on trypanotolerant draught cattle infected with *Trypanosoma congolense. Animal Science* **66**, 595–605.

BENNISON, J. J., SHERINGTON, J., WACHER, T. J., DEMPFLE, L. & LEAVER, J. D. (1998*b*). Effects of *Trypanosoma congolense* infection and groundnut (*Arachis hypogaea*) hay supplementation on ranging, activity and diet selection of N'Dama cows. *Applied Animal Behaviour Science* **58**, 1–12.

BLACK, S. & VANDERWEERD, V. (1989). Serum lipoproteins are required for multiplication of *Trypanosoma brucei brucei* under axenic culture conditions. *Molecular and Biochemical Parasitology* **37**, 65–72.

BOURN, D. & SCOTT, M. (1978). The successful use of work oxen in agricultural development of tsetse infected land in Ethiopia. *Tropical Animal Health and Production* **10**, 191–203.

CARROL, M. & McCROIRE, P. (1986). Lipid composition of blood stream forms of *Trypanosoma brucei brucei. Comparative Biochemical Physiology* **83B**, 647–651.

COOP, R. L. & HOLMES, P. H. (1996). Nutrition and parasite interaction. *International Journal for Parasitology* **26**, 951–962.

COPPENS, I., OPPERDOES, F. R., COURTOY, P. J. & BAUDHUIN, P. (1987). Receptor-mediated endocytosis in the bloodstream form of *Trypanosoma brucei. Journal of Protozoology* **34**, 465–473.

DWINGER, R. H., CLIFFORD, D. J., AGYEMANG, K., GETTINBY, G., GRIEVE, A. S., KORA, S. & BOJANG, M. A. (1992). Comparative studies on N'Dama and Zebu cattle following repeated infections with *Trypanosoma congolense. Research in Veterinary Science* **52**, 292–298.

FAGBEMI, B. O., OTESILE, E. B., MAKINDE, M. O. & AKINBOADE, O. A. (1990). The relationship between dietary energy levels and the severity of *Trypanosoma brucei* infection in growing pigs. *Veterinary Parasitology* **35**, 29–42.

GILLETT, M. P. T. & OWEN, J. S. (1987). *Trypanosoma brucei brucei* obtains cholesterol from plasma lipoproteins. *Biochemical Society Transactions* **15**, 258–259.

HECKER, P. A., COULIBALY, L., ROWLANDS, G. S., NAGDA, S. M. & d'ITEREN, G. D. M. (1991). Effect of plane of nutrition on trypanosome prevalence and mortality of Djallonke sheep exposed to high tsetse challenge. In *Proceedings of the 21st Meeting of the International Scientific Council for Trypanosomiasis Research and Control*, Cote d'Ivore, ISCTRC, Yamoussoukro, 21–25th October, 1991.

HOLMES, P. H. (1976). The use of radioisotopic tracer techniques in the study of the pathogenesis of trypanosomiasis. In *Nuclear Techniques in Animal Production and Health.* Vienna, International Atomic Energy Agency, pp. 663–674.

HOLMES, P. H. & BREMNER, K. C. (1971). The pathophysiology of ovine ostertagiaisis. Water balance and turnover studies. *Research in Veterinary Science* **12**, 381–383.

KATUNGUKA-RWAKISHAYA, E. (1992). *The Pathophysiology of Ovine Trypanosomiasis Caused by* Trypanosoma congolense. Ph.D. thesis, University of Glasgow, Scotland: UK.

KATUNGUKA-RWAKISHAYA, E., MURRAY, M. & HOLMES, P. H. (1992). Pathophysiology of ovine trypanosomiasis: Ferrokinetic and erythrocyte survival studies. *Research in Veterinary Science* **53**, 80–86.

KATUNGUKA-RWAKISHAYA, E., PARKINS, J. J., FISHWICK, G., MURRAY, M. & HOLMES, P. H. (1993). The pathophysiology of *Trypanosoma congolense* infection in Scottish Blackface sheep: Influence of dietary protein. *Veterinary Parasitology* **47**, 189–204.

KATUNGUKA-RWAKISHAYA, E., PARKINS, J. J., FISHWICK, G., MURRAY, M. & HOLMES, P. H. (1995). The influence of energy intake on the pathophysiology of *Trypanosoma congolense* infection in Scottish Blackface sheep. *Veterinary Parasitology* **59**, 207–218.

KYRIAZAKIS, I., OLDHAM, J. D., COOP, R. L. & JACKSON, F. (1994). The effect of subclinical intestinal nematode infection on the diet selection of growing sheep. *British Journal of Nutrition* **72**, 665–677.

LEEFLANG, P., IGE, K. & OLATUNDE, D. S. (1976). Studies on *Trypanosoma vivax*: The infectivity of cyclically and mechanically transmitted ruminant infections for mice and rats. *International Journal for Parasitology* **6**, 453–456.

LITTLE, D. A., DWINGER, R. H., CLIFFORD, D. J., GRIEVE, A. S., KORA, S. & BOJANG, M. (1990). Effect of nutritional level and body condition on susceptibility of N'Dama cattle to *Trypanosoma congolense* infection in The Gambia. *Proceedings of the Nutrition Society* **49**, 209A.

McCARTHY, D. O., KLUGER, M. J. & VANDER, A. J. (1985). Suppression of food intake during infection: is Interleukin-1 involved? *American Journal of Clinical Nutrition* **42**, 1179–1182.

MAKINDE, M. O., OTESILE, E. B. & FAGBEMI, B. O. (1991). Studies on the relationship between dietary energy levels and the severity of *Trypanosoma brucei* infection: The effects of diet and infection on blood and plasma volumes and the erythrocyte osmotic fragility of growing pigs. *Bulletin of Animal Health and Production in Africa* **39**, 161–166.

MURRAY, M. & BLACK, S. J. (1985). African trypanosomiasis in cattle: working with nature's solution. *Veterinary Parasitology* **18**, 167–182.

MURRAY, M., CLIFFORD, D. J., GETTINBY, G., SNOW, W. F. & McINTYRE, W. I. M. (1981). Susceptibility to African trypanosomiasis of N'Dama and Zebu cattle in an area of *Glossina morsitans submorsitans* challenge. *The Veterinary Record* **109**, 503–510.

MURRAY, M. & DEXTER, T. M. (1988). Anaemia in African trypanosomiasis. A review. *Acta Tropica* **45**, 389–432.

NANTULYA, V. M., MUSOKE, A. J., RURANGIRWA, F. R. & MOLOO, S. K. (1984). Resistance of cattle to tsetse-transmitted challenge with *Trypanosoma brucei* and *Trypanosoma congolense* after spontaneous recovery from syringe passaged infections. *Infection and Immunity* **43**, 735–738.

OTESILE, E. B., FAGBEMI, B. O. & ADEYEMO, O. (1991). The effect of infection on some serum biochemical parameters in boars on different planes of dietary energy. *Veterinary Parasitology* **40**, 207–216.

PALING, R. W., MOLOO, S. K., SCOTT, J. R., GETTINBY, G., McODIMBA, F. A. & MURRAY, M. (1991). Susceptibility of N'Dama and Boran cattle to sequential challenges with tsetse transmitted clones of *Trypanosoma congolense*. *Parasite Immunology* **13**, 427–445.

PARKINS, J. J., BAIRDEN, K. & ARMOUR, J. (1982). *Ostertagia ostertagi* in calves: a growth, nitrogen balance and digestibility study conducted during winter feeding following thiabendazole anthelmintic therapy. *Journal Comparative Pathology* **92**, 219–227.

PARKINS, J. J. & HOLMES, P. H. (1989). Effects of gastrointestinal parasites on ruminant nutrition. *Nutrition Research Reviews* **2**, 227–246.

REYNOLDS, L. & EKWURUKE, J. O. (1988). Effect of *Trypanosoma vivax* infection on West African Dwarf sheep at two planes of nutrition. *Small Ruminant Research* **1**, 175–188.

ROBERTS, C. J., TIZARD, I. R., MELLORS, A. & CLARKSON, M. J. (1977). Lysophospholipases, lipid metabolism and the pathogenesis of African trypanosomiasis. *The Lancet* December 3, 1187–1188.

ROMNEY, D. L., N'JIE, A., CLIFFORD, D. J., HOLMES, P. H., RICHARD, D. & GILL, M. (1997). The influence of nutrition on the effects of infection with *Trypanosoma congolense* in trypanotolerant cattle. *Journal of Agricultural Science* **129**, 83–89.

TIZARD, I., NIELSEN, K. H., SEED, J. R. & HALL, J. E. (1978). Biological active products from African trypanosomes. *Microbiological Reviews* **42**, 661–681.

TRAORE-LEROUX, T., FUMOUX, F. & PINDER, M. (1987). High density lipoprotein levels in the serum of trypanosensitive and trypanoresistant cattle. Changes during *Trypanosoma congolense* infection. *Acta Tropica* **44**, 315–323.

UDEN, P., COLUCCI, P. E. & VAN SOEST, P. J. (1980). Investigation of Chromium, Cerium and Cobalt as Markers of Digesta. Rate of Passage Studies. *Journal of the Science of Food and Agriculture* **31**, 625–632.

UDEN, P., ROUNSAVILLE, T. R., WIGGANS, G. R. & VAN SOEST, P. J. (1982). The measurement of liquid and solid digesta retention in ruminants, equines, and rabbits given timothy (*Phleum pratense*) hay. *British Journal of Nutrition* **48**, 329–339.

VAN DAM, J. T. P. (1996). *The Interaction Between Nutrition and Metabolism in West African Dwarf Goats, Infected with Trypanosomes*. Ph.D. thesis, Wageningen Agricultural University: The Netherlands.

VAN DAM, J. T. P., VAN DER HEIDE, D., VAN DER HEL, W., VAN DEN INGH, T. S. G. A. M., VERSTEGEN, M. W. A., WENSING, T. & ZWART, D. (1996). The effects of *Trypanosoma vivax* infection on energy and protein metabolisms and serum metabolites and hormones in West African Dwarf goats on different food intake levels. *Animal Science* **63**, 111–121.

VAN HOUTERT, M. F. J. & SYKES, A. R. (1996). Implications of nutrition for the ability of ruminants to withstand gastrointestinal nematode infections. *International Journal for Parasitology* **26**, 1151–1168.

VAN MIERT, A. S. J. P. A. M., VAN DUIN, C. T. M. & KOOT, M. (1990). Effects of *E. coli* endotoxin, some interferon inducers, recombinant interferon-a$_{2a}$ and *Trypanosoma brucei* infection on feed intake in dwarf goats. *Journal of Veterinary Pharmacology and Therapeutics* **13**, 327–331.

VENKATESAN, S. & ORMEROD, W. E. (1976). Lipid content of the slender and stumpy forms of *Trypanosoma brucei rhodesiense*: a comparative study. *Comparative Biochemistry and Physiology* **53B**, 481–487.

VERSTEGEN, M. W. A., ZWART, D., VAN DER HEL, W., BROUWER, B. O. & WENSING, T. (1991). Effect of *Trypanosoma vivax* infection on energy and nitrogen metabolism of West African dwarf goats. *Journal of Animal Science* **69**, 1667–1677.

VICKERMAN, K. & TETLEY, L. (1979). Biology and ultrastructure of trypanosomes in relation to pathogenesis. In *Pathogenicity of Trypanosomes* (ed. Losos, G. & Chouinard, Amy.), pp. 23–31. Ottawa: Canada, IDRC 132e.

WACHER, T. J. (1993). Interactions of season and trypanosome infection with ranging, activity and diet selection parameters in individual village managed N'Dama cattle in The Gambia. *Consultancy Report for the Natural Resources Institute* (*NRI*), January 1993.

WASSINK, G. J. (1997). *The Effect of Nutrition on the Pathophysiology of Trypanosomiasis in Scottish Blackface Sheep*. Ph.D. thesis, University of Glasgow: UK.

WASSINK, G. J., FISHWICK, G., PARKINS, J. J., GILL, M., ROMNEY, D. L., RICHARD, D. & HOLMES, P. H. (1997). The patho-physiology of *Trypanosoma congolense* in Scottish Blackface sheep: influence of diet on digestive function. *Animal Science* **64**, 127–137.

ZWART, D., BROUWER, B. O., VAN DER HEL, W., VAN DEN AKKER, H. N. & VERSTEGEN, M. W. A. (1991). Effect of *Trypanosoma vivax* on body temperature, feed intake and metabolic rate of West African dwarf goats. *Journal of Animal Science* **69**, 3780–3788.

Electrophysiological investigation of anthelmintic resistance

R. J. MARTIN[1]* and A. P. ROBERTSON[2]

[1] Department of Biomedical Sciences, College of Veterinary Medicine, Iowa State University, Ames, IA 50011, USA
[2] Department of Preclinical Veterinary Sciences, University of Edinburgh, Edinburgh EH9 1QH, UK

SUMMARY

It is pointed out that two of the three major groups of anthelmintic act by opening membrane ion-channels. It is appropriate, therefore, to use electrophysiological methods to study the properties of the sites of action of these drugs and the changes in the properties of these receptor sites associated with resistance. This paper describes the use of the patch-clamp technique to observe the currents that flow through the levamisole-activated channels as they open and close in levamisole-sensitive and levamisole-resistant isolates. It was found that, on average, the proportion of time the channels are open, is less in the resistant isolate. The patch-clamp technique also showed that the ion-channels are heterogeneous and that one of the subtypes is lost with the appearance of resistance. The use of the current clamp technique is illustrated to record a site of action of ivermectin in the pharyngeal muscle of *Ascaris*.

Key words: Levamisole, ivermectin, patch-clamp, anthelmintic, resistance, *Oesophagostomum dentatum*.

INTRODUCTION

The three major groups of anthelmintic drugs used for the treatment of nematode parasite infections are the benzimidazoles (e.g. thiabendazole and albendazole), the avermectins (e.g. ivermectin and doramectin), and the selective nictotinic agonists (e.g. levamisole and pyrantel). The mode of action of the avermectins and the nictotinic agonists is by means of selective opening of ligand-gated ion-channels found in the membranes of nerves and muscle of nematodes. If we wish to study the anthelmintic action of the avermectins and nicotinic agonists, then part of that investigation involves the use of electrophysiological techniques to study properties of the nematode ion-channel receptors. The action of benzimidazoles, by contrast, involves an exclusive binding to nematode β-tubulin and consequential inhibition of the formation of nematode microtubules. Electrophysiological techniques for the study of benzimidazoles may not be required.

Resistance to anthelmintic drugs is now recognized as a growing problem (Pritchard, 1994). Information relating to the mechanism of resistance to benzimidazoles is more advanced than the knowledge of the mechanisms of resistance to the two other groups of anthelmintic (Roos, 1990). We know comparatively little of the mechanisms of resistance to the avermectins and nicotinic anthelmintics. Information from studies on the model soil nematode, *C. elegans*, has given us a knowledge of some of the genes involved (Fleming *et al.* 1997; Dent, Davis & Avery, 1997). We know for example that the genes *avr-14* and *avr-15* are involved in the resistance to

ivermectin and that *lev-1, unc-38* and *unc-29* are involved in levamisole resistance. These genes encode subunits of ion-channels that are the target site of anthelmintics, so we know that resistance may involve modification of the target site of these drugs in the model nematode. However other *C. elegans* genes, whose function is not known (Fleming *et al.* 1997) are also involved in resistance to levamisole.

There are limitations to the information that is derived from *C. elegans* when it comes to applying it to parasitic nematodes. One problem is that the laboratory conditions under which *C. elegans* are kept are quite different from conditions required for a parasitic nematode to complete its life cycle. It means that *C. elegans* may survive in a paralysed condition on an agar plate when exposed to a particularly high concentration of anthelmintic but the same concentration of the anthlemintic would produce the demise of a parasitic nematode in the small intestine. Thus the genes identified for producing resistance in *C. elegans* may not always match the genes responsible for producing resistance in parasitic nematodes. There is another limitation and that is that the dynamic properties of the target ion-channel cannot be observed with molecular techniques. It may not be possible therefore to understand with just molecular techniques how the properties of the ion-channel are modified in the resistance. For these two reasons we have decided to look at the properties of ion-channels in parasitic nematodes and to see if and how they may be modified in resistance. We have started out study with *Oesophagostomum dentatum*, a nematode parasite of the pig that is about 1 cm in length. We have used the patch-clamp technique to compare the properties of levamisole receptor ion-channels in

* E-mail: rjmartin@iastate.edu

Parasitology (2000), **120**, S87–S94.   Printed in the United Kingdom   © 2000 Cambridge University Press

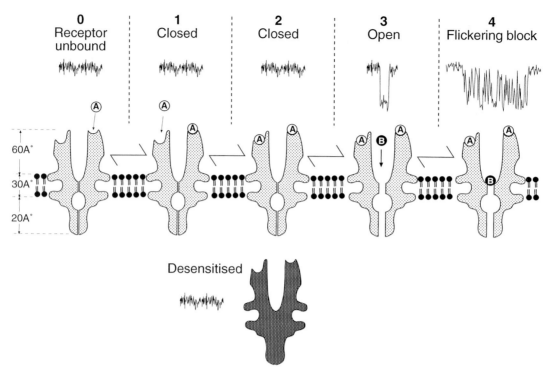

Fig. 1. Cartoon of a levamisole gated ion-channel (a nematode nicotinic acetylcholine channel) and its opening as a result of the binding of two molecules of agonist. Two agonist molecules (A e.g. levamisole) combine sequentially with non-equivalent binding sites on the extracellular surface of the channel and permit the opening of the channel. When the channel opens for a brief period (a few milliseconds) a current pulse of a few picoAmps flows in through the channel. A large inorganic cation (including the anthelmintic levamisole) (B) may enter but not pass through the channel pore and produces a 'flickering' channel block as it repeatedly binds and unbinds with a block site deep in the pore. The desensitised state is also shown as a closed non-conducting condition of the channel. Desensitisation is produced in a time-dependent manner following the addition of 'high' agonist drug concentrations and is a long closed state.

isolates that are sensitive (SENS) and isolates that are resistant (LEVR) to levamisole. This paper describes electrophysiological techniques we have used and some of our observations.

PROPERTIES OF LIGAND-GATED ION-CHANNELS

Fig. 1 illustrates how ligand-gated ion-channels are believed to operate. The channel is composed of five protein subunits that traverse the phospholipid membrane of the cell and that together form a tunnel through the membrane extending on both sides of the membrane. The ion-channel is normally closed in the absence of specific ligands but in the presence of specific ligands, it can open. It is usually taken that two molecules of the ligand bind to the extracellular domain of the ion-channel and this binding lowers the energy levels required for the channel to open. When the channel opens, it does so transiently giving rise to a pulse of current as ions move through the pore of the channel. The type of ions (charge and size) carried by the pore depends on the arrangement of amino acids that line the pore of the channel. Nicotinic ion-channels have negatively charged rings

of glutamate and aspartate amino acids in the pore, that allow the cations Na^+ and K^+ to pass through and carry the current: nicotinic channels are excitatory. Glutamate-gated Cl^- channels, that are the target site of avermectins, have a pore that permits the conduction of Cl^- ions: these channels are inhibitory. Both types of channel only open transiently for a period that lasts on average 1–20 msec, depending on the channel type: the mean open time, τ, of an ion-channel is a characteristic of that channel. As the channel opens, a pulse of current about 1 pA $(1 \times 10^{-12}$ A) flows. The channel opens and closes in the presence of the ligand and is only open for a small proportion of time, P_{open}, which is in the range 0·0001–0·01. An increase in the ligand concentration increases the proportion of time the channel is open and so increases the current that flows through the membrane. The amplitude of the current pulse is usually directly related to the voltage across the cell membrane; the relationship between the current carried by the pore and voltage across the membrane is then described by Ohm's law. The relationship between the current and the voltage across the ion-channel is described by its conductance (g: the reciprocal of resistance): g values of ion-channels are in the range 1–300 pS (pS $= 10^{-12}$ Ohms^{-1}).

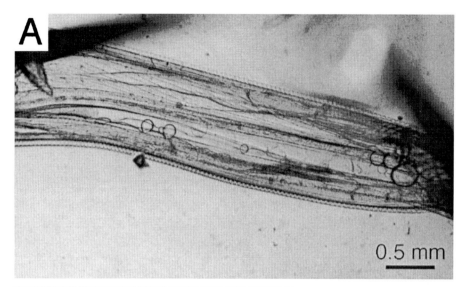

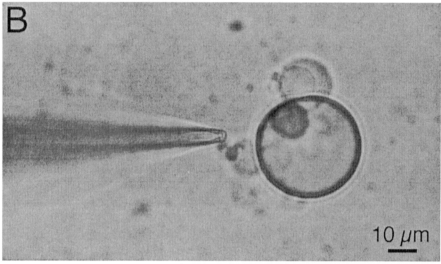

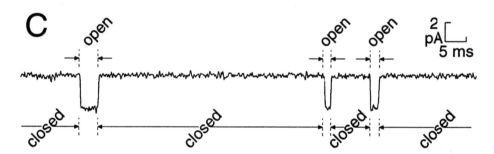

Fig. 2. (A) Muscle-flap preparation of *Oesophagostomum dentatum* showing the production of muscle membrane vesicles following collagenase treatment. (B) A single membrane vesicle and an approaching patch-pipette viewed under phase contrast. (C) Single channels currents activated by 30 μM levamisole in a cell-attached patch at -75 mV. The open- and closed-channel periods are indicated by the horizontal arrows. The scale bar showing the current amplitude and time duration is also shown.

Fig. 1 also illustrates the additional effects that a ligand may have on nicotinic ion-channels; desensitization and with some agonists, open-channel block. Desensitization is seen at some ligand-gated ion-channels as inactivation (no opening) following an active period of channel opening. It is most often produced by high concentrations of agonist. Open-channel block is produced when the ligand also enters the ion-channel pore and produces temporary block. The block usually occurs at high drug concentrations and gives rise to a 'flickering' appearance if the period of the block is short. Two mechanisms by which resistance could arise at ion-channels then is by an increase in ease with which an ion-channel receptor will desensitise (turn off) or by an increase in the binding of a drug like levamisole within the channel pore so that the receptor is blocked more frequently.

The mechanism by which ivermectin activates the glutamate-gated chloride channel presumably also involves a reduction in the potential energy level required for the channel to open. However, the binding site for ivermectin is not known; it is unlikely to be the same site on the ion-channel as the natural ligand, glutamate, because the chemical structures of glutamate (a dicarboxylic amino acid) and ivermectin (a macrocyclic lactone) are very different. Ivermectin is lipophilic, so the site it binds to may be within the membrane on the section of the ion-channel spanning the membrane.

THE PATCH-CLAMP TECHNIQUE FOR RECORDING OPENING OF LEVAMISOLE RECEPTOR ION-CHANNELS

Figs. 2A, B illustrate a muscle membrane vesicle preparation from *Oesophagostomum dentatum* that we have used to record levamisole activated single-channel currents from levamisole-sensitive and levamisole-resistant isolates. The currents that pass through the ion-channels as they open and close are collected with the patch-pipette and recorded with a high gain current-voltage converter (the technique referred to as the patch-clamp technique). The method involves preparing 'naked' membranes with the ion-channels present in them and pressing and sealing a blunted micropipette on the membrane to collect the ion-channels currents (Hamilton *et al.* 1981). The size of the pipette is about 2 μm across so it samples channels from only a small region of the membrane. The ligand required to activate the channels is included in the patch-pipette.

Fig. 2C illustrates a trace from a cell-attached muscle-vesicle patch recorded with 10 μM levamisole in the patch pipette at a membrane potential of −50 mV from a levamisole-sensitive isolate of *O. dentatum*. The openings are labelled and are down-ward. The closed-times are also marked. The electrical recording is processed using computer programmes that measure the individual open-times, the individual closed-times and the conductance of the openings. From this information we are able to calculate the mean open-times, τ, and the probability that the channel is in the open state, P_{open}, (= $\tau/[\tau +$ mean closed-time]). With the patch-clamp technique we are able to measure properties of individual receptors and to compare properties seen in sensitive (SENS) and resistant (LEVR) isolates. We can also count the number of active receptors that are present in the patch recorded at different concentrations of levamisole and compare sensitive and resistant isolates.

COMPARISON OF PROPERTIES OF LEVAMISOLE RECEPTORS IN SENSITIVE (SENS) AND RESISTANT (LEVR) ISOLATES

Number of active levamisole receptors is less in the resistant isolate

Fig. 3A shows a histogram of the proportion of patch pipettes that contained active channels in the sensitive and resistant isolates at the different concentration of levamisole. At 10 μM the proportion of patches that contained active channels was similar in both isolates; but at 30 μM, the concentration estimated to occur at therapeutic levels of the anthelmintic, we observed that there were less active channels in the resistant isolate than in the sensitive isolate. We interpreted this observation as indicating that the channels of the resistant isolate are likely to desensitise (turn off) more than in the sensitive isolate. This suggestion is consistent by the reduction in the proportion of patches that contained active receptors at 100 μM: at this concentration there was a decrease for both sensitive and resistant isolates.

The mean open-time is less in the resistant isolate

We also measured the mean open-times of the ion-channels in both isolates. Fig. 3B is a representative distribution histogram of the open-times of patches recorded at −50 mV with 30 μM levamisole as the ligand. Notice that the distributions are exponential with the brief openings being most frequent. The mean open-times are calculated by fitting an exponential curve with a maximum likelihood procedure: it is 1·01 msec in the sensitive (SENS) patch. In the resistant (LEVR) patch, the mean open-time is 0·61 msec. It is briefer and would carry less current on average during each channel-opening event. The mean open-times from all of our patch recordings observed at different concentrations are shown in Fig. 3C. As the concentration of levamisole increases the mean open-time of the channel decreases due to the open-channel block mentioned earlier. Despite the block, at all concentrations that the mean open-time of the resistant isolate is shorter than the sensitive isolate. We found at the estimated therapeutic concentration of 30 μM that the reduction in the number of active receptors and the averaged re-duction in the probability of the channel being open produced a 10-fold reduction in the levamisole current that would be carried across the membrane. In other words the receptors would produce a smaller contractile response to levamisole in the resistant isolates than the sensitive isolates.

Channel-conductance histograms show heterogeneity of receptors

Fig. 4A is a histogram of the conductances of individual ion-channel receptors activated by levamisole and recorded from levamisole sensitive

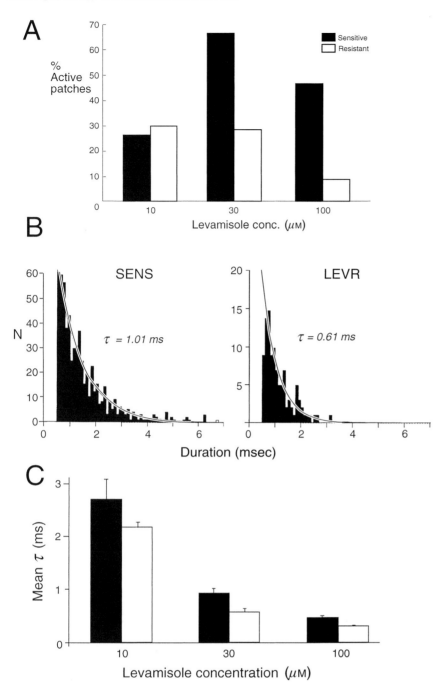

Fig. 3. Percentage of active patches at each levamisole concentration for SENS (■) and LEVR (□) parasites. Note that at 10 μM, the proportion of patches containing active patches was similar but at the estimated therapeutic concentration, there were more active receptors in the SENS isolate. (B) Examples of open-time distributions at −75 mV from experiments carried out on SENS and LEVR parasites (30 μM levamisole as agonist). Each distribution was fitted with a single exponential giving values for τ of 1·01 msec (SENS) and 0·61 msec (LEVR). (C) Histogram illustrating the mean ±s.e. τ values for −75 mV at the three levamisole concentrations tested; SENS (■) and LEVR (□). Note that at each of the concentrations, there is a lower shorter mean open-time for the LEVR isolate.

(SENS) *O. dentatum*. The histogram cannot be described by a single normal distribution but shows the presence of four peaks: *G25*, *G35*, *G40* and *G45*. The wide range of the conductances shown by the channels from 19 pS to 48 pS is not explained by any variation in experimental technique since the conductance of individual receptor channels may be measured to within ±1 pS. We have interpreted these observations (Robertson, Bjorn & Martin,

1999) as indicating that there are four subtypes of levamisole receptor present in the population of *O. dentatum*. The different subtypes may be produced by variation in the amino acid structure of any of the five subunits of the ion-channel pore, variation in the stoichiometry of the ion-channel, or by an alteration in the phosphorylation state of any of the constituent subunits of the receptor ion-channel. When we examined the conductance histogram of the channels

A

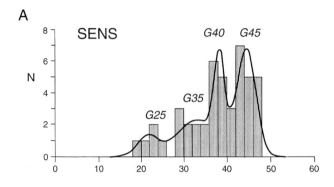

B

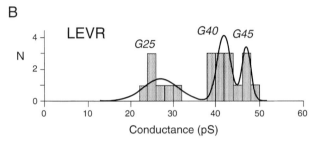

Fig. 4. Frequency histograms of single-channel conductances for levamisole-sensitive *O. dentatum* SENS (A) and levamisole-resistant *O. dentatum* LEVR (B) parasites. In the sensitive isolate, the conductance of the levamisole-activated channels ranged 18·1 pS–48·0 pS with a mean of 37·7 pS ± 1·1 pS, $n = 45$. In the resistant isolate the conductance values ranged 15·2 pS–47·8 pS with a mean of 36·6 pS ± 2·0 pS, $n = 22$. Note the large range between the maximum and minimum values of the histograms. Gaussian curves were fitted to each distribution using the maximum likelihood procedure. The peaks for the SENS isolate were 21·4 ± 2·3 pS (8 % area) labelled G25; 33·0 ± 4·8 pS (31 % area) labelled G35; 38·1 ± 1·2 pS (19 % area) labelled G40; and 44·3 ± 2·2 pS (42 % area) labelled G45. The peaks for the LEVR isolate were 25·2 ± 4·5 pS (21 % area) labelled G25; 41·2 ± 1·7 pS (49 % area) labelled G40; and 46·7 ± 1·1 pS (30 % area) labelled G45.

activated by levamisole in levamisole-resistant isolates (LEVR) we found that the *G35* peak was missing. This observation may be explained if the presence of the *G35* receptor is associated with susceptibility to levamisole. The development of resistance fits with the model whereby there are a number of allelic variations present in the nematode parasite, some of the alleles are sensitive to levamisole and some resistant. The selection pressure produced by the continued use of levamisole allows the expansion of the resistant population by preventing the survival of the sensitive isolates.

TWO-MICROPIPETTE CURRENT-CLAMP TECHNIQUE

The advantage of the patch-clamp technique is that it can determine the properties of individual receptors. The disadvantage is that it is not an easy technique to use and requires the preparation of membranes suitable for application of the patch-pipette. We have also used a two-micropipette technique to measure the change in the membrane resistance associated with the opening of ligand-gated ion-channels in nematode parasites (Martin, 1982, 1995). The principle of approach is that the two micropipettes are inserted into the cell of interest: one pipette is used to inject current and the other is used to measure the voltage response. Again Ohm's law is used to measure the conductance of the membrane. Opening of ion-channels is associated with an increase in the conductance of the membrane.

Ascaris *pharyngeal muscle*

Fig. 5A illustrates the use of the two-micropipette technique for measuring the input conductance of the pharynx of *Ascaris*. Also shown in Fig. 5 is the use of two additional micropipettes used for controlled application of drugs to the surface of the pharyngeal preparation.

Fig. 5B shows in the top two traces a diagram of the effect of injecting a 500 msec rectangular current pulse (I_{inj}) and its effect on the membrane potential (E_m). The membrane potential does not settle immediately to its final level because of the capacitance of the membrane. The input conductance of the membrane, g, may be measured by $I_{inj}/\Delta V$. The lower trace shows, on a lower time resolution, the effect of application of the amino acid L-glutamate. In the *Ascaris* pharyngeal preparation L-glutamate results in inhibition produced by a hyperpolarization with an increase in Cl⁻ conductance (Cl⁻ open). Fig. 5C shows that the application of ivermectin during the horizontal bar produces hyperpolarization along with an increase in conductance. During the continuous application of ivermectin, short pulses of L-glutamate are applied and they mimic the effect of ivermectin. These experiments demonstrate the presence of glutamate-gated Cl⁻ on the *Ascaris* pharynx. It will be interesting to compare the responses of sensitive and resistant isolates to controlled-dose application of glutamate and ivermectin.

Body muscle contains nicotinic receptors and GABA receptors revealed with current clamp

The two-micropipette technique may also be applied to examine receptors on the body muscle (Martin, 1980, 1982). The technique has identified the effect of piperazine (Martin, 1982) and levamisole (Harrow & Gration, 1985) on the GABA and nicotinic acetylcholine receptors, respectively. Again, controlled-dose application of levamisole or piperazine may be used to examine the properties of the receptors on somatic muscle of parasitic muscle and future

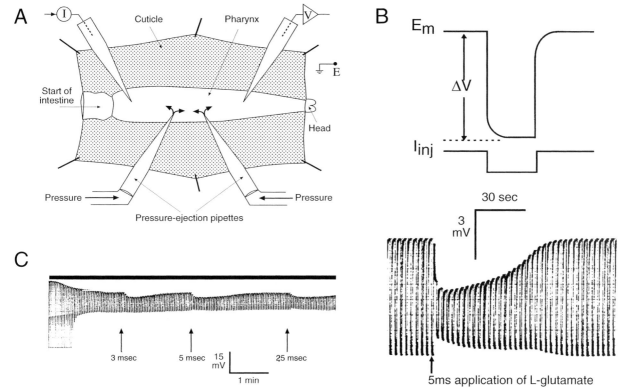

Fig. 5. Diagram of the two-microelectrode current-clamp recording technique that we have used to examine electrophysiological effects of focal application of glutamate and ivermectin to the pharyngeal muscle of *Ascaris suum*. One micropipette (labelled V) is placed in the pharyngeal muscle to record the membrane potential and a second micropipette (I) is also placed in the muscle to inject hyperpolarizing current pulses. Two additional micropipettes are shown which are used for local application of glutamate and ivermectin. A short burst or 'puff' of pressure is used to apply the drugs. (B) Diagram of the effect of injecting a rectangular current pulse (I_{inj}) on the membrane potential (E_m) using a two microelectrode clamp. The current produces an exponentially increasing hyperpolarization that settles after sufficient time to a change of ΔV mV. The input conductance of the pharynx can be determined from Ohms law because the current voltage relationship is linear. (C) The trace shows that brief application of glutamate using a 5 msec 'puff' or pressure application from a micropipette filled with 0·5 M L-glutamate leads a hyperpolarization of the membrane potential and an increase in the input conductance to 200 μS. As the ion-channels open up and carry Cl^- ions, the resistance of the membrane decreases and the membrane potential hyperpolarises. (D) Effects of glutamate and ivermectin on pharyngeal membrane potential and conductance. The trace shows the membrane potential and input conductance response to a continuous application of ivermectin (horizontal bar) and 3 msec, 5 msec and 25 msec 'puffs' of 0·5 M glutamate from a micropipette.

studies may determine changes associated with the development of resistance.

CONCLUSION

This paper has illustrated the use of electro-physiological techniques that may be used to observe changes in the properties of membrane ion-channels associated with the development of anthelmintic resistance. The use of these techniques relates to the fact that two major groups of anthelmintics, the avermectins and nicotinic anthelmintics, act by gating membrane ion-channels. One advantage of the techniques is that they allow a display of the activity of the channel in real time. Resolution of massed receptor responses is possible using the two-micropipette technique. Resolution at the single-channel level is possible with the patch-clamp technique. We are able to make biophysical obser-vations that tell us about the behaviour of the receptors we are examining. The techniques are likely to be combined with molecular techniques in the future to determine the effects of amino acid changes in the receptor and the development of anthelmintic resistance.

REFERENCES

DENT, J. A., DAVIS, M. W. & AVERY, L. (1997). *Avr-15* encodes a chloride channel subunit that mediates inhibitory glutamatergic neurotransmission and ivermectin sensitivity in *Caenorhabditis elegans*. *EMBO Journal* **16**, 5867–5879.

FLEMING, J. T., SQUIRE, M. D., BARNES, T. M., TORNOE, C., MATSUDA, K., AHNN, J., FIRE, A., SULSTON, J. E., BARNARD, E. A., SATTELLE, D. B. & LEWIS, J. A. (1997). *Caenorhabditis elegans* levamisole resistance genes *lev-1, unc-29*, and *unc-38* encode functional nicotinic acetylcholine receptor subunits. *Journal of Neuroscience* **17**, 5843–5857.

HAMILTON, O. P., MARTY, E., NEHER, B., SAKMANN, B. & SIGWORTH, F. J. (1981). Improved patch-clamp techniques for high-resolution current recording from the cells and cell-free membrane patches. *Pflugers Archives* **391**, 85–100.

HARROW, I. D. & GRATION, K. A. F. (1985). Mode of action of the anthelmintics morantel, pyrantel and levamisole on muscle-cell membrane of the nematode *Ascaris suum*. *Pesticide Science* **16**, 662–672.

MARTIN, R. J. (1980). The effect of γ-aminobutyric acid in the input conductance and membrane potential of *Ascaris* muscle. *British Journal of Pharmacology* **71**, 99–106.

MARTIN, R. J. (1982). Electrophysiological effects of piperazine and diethylcarbamazine on *Ascaris suum* somatic muscle. *British Journal of Pharmacology* **77**, 255–265.

MARTIN, R. J. (1995). An electrophysiological preparation of *Ascaris suum* pharyngeal muscle reveals a glutamate-gated chloride channel sensitive to the avermectin analogue milbemycin D. *Parasitology* **112**, 247–252.

PRICHARD, R. K. (1994). Anthelmintic resistance. *Veterinary Parasitology* **54**, 259–268.

ROBERTSON, A. P., BJORN, H. & MARTIN, R. J. (1999). Resistance to levamisole resolved at the single-channel level. *FASEB Journal* **13**, 749–760.

ROOS, M. H. (1990). The molecular nature of benzimidazole resistance in helminths. *Parasitology Today* **6**, 125–127.

The development of anthelmintic resistance in sheep nematodes

F. JACKSON* *and* R. L. COOP

Moredun Research Institute, International Research Centre, Pentlands Science Park, Bush Loan, Penicuik, Midlothian, Scotland EH26 0PZ, UK

SUMMARY

Anthelmintic resistance now poses problems to sheep farmers throughout the world. In some Southern hemisphere countries multiple resistance has reached levels which make sheep farming non-sustainable. Evidence from studies in the UK and Europe suggests (a) that the selection process occurs over a longer time frame than in Southern tropical/temperate regions and (b) that for some of the key ovine species little or no reversion to susceptibility may occur for many years after the withdrawal of the selecting agent. The dynamics of the selection process are influenced by a number of host, parasite, drug, management and environment-dependent factors. Recent mechanistic studies on resistance against avermectins and milbemycins (AM) suggest that there may be a number of mechanisms associated with resistance at the different target sites for these drugs. Within Europe endectocides within the AM drug group have now become the crucial element in strategies aimed at controlling important diseases such as sheep scab and nematodoses. Given that there is little likelihood of a series of novel action compounds emerging in the immediate future to replace this family the conservation of efficacy of the AM group should be accorded the highest priority for research in this area.

Key words: Anthelmintic resistance, nematodes, control.

INTRODUCTION

The emergence of resistance against antimicrobials (McKellar, 1998), antiprotozoals and against drugs and chemicals used to control arthropod pests and helminth parasites (Waller, 1997) provides a chilling reminder of the adaptive capacity of many invertebrate organisms associated with important diseases in man and animals. Given that resistance can either develop relatively simply through mutation or, as appears to be the case for many helminths, as a pre-adaptive phenomenon, then the emergence of resistant populations appears to be an inevitable consequence of intensive chemotherapy.

In the last 6 years there have been a number of review articles covering anthelmintic resistance in general (Craig, 1993; Jackson, 1993; Hazelby, Probert & Rowlands, 1994; Prichard, 1994; Waller, 1994, 1997; Condor & Campbell, 1995; Waller *et al.* 1995; Sangster, 1999) or its relationship to drug pharmacology (Sangster, 1996). Reported cases of anthelmintic resistance are now a global phenomenon and high levels of prevalence of nematodes exhibiting multiple resistance have recently been reported in South America (Echevarria *et al.* 1996; Eddi *et al.* 1996; Maciel *et al.* 1996; Nari *et al.* 1996) and South Africa (Van Wyk, personal communication). Table 1 summarizes the findings on the prevalence of resistance in *Haemonchus* from these surveys. In the absence of alternative control

measures such high levels of resistance, involving a highly pathogenic parasite such as *Haemonchus*, may make it impossible to sustain sheep production on certain farms.

This paper reviews the current situation in the UK and our understanding of some of the key processes that affect the development of anthelmintic resistance in sheep as well as the strategic importance of chemotherapy and chemoprophylaxis in many of our production systems.

PREVALENCE OF ANTHELMINTIC RESISTANCE IN UK AND EUROPE

The first case of BZ resistance (*Teladorsagia* (*Ostertagia*) *circumcincta*) in sheep was recorded on a farm in Cheshire (Britt, 1982) about 20 years after thiabendazole (TBZ) was first introduced into the UK. As a consequence of this report studies were initiated in the south of England (Cawthorne & Whitehead, 1983) to screen both a research flock and a commercial enterprise for the presence of BZ resistant isolates. These studies identified *T. circumcincta* populations which were resistant to TBZ, fenbendazole (FBZ), oxfendazole (OFZ) and albendazole (ABZ) demonstrating the presence of side-resistance. Subsequently, Cawthorne & Cheong (1984) applied both *in vitro* and *in vivo* tests to investigate 52 commercial sheep flocks in the south east of England and evidence of resistance to TBZ was found on 13·5% of the farms with *Haemonchus contortus* being the predominant resistant species

* Corresponding author. Tel: 0131 445 5111 Ext. 144. Fax: 0131 445 6111. E-mail: jackf@mri.sari.ac.uk

Parasitology (2000), **120**, S95–S107.   Printed in the United Kingdom   © 2000 Cambridge University Press

Table 1. *Prevalence of resistance of* Haemonchus *against various anthelmintics in South Africa and South America*

Country	Farm	Percentage of resistant farms for given anthelmintics					
		Bz	Iv	Lev	Bz + Lev	Rfx	Clos
S.Africa	80	79	73	23	—	89	—
Paraguay	37	70	67	47	—	—	—
Uraguay	242	61	1	29	—	—	—
Brazil	182	68	7	19	15	—	20
Argentina	65	37	2	8	5	—	—

Key: BZ, benzimidazoles; Iv, Ivermectin; Lev, Levamisole; Rfx, rafoxanide; Clos, Closantel.

(Table 2). These findings were supported by a further investigation in the same region of the country (Taylor & Hunt, 1988).

These early investigations primarily focused on farms where a drug resistance problem might be anticipated from information on the frequency of drenching and the animal husbandry system in operation. More recently several detailed surveys have assessed the prevalence of anthelmintic resistance in a random sample of sheep and goat flocks from several different geographical regions of England. Hong *et al.* (1992) investigated 209 sheep farms in southern England and detected BZ resistance on 35, 61 and 44% of farms tested in East and West Sussex and Oxfordshire, respectively. The predominant resistant species was *T. circumcincta* with smaller numbers of resistant *H. contortus* detected. A further survey of 138 sheep farms conducted by these workers in 1992 (Hong, Hunt & Coles, 1996) revealed the presence of BZ resistant strains on 44% of farms in the south west and on 15% of farms from the north east of England (Table 2) with *T. circumcincta* again being the predominant species. Limited data are available for other regions of the UK. A non-random survey conducted in Scotland in 1990 using a FECRT and an EHA found BZ resistant *T. circumcincta* on 24% of the farms examined (Mitchell, Jackson & Coop, 1991).

These investigations suggest that the prevalence of resistant isolates in sheep is lower in the northern areas of the UK. These differences probably reflect the management systems and topography of the land, southern counties mainly comprising intensively stocked lowland farms with a high frequency of anthelmintic usage whereas in northern England and in Scotland a larger proportion of flocks would be grazed more extensively on upland and hill areas, and receive less anthelmintic input. Both the selective and random surveys indicate that anthelmintic resistance in sheep in the UK is predominantly to the BZ group of drugs and that the main species implicated is *T. circumcincta* with some cases of *H. contortus*, particularly in southern Britain. There have been isolated reports of other species of ovine

nematodes showing resistance to the BZs such as *Cooperia curticei* (Hunt *et al.* 1992). This pattern is in general agreement with the findings from other areas of Europe (Coles, Borgsteede & Geerts, 1993).

Levamisole resistance in sheep was first reported on a farm in south west England (Hong *et al.* 1996). Recently resistance to LEV was confirmed in an isolate of *Teladorsagia* (*Ostertagia*) spp. on a farm in Buckinghamshire (Coles & Simkins, 1996) and in Wales (Coles *et al.* 1998) where resistance was evident in both *Teladorsagia* and *Trichostrongylus vitrinus*. Caution is necessary when examining flocks for suspected resistance to LEV as variable efficacy of LEV against inhibited or immature nematodes can occur. Although there have been relatively few reported cases of levamisole resistance current studies in England suggest that the prevalence of resistance against this drug may have increased substantially in recent years (Hunt, personal communication).

The majority of reported cases of anthelmintic resistance in sheep have been trichostrongylid nematodes, however Mitchell, Maris & Bonniwell (1998) reported evidence of resistance in *Fasciola hepatica* to the commonly used flukicide, triclabendazole.

Although this paper primarily focuses on sheep one cannot exclude goats as both ruminants are often reared or grazed on common pasture and they harbour the same nematode species and thus have the potential to select and disseminate anthelmintic resistant populations. Surveys have shown that the prevalence of BZ resistance in goats in the UK is higher (65–70%) than in sheep (Jackson *et al.* 1992*a*; Hong *et al.* 1996) and that *Haemonchus* is involved as well as *Teladorsagia* (*Ostertagia*). The higher prevalence is considered to result partly from the requirement for more frequent treatment of both young and adult goats as they are more susceptible to gastrointestinal nematodes than sheep, and partly due to the more rapid metabolism of anthelmintics which can lead to the equivalent of underdosing. Interestingly, the first case of ivermectin resistance in the UK was reported in cashmere goats which were already carrying worm populations that were

Table 2. *Prevalence of anthelmintic (BZ) resistant nematodes in the UK*

Host/ region	Type of survey	No. farms tested	Percentage positive	Dominant species	Reference
Sheep					
SE England	S	52	14	H.c.	Cawthorne & Cheong, 1984
Scotland	S	37	24	T.c.	Mitchell *et al.* 1991
S England	R	209	47	T.c.	Hong *et al.* 1992
SW England	R	84	44	T.c	Hong *et al.* 1996
NE England	R	54	15	T.c	Hong *et al.* 1996
Goats					
Scotland	S	10	70	T.c.	Jackson *et al.* 1992*a*
England & Wales	R	63	65	H.c.	Hong *et al.* 1996

H.c., *Haemonchus contortus*; T.c., *Teladorsagia (Ostertagia) circumcincta*; S, selective; R, random.

resistant to the BZs and in which only a limited number of parasite generations had been exposed to the drug over a 2 year period (Jackson, Jackson & Coop, 1992*b*). Recently, ivermectin resistant *Teladorsagia circumcincta* has been confirmed in two commercial Angora goat herds using a FECRT where the isolate was also resistant to BZ and LEV (Coles, Warren & Best, 1996). This triple resistance to all three broad-spectrum anthelmintics emphasizes the ability of goats to select resistant isolates.

There is less sequential information on the prevalence of anthelmintic resistance in small ruminants in Europe although a European Community symposium held in 1993 (Coles *et al.* 1993) highlighted occurrence of resistance to BZs in several central and western European countries. Since this meeting reports from these countries (Borgsteede *et al.* 1995; Borgsteede, Pekelder & Dercksen, 1996; Maingi *et al.* 1996, 1997; Requejo-Fernandez *et al.* 1997; Chartier *et al.* 1998) and also from eastern Europe (Praslicka & Corba, 1995; Corba *et al.* 1998) suggest that the number of reported cases of anthelmintic resistance in both sheep and goats are increasing and that multiple resistant isolates are beginning to feature in Europe.

SELECTION OF ANTHELMINTIC RESISTANCE

Anthelmintic resistance is now accepted as a pre-adaptive phenomenon, i.e. the gene or genes that confer resistance may already exist within the existing phenotypic range of a species. Under these circumstances the introduction and continued use of an anthelmintic confers some survival advantage to individuals carrying a resistance gene or genes. The rate at which the selection process occurs is influenced by a number of factors that act either independently or in an additive fashion. These factors are either parasite, host, host management, or drug dependent, and/or are dependent on environmental parameters which, through their effects upon suprapopulation size, influence the proportion of the total population exposed to treatment (Taylor & Hunt, 1989; Jackson, 1993). The size of the population *in refugia* at the time of treatment, the frequency of treatment and extent of underdosing are foremost amongst the key factors that affect selection dynamics and help to explain some of the very apparent geographical differences in distribution of anthelmintic resistance (Waller, 1997). In many countries treatments administered during seasonal droughts, when the size of the suprapopulation may be drastically reduced, increase the selection pressure since resistant surviving worms then have the opportunity to re-populate pastures/grazing that are minimally populated.

The marked differences in prevalence of resistance in and between small and large ruminant species may, at least in part, be explained by differences in the exposure of adult small and large ruminants to chemoprophylaxis. In many cattle enterprises adult animals remain untreated and thus will not contribute to the selection process. The prevalence of resistance is highest and the selection process appears to occur most intensively in ruminants such as goats where, because of their relatively poor ability to regulate gastrointestinal nematodes, it is common practice to treat all age classes of non-dairy breeds on a frequent basis. Bioavailability may also be limited in goats by the high incidence of rumen by-pass (Sangster *et al.* 1991) and the relatively short half-life enjoyed by drugs in the broad-spectrum families (Galtier *et al.* 1981; Bogan, Benoit & Delatour, 1987).

In host species which are treated regularly such as goats (Jackson *et al.* 1992*b*) and where hypobiotic stages may form a large part of the infrapopulation, hypobiotic larvae may also make a significant contribution to the selection process. The relative longevity of these stages can result in them being exposed to several drug treatments which may result in the accumulation of a resistant hypobiotic population.

Underdosing is thought to enable the survival of heterozygous resistant individuals which, in turn, allows the accumulation of resistance alleles within the population. Underdosing may occur if animal liveweights are underestimated or animals are dosed at the mean weight for the group. In countries where fake or adulterated anthelmintics find their way onto the market then underdosing may also contribute to the acceleration of selection (Monteiro *et al.* 1998).

The rate at which the selection process advances may also be influenced by the numbers of genes involved in the resistance mechanisms and whether they are inherited in a dominant fashion. Resistance to a compound is likely to arise more rapidly if it is monogenic and inherited as a dominant trait. Monogenic resistance has been reported for avermectin resistance in *H. contortus* (Le Jambre, 1993; Dobson, Le Jambre & Gill, 1996) and for levamisole resistance in *T. colubriformis* (Martin & McKenzie, 1990) where it was also a sex-linked trait. However, the results from a back-cross study using a benzimidazole- and levamisole-resistant isolate of *H. contortus* (Sangster, Redwin & Bjorn, 1998*a*) suggest that resistance to these two drugs in this species is an incompletely recessive characteristic attributable to more than one gene and moreover is not sex linked. Earlier studies on the inheritance of thiabendazole resistance in *Haemonchus* (Le Jambre, Royal & Martin, 1979; Herlich, Rew & Colglazier, 1981) have also suggested that it is multigenic. The genetic mechanisms underpinning anthelmintic resistance have the potential to influence the rate at which resistance genes accumulate within a population; however the multifactorial nature of the selection process ensures that even those types of resistance with an apparently unpromising genetic background may, as a result of intense selection pressure, come to achieve a position of predominance within a population.

Mechanistic studies using *C. elegans* and trichostrongyloid parasites of sheep have gone some way in defining the mechanisms that are associated with resistance against levamisole (Sangster, 1999), the benzimidazoles (Roos, Kwa & Grant, 1995) and more recently the avermectins/milbemycins (Gill *et al.* 1998). Resistance against levamisole in *H. contortus* and *T. colubriformis* appears to be due to alterations in drug pharmacokinetics at the ACh receptor. Levamisole-resistant *H. contortus* have more binding sites and different drug affinities at the low affinity binding sites than susceptible *H. contortus* and are thus desensitized against the action of levamisole (Sangster, Riley and Wiley, 1998). Most cases of resistance against the benzimidazoles appears to be due to changes in the β-tubulin isotype pattern which results in the loss of high affinity receptor binding sites (Lacey & Gill, 1994). Studies using *H. contortus* and *T. colubriformis* (Kwa, Veenstra & Roos, 1994) and *T. circumcincta* (Elard, Comes & Humbert, 1996) have shown a common point mutation at amino acid 200 in isotype 1 β-tubulin from resistant isolates. Resistant isolates have tyrosine at this locus whereas susceptibles have phenylalanine. It has been suggested that there may be two mechanisms that associate with resistance to the BZs, the selection of an isotype 1 β-tubulin with a reduced affinity for the BZs and the elimination of isotype 2 β-tubulin genes from highly resistant individuals (Condor & Campbell, 1995).

Whilst there appear to be one or two resistance mechanisms for the imidazothiazoles and benzimidazoles recent studies using avermectin/milbemycin-susceptible and -resistant isolates suggest that resistance in nematodes may be similar to that seen in insects, in that it can manifest itself in different ways, depending upon the selection processes involved in its development (Gill & Lacey, 1998; Gill *et al.* 1998). *In vitro* studies using resistant isolates of *H. contortus*, *T. colubriformis* and *T. circumcincta* enabled them to be classified into five distinct types according to their sensitivity to avermectins/milbemycins in larval development assays, L_3 motility assays or their sensitivity to paraherquamide. Selection via suboptimal dosing or through exposure to the full recommended dose rate appears to result in different resistant phenotypes (Gill & Lacey, 1998). Trichostrongyloid resistance mechanisms for drugs within the macrocyclic lactone family remain to be elucidated although it is known that avermectins/milbemycins can affect both somatic (Gill *et al.* 1991; Geary *et al.* 1993) and pharyngeal musculature (Geary *et al.* 1993). Interspecific differences in sensitivity at these two sites of action have been demonstrated in *in vivo* studies on the dynamics of expulsion of *H. contortus*, *T. colubriformis* and *T. circumcincta* following ivermectin treatment (Gill & Lacey, 1998). These studies show a reduced rate of expulsion for *T. circumcincta* and led Gill & Lacey (1998) to suggest that the key event leading to expulsion of this species was the availability of energy which was reduced over time by the effects of ivermectin on the pharynx. The more rapid expulsion of *H. contortus* and *T. colubriformis* was thought to be associated with a direct effect upon motility. Generally, AM-resistant isolates that had been selected using 'full' doses demonstrated resistance at both sites of activity whereas those selected via underdosing in the laboratory or in the field tended not to affect larval motility or only affected motility to a lesser extent (Gill & Lacey, 1998).

The existence of phenotypic differences arising from the different selection processes involved in AM resistance has profound consequences for both mechanistic and genetic studies and also for research into the diagnosis of AM resistance in the field.

The potential of ML drug pharmacokinetics to influence the selection process has been the source of considerable debate. It has been suggested that there may be a 'tail effect', with increased selection occurring during the slow decline in drug bio-availability that is a characteristic of some persistent macrocyclic lactones. An early attempt to examine the problem using a *T. colubriformis* model (Dobson *et al.* 1996) suggested persistence and initial efficacy were far more important in determining the rate of selection of resistance than the 'tail' effect which allowed selection amongst incoming infective larvae. However, subsequent studies using resistant *H. contortus* (Le Jambre, Dobson & Lenane, 1998) and *T. circumcincta* (Sutherland *et al.* 1997) have shown that some selection amongst incoming larvae does occur during the period of drug persistence. Using morphologically distinct *Haemonchus* isolates in a cross-over study, groups of adult (immune) and immature (non-immune) sheep were trickle infected with either a mixture of resistant/susceptible or susceptible *Haemonchus* larvae prior to treatment with a non-persistent (ivermectin) or a persistent (moxidectin) macrocyclic lactone. Worm counts showed a significantly higher proportion of resistant worms in the moxidectin treated group infected post-treatment with the resistant/susceptible larvae. Similar findings have been reported in studies using *T. circumcincta* where some establishment of resistant worms occurred in the presence of moxidectin at levels which prevented establishment of susceptible worms (Sutherland *et al.* 1997).

Reversion

Reversion is the return towards susceptibility of a resistant nematode isolate in the absence of the selecting drug. Reversion will only occur under these circumstances if resistant alleles are selected against. Normally in unselected populations of worms resistant alleles are present at very low frequencies and it is assumed that these alleles have a selective disadvantage for fitness.

Early laboratory studies which assessed reversion in *T. colubriformis*, *T. circumcincta* or *H. contortus* over several generations (Prichard *et al.* 1980; Hall, Ritchie & Kelly, 1982; Le Lambre, Martin & Jarrett, 1982; Martin, 1987) were unable to demonstrate any significant return to susceptibility and in one report (Le Jambre *et al.* 1982) the level of resistance actually increased following drug withdrawal in an isolate which initially showed an intermediate level of resistance. In contrast, Simpkin & Coles (1978) showed that the resistance of eggs from BZ-resistant *H. contortus* and *T. colubriformis* decreased in the absence of benzimidazole treatment and Waller *et al.* (1985*a*) showed that a BZ-resistant strain of *T. colubriformis* reverted to susceptibility to thiabendazole after use of levamisole for 6 years. However, levamisole resistance developed during this period and on withdrawal of the drug there was some reversion to susceptibility which was more rapid in the presence of selection with a

benzimidazole. This confirms earlier evidence (Donald *et al.* 1980) that the introduction of another drug family might select positively against resistance to the primary anthelmintic which is at a faster rate than the process of reversion. Contrasting findings were reported by Martin *et al.* (1988) who, in a laboratory study, found no significant reversion occurred in a BZ resistant strain of *Teladorsagia* (*Ostertagia*) following counter-selection over 4 years with levamisole.

Whether or not reversion occurs will depend upon the level of resistant genes in the parasite population. Once the population is comprised of predominantly homozygous resistant worms then little reversion, if any, will occur (Le Jambre *et al.* 1981) whereas withdrawal of the selecting drug where the population is comprised mainly of heterozygotes will allow a measure of reversion to susceptibility (Waller, Dobson & Axelson, 1988). However, re-introduction of the selecting drug can result in a rapid return to a resistant state. In these laboratory studies the selected resistant nematodes were not diluted with susceptible or hybrid larval genotypes as would occur on pasture.

The situation in the field is less controlled than in the laboratory and limited data are available. Kelly & Hall (1979) reported a reduction in resistance to thiabendazole in an isolate of *H. contortus* after 5 years of exposure to levamisole. As in other reports resistance to BZ rapidly returned when the drug was reintroduced.

Waller *et al.* (1988) studied the changes in anthelmintic resistance in nematode populations in grazing sheep on two farms in Australia over 16 years. After 9 years of continuous use of BZs a high level of BZ resistance was apparent in *Teladorsagia* (*Ostertagia*) and *Trichostrongylus* spp. populations. A change to levamisole usage over 2 years resulted in significant reversion towards partial susceptibility to the benzimidazoles, although resistance to levamisole also developed. Interestingly a high level of resistance to BZ returned rapidly following the reintroduction of oxfendazole. Following the first field case of BZ resistance in the Netherlands, Borgsteede & Duyn (1989) assessed the level of resistance of the *H. contortus* isolate after 6 years of levamisole usage. The results showed that resistance of *H. contortus* to benzimidazoles was still present and that there was no evidence that resistance to levamisole had developed over this period. Recently, a long-term study of anthelmintic reversion over 15 years has been completed in the UK (Jackson, Jackson & Coop, 1998). Benzimidazole resistance was first diagnosed in 1983 in an isolate of *T. circumcincta* on a research farm using a controlled efficacy test (CET) in naturally infected lambs (43 % efficacy using fenbendazole). Since 1983 benzimidazoles have not been used on the farm, control of gastrointestinal parasites being achieved by an

Table 3. Efficacies against *T. circumcincta* in CETs conducted on naturally infected lambs between 1983 and 1998

Anthelmintic	1983 % efficacy	1993 % efficacy	1998 % efficacy
BZ (5 mg kg^{-1})	43	59·1	27
LEV (7·5 mg kg^{-1})	—	82·1	—
IVM (0·2 mg kg^{-1})	—	99·8	—
BZ/LEV (5+7·5 mg kg^{-1})	—	96·5	—

Table 4. *In vivo* and *in vitro* assays used in the detection of anthelmintic resistance

Assay	Spectrum	Type	Author(s)
Controlled test	All drugs	*In vivo* BA	Powers *et al.* 1982
Egg count reduction	All drugs	*In vivo* BA	Presidente, 1985
Egg hatch assay	BZ	*In vitro* BA	Le Jambre, 1976
			Coles & Simpkins, 1977
			Hunt & Taylor, 1989
Egg hatch assay (larval paralysis)	LEV	*In vitro* BA	Dobson *et al.* 1986
Larval paralysis	LEV	*In vitro* BA	Martin & Le Jambre, 1979
	IV	*In vitro* BA	Gill *et al.* 1991
	IV	*In vitro* BA	D'Assonville *et al.* 1996
Larval development	BZ, IV	*In vitro* BA	Coles *et al.* 1988
	BZ, LEV	*In vitro* BA	Taylor, 1990
	BZ, IV, LEV	*In vitro* BA	Lacey *et al.* 1990
	BZ, IV, LEV	*In vitro* BA	Hubert & Kerboeuf, 1992
Tubulin binding	BZ	*In vitro* BC	Lacey & Snowden, 1988
Esterase activity	BZ	*In vitro* BC	Sutherland *et al.* 1989
Tubulin probe	BZ	*In vitro* G	Roos & Boersma, 1990
	BZ	*In vitro* G	Le Jambre, 1990
	BZ	*In vitro* G	Beech *et al.* 1994
	BZ	*In vitro* G	Elard *et al.* 1999

BA, Bioassay; BC, biochemical assay; G, genetic assay.

annual rotation between levamisole and ivermectin. Table 3 contains details of the efficacies against *T. circumcincta* in CETs conducted on naturally infected lambs between 1983 and 1998.

A CET conducted in 1993 showed that the isolate was still resistant to benzimidazoles (Table 3) and a further reversion test conducted at the end of the 1998 grazing season confirmed that benzimidazole resistance was still present (27% efficacy). The results also suggest that a low level of levamisole resistance has developed despite the use of ivermectin in the intervening year. It is interesting that a combination product (benzimidazole plus levamisole) showed a high level of efficacy (96·5%) on the farm. These findings support the laboratory studies above and suggest that little or no reversion back to a susceptible state is likely to occur during the period when the selecting drug class is withdrawn.

During several generations of selection re-segregation of genes may result in resistant phenotypes which have similar or greater general fitness than the susceptible phenotypes in the absence of the drug.

Earlier studies which attempted to compare the fitness of resistant or susceptible strains have yielded contradictory results (Kelly *et al.* 1978; Maingi, Scott & Prichard, 1990) but invariably the comparisons used different isolates which could account for some of the variation. Recently Elard, Sauve & Humbert (1998) showed no differences in a range of fitness-related traits between BZ-resistant and -susceptible isolates derived from the same strain of *T. circumcincta*. These data would suggest that once a high level of anthelmintic resistance is established on a farm then for all practical purposes there will be little reversion to susceptibility.

Restoration of anthelmintic efficacy has been attempted in South Africa by the reintroduction of susceptible nematode larvae into a trial site where a benzimidazole-resistant strain of *Haemonchus* was established (Van Wyk & van Schalkwyk, 1990). In theory the dosing of sheep with susceptible *Haemonchus* larvae would dilute the existing resistant isolates but balancing the pathogenic effects of the parasite, in the absence of anthelmintic treatment, may be a difficult exercise in the field.

Detection of resistance

The accepted definitions of anthelmintic resistance inevitably centre around the relative effectiveness of an anthelmintic, resistance becoming apparent when a previously effective drug ceases to be so. The presence of anthelmintic resistance within a population can be determined using *in vivo* bioassays such as controlled efficacy tests (CET) and faecal egg count reduction tests (FECRT) or *in vitro* using bioassays, biochemical assays or genetic assays (Table 4).

In the field resistance is usually detected using an *in vivo* assay such as the FECRT (Coles *et al.* 1992) or by the use of a bioassay which measures the degree of disruption of developmental or behavioural function in the free-living stages of nematode parasites such as the egg hatch assay (EHA) or larval development assay (LDA). Although the early detection of emerging resistance is accepted as vital to strategies designed to delay the onset of resistance in nematodes (Coles & Roush, 1992), regular on-farm testing for resistance is rarely, if ever practised. In Europe and the UK most testing is reactive and, since in most positive cases the selection processes will have run their full course, little can be done in the way of management to conserve efficacy against the resistant species.

All of the tests that are used in the detection of resistance have drawbacks either in terms of cost, applicability or sensitivity and, in the case of genetic assays, availability at the right time. Molecular genetic assays of the type described for benzimidazole resistance in *Teladorsagia* (Elard, Cabaret & Humbert, 1999) offer the potential to study the dynamics of the selection and spread of resistance. Since their utility is greatest when resistance alleles are at a low frequency within the population it has, quite rightly, been suggested that the development of suitable molecular probes for detecting macrocyclic lactone resistance should be given the highest research priority (Geary, Sangster & Thompson, 1999). However, given the paucity of research funding in this area and the relative complexity of resistance mechanisms for AM drugs (Gill & Lacey, 1998) whether these probes will become available before AM resistance alleles become relatively common in the field remains a moot point. In the northern hemisphere AM resistance in small ruminants is at present relatively rare, however this enviable situation may not persist since this family is used intensively in the chemoprophylactic control of our economically important ovine endo- and ecto-parasitoses. Clearly in those areas where AM resistance is at a low level attention needs to be focused on monitoring changes in susceptibility. Specific studies charting these changes and surveillance programmes not only require economic and political support but may also require the development of new assays and means of improving the sensitivity of our existing tests that are used to detect resistance against this family of anthelmintics.

The faecal egg count reduction test, using an established efficacy threshold of 95 %, is arguably our most traditional field test for anthelmintic resistance but has long been recognised as insensitive (Martin, Anderson & Jarrett, 1989), particularly when used as an undifferentiated test. Studies in New Zealand (McKenna, 1996, 1997) have shown that the sensitivity of the FECRT can be significantly improved by the use of pre- and post-treatment coprocultures, particularly when any sensitive species are highly fecund. Improved methods of speciation would be of great value to the FECRT particularly since there is evidence that this test may need to be tailored to suit the drug and parasite species under investigation. Survival of non-resistant histotrophic stages may occur when using the FECRT to detect levamisole resistance (Grimshaw, Hong & Hunt, 1996) and for some AM resistant species such as *Teladorsagia* a temporary suppression of ovulation appears to occur post-treatment (Jackson, 1993; Sutherland, Leathwick & Brown, 1999). These characteristics which affect the timing of re-sampling obviously need to be brought into consideration in field studies on anthelmintic resistance. Speciation techniques will also be critical within AM FECRTs if the loss of the persistent effects that are inherent in some of our current endectocides proves to be a characteristic of developing resistance against these drugs.

International non-parochial research on the development of AM resistance not only requires inter-institutional collaborations, which must be adequately funded, but must be underpinned by a range of suitable, robust techniques. Robust techniques for field investigations have simple criteria/thresholds for resistance which are applicable across the species range. Evidence from the literature sadly suggests that, at present, we lack some of the necessary panel of investigative techniques.

Studies comparing *in vivo* FECRT and *in vitro* methods such as the egg hatch assay, egg hatch paralysis assay and various forms of the larval development assay have generally shown a good agreement between these tests for the benzimidazoles but have reported difficulties of interpretation when these *in vitro* tests are used in field investigations into levamisole resistance (Maingi, Bjørn & Dangolla, 1998; Varady & Corba, 1999). Studies using the larval development assay with *Teladorsagia (Ostertagia) circumcincta* demonstrated that the LDA was able to detect resistance against the benzimidazoles and levamisole but was less useful in detecting resistance against ivermectin (Amarante *et al.* 1997). The same study also demonstrated considerable variation in LD_{50} values which increased to peak values for all three drugs around 50–60 days after infection. A similar pattern has also been reported previously with benzimidazole-resistant *Haemonchus* (Borgesteede & Couwenberg, 1987).

The larval development assay has the unique ability to be used with drugs in the three broad-spectrum families and has been used to great effect with monospecific isolates to identify AM resistance mechanisms (Gill & Lacey, 1998). However, specific differences in AM susceptibility (Shoop, Mrozik & Fisher, 1995) that relate to the sensitivity of the different target sites and the existence of different resistance mechanisms (Gill & Lacey, 1998) create problems for the use of the LDA in detecting AM resistance in the multi-specific infections that are common in the field. Alternatively it may be possible to assay for IVM resistance using the most susceptible target, the pharynx. Preliminary studies at the Moredun Research Institute using L_1 larvae of resistant and susceptible isolates of *H. contortus* and *T. circumcincta* in an IVM B_1 larval feeding inhibition assay based upon that described for adults by Geary *et al.* (1993) provided resistance factors of 8 (*H. contortus*) and 5 (*T. circumcincta*). Examination of other drug-susceptible gastrointestinal species held in the laboratory (*T. colubriformis*, *T. vitrinus*, *T. axei* and *C. oncophora*) where feeding was inhibited at concentrations below 2 nM IVM provides the tantalizing suggestion of a defined upper limit beyond which susceptible individuals do not feed. However, further studies are required using different species, resistant isolates and other drug analogues to confirm that this is indeed the case and to determine the potential of this method as a means of detecting macrocyclic lactone resistance in field material.

The development of molecular and genetic probes capable of determining individual susceptibility to a drug have the potential to provide powerful tools for examining adaptation in populations. However, for ML drugs these tools may be highly specific given the existence of different resistance mechanisms that may map to different sites. The specificity of these probes together with their costs will almost certainly limit their application to research rather than routine diagnosis.

Management of resistance

Management practices to delay the development of resistance such as reducing the frequency of anthelmintic treatment, avoiding underdosing and rotation between drug families have been well documented (Coles & Roush, 1992). However, uptake of these recommendations has been slow in the UK despite the increasing number of reports of resistance to the benzimidazole drugs (Coles, 1997).

Most of the strategies that have been developed for managing anthelmintic resistance in sheep have the twin linked aims of maintaining susceptible alleles within the parasite population by reducing selection pressure and conserving the efficacy of existing drug families. The central tenet of most recommendations to combat anthelmintic resistance has been the adoption of measures which reduce treatment frequency (Prichard *et al.* 1980; Waller *et al.* 1985*b*; Besier, 1997). Regionally-based management systems developed in Australia (Dash, Newman & Hall, 1985; Waller *et al.* 1995) utilize a small number of epidemiologically-based strategic treatments in order to reduce selection pressure. The use of relatively few treatments over an extended period in these schemes obviously enables survival of susceptible alleles. Susceptible alleles may also be conserved by withholding treatment from a proportion of the flock (Barnes, Dobson & Barger, 1995) or by identifying and treating the most susceptible individuals. The latter approach has been investigated in South Africa using an assessment of the colour of the conjunctiva as a clinical indicator of anaemia to identify individuals most affected by *H. contortus*. In an irrigated pasture trial conducted over 4 months 69% of the sheep required no anthelmintic treatments and 21% required only a single treatment (Van Wyk, Malan & Bath, 1997). Maintaining susceptible alleles in this way may impose costs in terms of productivity which must be set against the less easily defined benefits of reducing selection pressure.

Although the use of an annual rotation between different mode-of-action families has been proposed as a means of conserving efficacy in the long term, no studies have demonstrated its value in the field. Modelling studies covering a 20-year period suggest that the use of combination products is the best way to delay the selection process, with an annual rotation being the second option (Barnes *et al.* 1995). Although combination products containing current individually registered anthelmintics have not been registered for use in the UK they are available for use in Australia and New Zealand. However, the registration process and both commercial and marketing pressures are such that combination products containing a novel mode-of-action product are unlikely to appear on the market. Introducing combination products at a time when resistance alleles are common within the population may, in the short term, provide high levels of efficacy but may do little to slow the advance of resistance (Hennessy, 1997).

Conserving the efficacy of existing compounds against resistant populations provides a means of extending not only the life expectancy of resistance-selected drugs but, by maintaining their use, may provide some benefit for those drugs against which resistance has not emerged. Research on optimising efficacy against resistant populations has shown that efficacy may be enhanced in a number of different ways which maximize and/or extend anthelmintic bioavailability (Hennessy, 1997). Reductions in feed intake which reduce gastrointestinal digesta flow rates and provide an increased period for drug absorption and recycling have been shown to significantly improve efficacy against benzimidazole-resistant isolates of *H. contortus*, *T. colubriformis* (Ali & Hennessy, 1995) and *T. circumcincta* (Barrett *et al.* 1998). Administering divided doses has also been shown to increase efficacy against benzimidazole-resistant populations of *Haemonchus*, *Teladorsagia* (*Ostertagia*) and *Trichostrongylus* (Sangster *et al.* 1991). Co-administration of benzimidazole synergists such as piperonyl butoxide which alter drug metabolism has been shown in the laboratory to increase efficacy against resistant isolates of *T. circumcincta* (Benchaoui & McKellar, 1994). Unfortunately the field application of the various methods which can improve the uptake and retention of the benzimidazoles may be limited not only by the prevalence and extent of resistance but also by ease of application and the difficulty of registering combination products.

Given the potential for rapid development of resistance within a few parasite generations (Besier, 1997), the relative insensitivity of current diagnostic methods, their limited application in the field and the time required to develop and institute strategies it is hardly surprising that defined approaches are usually instituted when resistance alleles are common within the region. The new classes of anthelmintics which have potential for veterinary application, the diketopiperazines, cyclic depsipeptides and nitazoxanide are still at the earliest stages of development and commercial evaluation (Geary *et al.* 1999). Since we cannot expect novel mode-of-action families to appear regularly on the market the conservation of the macrocyclic lactone family, which provides endectocides that are crucial to the sheep industry, should clearly be given the highest priority as far as the UK and Europe are concerned. Although in temperate areas of Europe our, at present, largely susceptible populations *in refugia* may appear to provide a formidable obstacle against developing ML resistance, there is ample evidence supporting the view that this may only provide temporary protection. Europe clearly has much of the scientific expertise, both in academia and in industry, to begin tackling the problem of ML resistance but what it evidently lacks is the best means of exploiting this potential.

The Australian experience highlights the importance of a multidisciplinary approach and the necessity for surveillance and advisory/support services to provide information and advice. Unless some means is found to establish, promote and maintain the required scientific forum then, at some point in future, we face the dismal prospect of simply charting the establishment of ML resistance and reactively attempting to combat the problems that it poses to an already beleaguered sheep industry. The findings from research on anthelmintic resistance are not only important to the livestock industry whose productivity is currently dependent to a very great extent upon chemotherapy and prophylaxis but may also provide important data for developing models of metazoan adaptation.

REFERENCES

ALI, D. N. & HENNESSY, D. R. (1995). The effect of reduced feed intake on the efficacy of oxfendazole against benzimidazole resistant *Haemonchus contortus* and *Trichostrongylus colubriformis* in sheep. *International Journal for Parasitology* **25**, 71–74.

AMARANTE, A. F. T., POMROY, W. E., CHARLESTON, W. A. G., LEATHWICH, D. M. & TORNERO, M. T. T. (1997). Evaluation of a larval development assay for the detection of anthelmintic resistance in *Teladorsagia* (*Ostertagia*) *circumcincta*. *International Journal for Parasitology* **27**, 305–311.

BARNES, E. H., DOBSON, R. J. & BARGER, I. A. (1995). Worm control and anthelmintic resistance: adventures with a model. *Parasitology Today* **11**, 56–63.

BARRETT, M., JACKSON, F., PATTERSON, M., JACKSON, E. & MCKELLAR, Q. A. (1998). Comparative field evaluation of divided-dosing and reduced feed intake upon treatment efficacy against resistant isolates of *Teladorsagia circumcincta* in sheep and goats. *Research in Veterinary Science* **64**, 101–104.

BEECH, R. N., PRICHARD, R. K. & SCOTT, M. E. (1994). Genetic variability of the beta-tubulin genes in benzimidazole susceptible and resistant strains of *Haemonchus contortus*. *Genetics* **138**, 103–110.

BENCHAOUI, H. A. & MCKELLAR, Q. A. (1994). Potentiation of fenbendazole: pharmacokinetic and efficacy assessments of a drug combination in sheep. *Proceedings of the 6th International Congress of the European Association for Veterinary Pharmacology and Toxicology*, Edinburgh, Session 10, O7, p. 252.

BESIER, R. B. (1997). Ecological selection for anthelmintic resistance: Re-evaluation of sheep worm control programmes. In *Managing Anthelmintic Resistance in Endoparasites. Workshop at the 16th International Conference of the World Association for the Advancement of Veterinary Parasitology* (ed. Van Wyk, J. A. & van Schalkwyk, P. C.), pp. 30–38. Sun City: South Africa.

BOGAN, J., BENOIT, E. & DELATOUR, P. (1987). Pharmacokinetics of oxfendazole in goats: a comparison with sheep. *Journal of Veterinary Pharmacology and Therapeutics* **10**, 305–309.

BORGESTEEDE, F. M. M. & COUWENBURG, T. (1987). Changes in LC50 in an *in vitro* egg development assay during the patent period of *Haemonchus contortus* in sheep. *Research in Veterinary Science* **42**, 413–414.

BORGSTEEDE, F. H. M. & DUYN, S. P. J. (1989). Lack of reversion of a benzimidazole resistant strain of *Haemonchus contortus* after six years of levamisole usage. *Research in Veterinary Science* **47**, 270–272.

BORGSTEEDE, F. H. M., PEKELDER, J. J., DERCKSEN, D. P., SOL, J., VELLEMA, P., GAASENBEEK, C. P. H. & VANDERLINDEN, J. N. (1995). Anthelmintic resistance in nematodes of sheep in the Netherlands. *Tijdschrift voor Diergeneeskunde* **120**, 173–176.

BORGSTEEDE, F. H. M., PEKELDER, J. J. & DERCKSEN, D. P. (1996). Anthelmintic resistant nematodes in goats in the Netherlands. *Veterinary Parasitology* **65**, 83–87.

BRITT, D. P. (1982). Benzimidazole-resistant nematodes in Britain. *Veterinary Record* **110**, 343–344.

CAWTHORNE, R. J. G. & CHEONG, F. H. (1984). Prevalence of anthelmintic resistant nematodes in sheep in south-east England. *Veterinary Record* **114**, 562–564.

CAWTHORNE, R. J. G. & WHITEHAD, J. D. (1983). Isolation of benzimidazole resistant strains of *Teladorsagia* (*Ostertagia*) *circumcincta* from British sheep. *Veterinary Record* **112**, 274–277.

CHARTIER, C., PORS, I., HUBERT, J., ROCHETEAU, D., BENOIT, C. & BERNARD, N. (1998). Prevalence of anthelmintic resistant nematodes in sheep and goats in western France. *Small Ruminant Research* **29**, 33–41.

COLES, G. C. (1997). Nematode control practices and anthelmintic resistance on British sheep farms. *Veterinary Record* **141**, 91–93.

COLES, G. C., BAUER, C., BORGSTEEDE, F. H. M., GEERTS, S., KLEI, T. R., TAYLOR, M. A. & WALLER, P. J. (1992). World Association for the Advancement of Veterinary Parasitology (WAAVP) methods for the detection of anthelmintic resistance in nematodes of veterinary importance. *Veterinary Parasitology* **44**, 35–44.

COLES, G. C., BORGSTEEDE, F. H. M. & GEERTS, S. (1993). Anthelmintic resistance in nematodes of farm animals. *CEC Publication*, Brussels.

COLES, G. C., RHODES, A. C., GLOVER, M. G., PRESTON, G. D. & COLES, E. M. (1998). Avoiding introduction of levamisole-resistant nematodes. *Veterinary Record* **143**, 667.

COLES, G. C. & ROUSH, R. T. (1992). Slowing the spread of anthelmintic resistance nematodes of sheep and goats in the United Kingdom. *Veterinary Record* **130**, 505–510.

COLES, G. C. & SIMKINS, K. (1977). Resistance of nematode eggs to the ovicidal activity of benzimidazoles. *Research in Veterinary Science* **22**, 386–389.

COLES, G. C. & SIMKINS, K. (1996). Resistance to levamisole. *Veterinary Record* **139**, 124.

COLES, G. C., TRITSCHLER, J. P. II, GIORDANO, D. J., LASTE, N. J. & SCHMIDT, A. L. (1988). Larval development test for detection of anthelmintic resistant nematodes. *Research in Veterinary Science* **45**, 50–53.

COLES, G. C., WARREN, A. K. & BEST, J. R. (1996). Triple resistant *Teladorsagia* (*Ostertagia*) from Angora goats. *Veterinary Record* **139**, 299–300.

CONDOR, G. A. & CAMPBELL, W. C. (1995). Chemotherapy of nematode infections of veterinary importance, with special reference to drug resistance. *Advances in Parasitology* **35**, 1–84.

CORBA, J., VARADY, M., PRASLICKA, J., TOMASOVICOVA, O. & KONIGOVA, A. (1998). Anthelmintic resistance in nematodes of domestic animals in Slovakia. *Slovensky Veterinarsky Casopis* **23**, 61–66.

CRAIG, T. M. (1993). Anthelmintic resistance. *Veterinary Parasitology* **46**, 121–131.

DASH, K. M., NEWMAN, R. J. & HALL, E. (1985). Recommendations to minimise selection for anthelmintic resistance in nematode control programmes. In *Resistance in Nematodes to Anthelmintic Drugs* (ed. Anderson, N. & Waller, P. J.),

pp. 161–169. Melbourne: Australian Wool Corporation Technical Publication.

D'ASSONVILLE, J. A., JANOVSKY, E. & VERSTER, A. (1996). *In vitro* screening of *Haemonchus contortus* third stage larvae for Ivermectin resistance. *Veterinary Parasitology* **61**, 73–80.

DOBSON, R. J., DONALD, A. D., WALLER, P. J. & SNOWDEN, K. L. (1986). An egg hatch assay for resistance to levamisole in trichostronglyoid nematode parasites. *Veterinary Parasitology* **19**, 77–84.

DOBSON, R. J., LE JAMBRE, L. F. & GILL, J. H. (1996). Management of anthelmintic resistance: inheritance of resistance and selection with persistent drugs. *International Journal for Parasitology* **26**, 993–1000.

DONALD, A. D., WALLER, P. J., DOBSON, R. J. & AXELSON, A. (1980). The effect of selection with levamisole on benzimidazole resistance in *Teladorsagia* (*Ostertagia*) spp. of sheep. *International Journal for Parasitology* **10**, 381–389.

ECHEVARRIA, F., BORBA, M. F. S., PINHEIRO, A. C., WALLER, P. J. & HANSEN, J. W. (1996). The prevalence of anthelmintic resistance in nematode parasites of sheep in Southern Latin America. *Veterinary Parasitology* **62**, 199–206.

EDDI, C., CARACOSTANTOGOLO, J., PEÑA, M., SCHAPIRO, L., MARANGUNICH, L., WALLER, P. J. & HANSEN, J. W. (1996). The prevalence of anthelmintic resistance in nematode parasites of sheep in Southern Latin America: Argentina. *Veterinary Parasitology* **62**, 189–197.

ELARD, L., CABARET, J. & HUMBERT, J. F. (1999). PCR diagnosis of benzimidazole-susceptibility or -resistance in natural populations of the small ruminant parasite, *Teladorsagia circumcincta*. *Veterinary Parasitology* **80**, 231–237.

ELARD, L., COMES, A. M. & HUMBERT, J. F. (1996). Sequences of beta-tubulin cDNA from benzimidazole-susceptible and -resistant strains of *Teladorsagia circumcincta*, a nematode parasite of small ruminants. *Molecular and Biochemical Parasitology* **79**, 249–253.

ELARD, L., SAUVE, C. & HUMBERT, J. F. (1998). Fitness of benzimidazole-resistant and -susceptible worms of *Teladorsagia circumcincta*, a nematode parasite of small ruminants. *Parasitology* **117**, 571–578.

GALTIER, P., ESCOULA, L., CAMGUILHEM, R. & ALVINIERIE, M. (1981). Comparative bioavailability of levamizole in non lactating ewes and goats. *Annales Recherches Veterinarie* **12**, 109–115.

GEARY, T. G., SANGSTER, N. C. & THOMPSON, D. P. (1999). Frontiers in anthelmintic pharmacology. *Veterinary Parasitology* **84**, 275–295.

GEARY, T. G., SIMS, S. M., THOMAS, E. M., VANOVER, L., DAVIS, J. P., WINTERROWD, C. A., KLEIN, R. D., HO, N. F. H. & THOMPSON, D. P. (1993). *Haemonchus contortus* – Ivermectin-induced paralysis of the pharynx. *Experimental Parasitology* **77**, 88–96.

GILL, J. H. & LACEY, E. (1998). Avermectin/milbemycin resistance in trichostrongyloid nematodes. *International Journal for Parasitology* **28**, 863–877.

GILL, J. H., KERR, C. A., SHOOP, W. L. & LACEY, E. (1998). Evidence of multiple mechanisms of avermectin resistance in *Haemonchus contortus* – comparison of selection protocols. *International Journal for Parasitology* **28**, 783–789.

GILL, J. H., REDWIN, J. M., VAN WYK, J. A. & LACEY, E. (1991). Detection of resistance to Ivermectin in *Haemonchus contortus*. *International Journal for Parasitology* **21**, 771–776.

GRIMSHAW, W. T. R., HONG, C. & HUNT, K. R. (1996). Potential for misinterpretation of the faecal egg count reduction test for levamisole resistance in gastrointestinal nematodes of sheep. *Veterinary Parasitology* **62**, 267–273.

HALL, C. A., RITCHIE, L. & KELLY, J. D. (1982). Effect of removing anthelmintic selection pressure on the benzimidazole resistance status of *Haemonchus contortus* and *Trichostrongylus colubriformis* in sheep. *Research in Veterinary Sciences* **33**, 54–57.

HAZELBY, C. A., PROBERT, A. J. & ROWLANDS, D. A. T. (1994). Anthelmintic resistance in nematodes causing parasitic gastroenteritis of sheep in the UK. *Journal of Veterinary Pharmacology and Therapeutics* **17**, 245–252.

HERLICH, H., REW, R. S. & COLGLAZIER, M. L. (1981). Inheritance of cambendazole resistance in *Haemonchus contortus*. *American Journal of Veterinary Research* **42**, 1342–1344.

HENNESSY, D. R. (1997). Practical aspects of parasite treatment. In *Managing Anthelmintic Resistance in Endoparasites. Workshop at the 16th International Conference of the World Association for the Advancement of Veterinary Parasitology* (ed. Van Wyk, J. A. & van Schalkwyk, P. C.), pp. 40–49.

HONG, C., HUNT, K. & COLES, G. C. (1996). Occurrence of anthelmintic resistant nematodes on sheep farms in England and goat farms in England and Wales. *Veterinary Record* **139**, 83–86.

HONG, C., HUNT, K. R., HARRIS, T. J., COLES, G. C., GRIMSHAW, W. T. R. & MCMULLIN, P. F. (1992). A survey of benzimidazole resistant nematodes in sheep in three counties of southern England. *Veterinary Record* **131**, 5–7.

HUBERT, J. & KERBOEUF, D. (1992). A microlarval development assay for the detection of anthelmintic resistance in sheep nematodes. *Veterinary Record* **130**, 442–446.

HUNT, K. R., HONG, C., COLES, G. C., SIMPSON, V. R. & NEAL, C. (1992). Benzimidazole-resistant *Cooperia curticei* from Cornwall, England. *Veterinary Record* **130**, 164.

HUNT, K. R. & TAYLOR, M. A. (1989). Use of the egg hatch assay on sheep faecal samples for the detection of benzimidazole resistant nematodes. *Veterinary Record* **125**, 153–154.

JACKSON, F. (1993). Anthelmintic resistance – The state of play. *British Veterinary Journal* **149**, 123–138.

JACKSON, F., JACKSON, E. & COOP, R. L. (1992*b*). Evidence of multiple anthelmintic resistance in a strain of *Teladorsagia circumcincta* (*Teladorsagia* (*Ostertagia*) *circumcincta*) isolated from goats in Scotland. *Research in Veterinary Science* **53**, 371–374.

JACKSON, F., JACKSON, E. & COOP, R. L. (1998). Reversion and susceptibility studies at Moredun Research Institute's Firth Mains Farm. *Proceedings of the Sheep Veterinary Society* **22**, 149–150.

JACKSON, F., JACKSON, E., LITTLE, S., COOP, R. L. & RUSSEL,

A. J. F. (1992*a*). Prevalence of anthelmintic-resistant nematodes in fibre-producing goats in Scotland. *Veterinary Record* **131**, 282–285.

KELLY, J. D. & HALL, C. A. (1979). Resistance of animal helminths to anthelmintics. *Advances in Pharmacology and Chemotherapy* **16**, 89–128.

KELLY, J. D., WHITLOCK, H. V., THOMPSON, H. G., HALL, C. A., CAMPBELL, N. J., MARTIN, I. C. A. & LE JAMBRE, L. F. (1978). Physiological characteristics of the free-living and parasitic stages of strains of *Haemonchus contortus* susceptible or resistant to benzimidazole anthelmintics. *Research in Veterinary Science* **25**, 376–385.

KWA, M. S. G., VEENSTRA, J. G. & ROOS, M. H. (1994). Benzimidazole resistance in *Haemonchus contortus* is correlated with a conserved mutation at amino-acid-200 in beta-tubulin isotype-1. *Molecular and Biochemical Parasitology* **63**, 299–303.

LACEY, E. & GILL, J. H. (1994). Biochemistry of benzimidazole resistance. *Acta Tropica* **56**, 245–262.

LACEY, E., REDWIN, J. M., GILL J. H., DEMARGHERITI, V. M. & WALLER, P. J. (1990). In *Resistance of Parasites to Antiparasitic Drugs* (ed. Boray, J. C., Martin, P. J. & Roush, R. T.), pp. 177–184. Rahway: MSD Agvet.

LACEY, E. & SNOWDON, K. L. (1988). A routine diagnostic assay for the detection of benzimidazole resistance in parasitic nematodes using tritiated benzimidazole carbamates. *Veterinary Parasitology* **27**, 309–324.

LE JAMBRE, L. F. (1976). Egg hatch as an *in vitro* assay of thiabendazole resistance in nematodes. *Veterinary Parasitology* **2**, 385–391.

LE JAMBRE, L. F. (1990). Molecular biology and anthelmintic resistance in parasitic nematodes. In *Resistance of Parasites to Antiparasitic Drugs* (ed. Boray, J. C., Martin, P. J. & Roush, R. T.), pp. 155–164. Rahway: MSD Agvet.

LE JAMBRE, L. F. (1993). Molecular variation in trichostrongylid nematodes from sheep and cattle. *Acta Tropica* **53**, 331–343.

LE JAMBRE, L. F., DOBSON, R. J. & LENANE, I. J. (1998). Selection for anthelmintic resistance by macrocyclic lactones in adult and young sheep. *Abstracts of the Second International Conference ; Novel Approaches to the Control of Helminth Parasites of Livestock* **55**. Armidale, NSW, Australia

LE JAMBRE, L. F., MARTIN, P. J. & JARRETT, R. G. (1982). Comparison of changes in resistance of *Haemonchus contortus* eggs following withdrawal of thiabendazole selection. *Research in Veterinary Science* **32**, 39–43.

LE JAMBRE, L. F., PRICHARD, R. K., HENNESSY, D. R. & LABY, R. H. (1981). Efficiency of oxfendazole administered as a single dose or in a controlled release capsule against benzimidazole-resistant *Haemonchus contortus*, *Teladorsagia* (*Ostertagia*) *circumcincta* and *Trichostrongylus colubriformis*. *Research in Veterinary Science* **31**, 289–294.

LE JAMBRE, L. F., ROYAL, W. M. & MARTIN, P. J. (1979). The inheritance of thiabendazole resistance in *Haemonchus contortus*. *Parasitology* **78**, 107–119.

MACIEL, S., GIMÉNEZ, A. M., GAONA, C., WALLER, P. J. & HANSEN, J. W. (1996). The prevalence of anthelmintic resistance of sheep in Southern Latin America: Paraguay. *Veterinary Parasitology* **62**, 207–212.

MAINGI, N., BJØRN, H. & DANGOLLA, A. (1998). The

relationship between faecal egg count reduction and the lethal dose 50 % in the egg hatch assay and larval development assay. *Veterinary Parasitology* **77**, 133–145.

MAINGI, N., BJORN, H., THAMSBORG, S. M., BOGH, H. O. & NANSEN, P. (1997). Anthelmintic resistance in nematode parasites of sheep in Denmark. *Small Ruminant Research* **23**, 171–181.

MAINGI, N., BJORN, H., THAMSBORG, S. M., BOGH, H., THAMSBORG, S. M., BOGH, H. O. & NANSEN, P. (1996). A survey of anthelmintic resistance in nematode parasites of goats in Denmark. *Veterinary Parasitology* **66**, 53–66.

MAINGI, N., SCOTT, M. E. & PRICHARD, R. K. (1990). Effect of selection pressure for thiabendazole resistance on fitness of *Haemonchus contortus* in sheep. *Parasitology* **100**, 327–335.

MARTIN, P. J. (1987). Development and control of resistance to anthelmintics. *International Journal for Parasitology* **17**, 493–501.

MARTIN, P. J., ANDERSON, N., BROWN, T. H. & MILLER, D. W. (1988). Changes in resistance of *Teladorsagia* (*Ostertagia*) spp. to thiabendzole following natural selection or treatment with levamisole. *International Journal for Parasitology* **18**, 333–340.

MARTIN, P. J., ANDERSON, N. & JARRETT, R. G. (1989). Detecting benzimidazole resistance with faecal egg count reduction tests and *in vitro* assays. *Australian Veterinary Journal* **66**, 236–240.

MARTIN, P. J. & LE JAMBRE, L. F. (1979). Larval paralysis as an *in vitro* assay of levamisole and morantel tartrate resistance in *Teladorsagia* (*Ostertagia*). *Veterinary Scientific Communication* **3**, 159–164.

MARTIN, P. J. & MCKENZIE, J. A. (1990). Levamisole resistance in *Trichostrongylus colubriformis*: A sex linked recessive character. *International Journal for Parasitology* **20**, 867–872.

MCKELLAR, Q. A. (1998). Antimicrobial resistance: a veterinary perspective. *British Medical Journal* **317**, 610–611.

MCKENNA, P. B. (1996). Potential limitations of the undifferentiated faecal egg count reduction test for the detection of anthelmintic resistance. *New Zealand Veterinary Journal* **44**, 73–75.

MCKENNA, P. B. (1997). Further potential limitations of the undifferentiated faecal egg count reduction test for the detection of anthelmintic resistance. *New Zealand Veterinary Journal* **45**, 244–246.

MITCHELL, G. B. B., JACKSON, F. & COOP, R. L. (1991). Anthelmintic resistance in Scotland. *Veterinary Record* **129**, 58.

MITCHELL, G. B. B., MARIS, L. & BONNIWELL, M. A. (1998). Triclabendazole-resistant liver fluke in Scottish sheep. *Veterinary Record* **143**, 399.

MONTEIRO, A. M., WANYANGU, S. W., KARIUKI, D. P., BAIN, R., JACKSON, F. & MCKELLAR, Q. A. (1998). Pharmaceutical quality of anthelmintics sold in Kenya. *Veterinary Record* **142**, 396–398.

NARI, A., SALLES, J., GIL, A., WALLER, P. J. & HANSEN, J. W. (1996). The prevalence of anthelmintic resistance of sheep in Southern Latin America: Uruguay. *Veterinary Parasitology* **62**, 213–222.

POWERS, K. G., WOOD, I. B., ECKERT, J., GIBSON, T. &

SMITH, H. J. (1982). World Association for the Advancement of Veterinary Parasitology (WAAVP). Guidelines for evaluating the efficacy of anthelmintics in ruminants (Bovine and Ovine). *Veterinary Parasitology* **10**, 265–284.

PRASLICKA, J. & CORBA, J. (1995). Resistance to anthelmintics in nematodes of sheep and goats. *Veterinarni Medicina* **40**, 257–260.

PRESIDENTE, P. J. A. (1985). Methods for detection of resistance to anthelmintics. In *Resistance in Nematodes to Anthelmintic Drugs* (ed. Anderson, N. & Waller, P. J.), pp. 13–28. Melbourne: CSIRO and Australian Wool Corporation Technical Publication.

PRICHARD, R. (1994). Anthelmintic resistance. *Veterinary Parasitology* **54**, 259–268.

PRICHARD, R. K., HALL, C. A., KELLY, J. D., MARTIN, I. C. A. & DONALD, A. D. (1980). The problem of anthelmintic resistance in nematodes. *Australian Veterinary Journal* **56**, 239–251.

REQUEJO-FERNANDEZ, J. A., MARTINEZ, A., MEANA, A., ROJO VAZQUEZ, F. A., OSORO, K. & ORTEGA MORA, L. M. (1997). Anthelmintic resistance in nematode parasites from goats in Spain. *Veterinary Parasitology* **73**, 83–88.

ROOS, M. H. & BOERSMA, J. H. (1990). Comparison of 4 benzimidazole susceptible and 9 resistant *Haemonchus contortus* populations by restriction fragment length polymorphism. In *Resistance of Parasites to Antiparasitic Drugs* (ed. Boray, J. C., Martin, P. J. & Roush, R. T.), pp. 165–169. Rahway: MSD Agvet.

ROOS, M. H., KWA, M. S. G. & GRANT, W. N. (1995). New genetic and practical implications of selection for anthelmintic resistance in parasitic nematodes. *Parasitology Today* **11**, 148–150.

SANGSTER, N. (1996). Pharmacology of anthelmintic resistance. *Parasitology* **113**, S201–S216.

SANGSTER, N. C. (1999). Anthelmintic resistance: past, present and future. *International Journal for Parasitology* **29**, 115–124.

SANGSTER, N. C., REDWIN, J. M. & BJORN, H. (1998*a*). Inheritance of levamisole and benzimidazole resistance in an isolate of *Haemonchus contortus*. *International Journal for Parasitology* **28**, 503–510.

SANGSTER, N. C., RICKARD, J. M., HENNESSY, D. R., STEEL, J. N. & COLLINS, G. H. (1991). Disposition of oxfendazole in goats and efficacy compared to sheep. *Research in Veterinary Science* **51**, 258–263.

SANGSTER, N. C., RILEY, F. L. & WILEY, L. J. (1998*b*). Binding of [H-3]m-aminolevamisole to receptors in levamisole-susceptible and -resistant *Haemonchus contortus*. *International Journal for Parasitology* **28**, 707–717.

SHOOP, W. L., MROZIK, H. & FISHER, M. H. (1995). Structure and activity of avermectins and milbemycins in animal health. *Veterinary Parasitology* **59**, 139–156.

SIMPKIN, K. G. & COLES, G. C. (1978). Instability of benzimidazole resistance in nematode eggs. *Research in Veterinary Science* **25**, 249–250.

SUTHERLAND, I. A., LEATHWICK, D. M. & BROWN, A. E. (1999). Moxidectin: persistence and efficacy against drug-resistant *Teladorsagia* (*Ostertagia*) *circumcincta*. *Journal of Veterinary Pharmacology and Therapeutics* **22**, 2–5.

SUTHERLAND, I. A., LEATHWICK, D. M., BROWN, A. E. &

MILLER, C. M. (1997). Prophylactic efficacy of persistent anthelmintics against challenge with drug-resistant and susceptible *Teladorsagia* (*Ostertagia*) *circumcincta*. *Veterinary Record* **141**, 120–123.

SUTHERLAND, I. A., LEE, D. L. & LEWIS, D. (1989). Colorimetric assay for the detection of benzimidazole resistance in trichostrongylids. *Research in Veterinary Science* **46**, 363–366.

TAYLOR, M. A. (1990). A larval development test for the detection of anthelmintic resistant nematodes in sheep. *Research in Veterinary Science* **49**, 198–202.

TAYLOR, M. A. & HUNT, K. R. (1988). Field observations on the control of ovine parasitic gastroenteritis in south-east England. *Veterinary Record* **123**, 241–245.

TAYLOR, M. A. & HUNT, K. R. (1989). Anthelmintic drug resistance in the UK. *Veterinary Record* **125**, 143–147.

VARADY, M. & CORBA, J. (1999). Comparison of six *in vitro* tests in determining benzimidazole and levamisole resistance in *Haemonchus contortus* and *Teladorsagia* (*Ostertagia*) *circumcincta* of sheep. *Veterinary Parasitology* **80**, 239–249.

VAN WYK, J. A., MALAN, F. S. & BATH, G. F. (1997). Rampant anthelmintic resistance in sheep in South Africa – what are the options? In *Workshop at the 16th International Conference of the World Association for the Advancement of Veterinary Parasitology* (ed. Van Wyk, J. A. & van Schalkwyk, P. C.), pp. 51–63.

VAN WYK, J. A. & VAN SCHALKWYK, P. C. (1990). A novel approach to the control of anthelmintic-resistant *Haemonchus contortus* in sheep. *Veterinary Parasitology* **35**, 61–69.

WALLER, P. J. (1994). The development of anthelmintic resistance in ruminant livestock. *Acta Tropica* **56**, 233–243.

WALLER, P. J. (1997). Anthelmintic resistance. *Veterinary Parasitology* **72**, 391–412.

WALLER, P. J., DASH, K. M., BARGER, I. A., LE JAMBRE, L. F. & PLANT, J. (1995). Anthelmintic resistance in nematode parasites of sheep – learning from the Australian experience. *Veterinary Record* **136**, 411–413.

WALLER, P. J., DOBSON, R. J. & AXELSON, A. (1988). Anthelmintic resistance in the field: changes in resistance status of parasitic populations in response to anthelmintic treatment. *Australian Veterinary Journal* **65**, 376–379.

WALLER, P. J., DOBSON, R. J., DONALD, A. D., GRIFFITHS, D. A. & SMITH, E. F. (1985*a*). Selection studies on anthelmintic resistance and susceptible populations of *Trichostrongylus colubriformis* in sheep. *International Journal for Parasitology* **15**, 669–676.

WALLER, P. J., DONALD, A. D., DOBSON, R. J., LACEY, E., HENESEY, D. R., ALLERTON, G. R. & PRICHARD, R. K. (1985*b*). Changes in anthelmintic resistance status of *Haemonchus contortus* and *Trichostrongylus colubriformis* exposed to different anthelmintic selection pressures in grazing sheep. *International Journal for Parasitology* **19**, 99–110.

Value of present diagnostic methods for gastrointestinal nematode infections in ruminants

M. EYSKER* *and* H. W. PLOEGER

Division of Parasitology and Tropical Veterinary Medicine, Utrecht University, P.O. Box 80.165,
3508 TD Utrecht, The Netherlands

SUMMARY

In this paper the different options for the diagnosis of gastrointestinal nematode infections are discussed. Diagnostic tests have a role in confirming the clinical diagnosis of parasitic gastroenteritis, but are more important for herd health monitoring of nematode infections, in particular for cattle. Therefore, emphasis is placed on discussing the available diagnostic parameters on their usefulness for that purpose. For clinical diagnosis the clinical signs, combined with the history of the animals is usually sufficient and a laboratory confirmation is not required. Faecal egg counts are, with two exceptions, not suitable for confirmation of the clinical diagnosis, because correlation between faecal egg counts and infection levels is usually low. These exceptions are the diagnosis of haemonchosis in small ruminants and the detection of anthelmintic resistance. This also limits the value of DNA-based tests of faecal material; even quantitative tests of nematode species specific DNA will have little value for diagnosis and monitoring. Pasture larval counts and worm counts are useful parameters for basic epidemiological studies on nematode infections. However, they are too laborious to be used for either routine diagnosis or monitoring. Blood parameters, such as gastrin and pepsinogen and serology are valuable tools for diagnosis. Pepsinogen and ELISAs based on recombinant proteins show most promise as parameters for herd health monitoring. However, extensive epidemiological studies are still needed before these parameters can be implemented in routine herd health monitoring schemes for parasitic gastroenteritis.

Key words: Ruminants, gastrointestinal nematodes, diagnostic methods, pepsinogen, ELISA, herd health monitoring.

INTRODUCTION

Worldwide parasitic gastroenteritis is an important cause of production losses in cattle and small ruminants. These losses include the direct effects of severe clinical signs such as anaemia and associated oedema, diarrhoea and anorexia. These can easily result in poor general performance and even mortality, particularly in young animals. More important are the less visible subclinical losses, such as decreased weight gain (Ploeger *et al.* 1990*b*, *c*; Ploeger & Kloosterman, 1993), decreased milk yield (reviewed by Gross, Ryan & Ploeger, 1999) and decreased fertility (Ankers *et al.* 1998; Osaer *et al.* 1999). Subclinical losses in older cattle are becoming more important because highly effective nematode control schemes using anthelmintics are available now for young cattle. Application of these schemes in dairy replacement heifers, in particular when combined with evasive grazing management, can result in insufficient exposure to nematodes in the first grazing season and as a consequence subclinical production losses are delayed to later grazing seasons (Ploeger *et al.* 1990*b*). This implies that diagnostic tools are needed to monitor exposure to nematode infections or, even better, the level of immunity against these nematodes in young cattle. Unfortunately, current diagnostic techniques do not permit monitoring the level of immunity directly. Therefore, in discussing the available diagnostic techniques, emphasis will be placed on their usefulness for estimating nematode exposure.

HERD HEALTH MONITORING

Nowadays veterinary care of domestic ruminants more and more implies a herd-based approach to preventive disease control. This implies contracts between farmers and veterinary practices for regular visits to monitor all veterinary aspects of the herd. Good records of all animals in a herd and software to analyse these are essential. Furthermore, suitable parameters have to be available for decision making. At the moment parameters suitable for monitoring parasitic gastroenteritis in dairy cattle in such herd health monitoring programmes have not yet been sufficiently validated. The main requirements of such parameters should be that: (1) they enable an estimate of nematode exposure; (2) values reflect production losses from these infections; (3) they can be used to predict the risks of future production losses and allow recommendation of appropriate preventive measures; (4) they are easy to assess and (5) they are inexpensive. Tests should be quantitative because grazing ruminants will always be infected with gastrointestinal nematodes. Furthermore, they should be used to evaluate the status of herds/flocks rather than of individual animals. In dairy cattle parameters for monitoring nematode infections are

* Corresponding author: Tel: +31 30 2531223. Fax: +31 30 2540784.
E-mail: m.eysker@vet.uu.nl

Parasitology (2000), **120**, S109–S119.   Printed in the United Kingdom   © 2000 Cambridge University Press

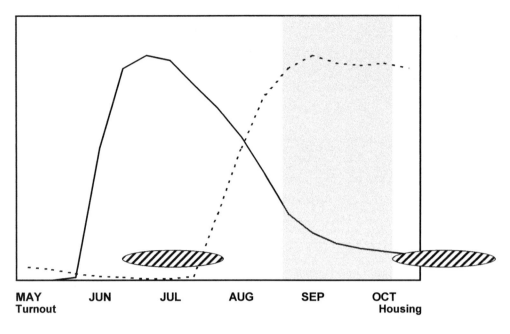

| Monitoring at 5-10 weeks after turnout or around housing |
| Period when clinical disease is usually observed |

Fig. 1. Diagram indicating the epidemiological pattern of gastrointestinal nematode infections in first-season grazing calves and the periods when herd health monitoring of these infections could be applied. Solid line indicates faecal egg excretion and the dotted line the pasture infectivity.

needed most for replacement heifers, though it would also be an advantage to have a test that could predict whether or not anthelmintic treatment of lactating animals would be profitable on a certain farm or not (Gross *et al.* 1999). Timing of monitoring depends on the epidemiological pattern of nematode infections and the ease of collecting samples. The best periods for sampling first-year calves are indicated in Fig. 1 which also shows the normal pattern of faecal egg counts and pasture larval counts in set-stocked calves. Good opportunities for monitoring would be 5–10 weeks after turnout and after housing. The first period would allow an adaptation of nematode control when initial infections are higher or lower than expected. The second period is convenient because animals can be sampled easily. Results can be used to decide whether anthelmintic treatment is required during housing and to predict whether problems with nematodes may be expected in the second grazing season. It also can be used to re-evaluate applied herd management so that this can be adapted, if necessary, for the next crop of first-season grazing calves.

CLINICAL DIAGNOSIS

The diagnosis of the different forms of parasitic gastroenteritis in domestic ruminants always includes a combination of the clinical signs and the history of the animals. Clinical parasitic gastro-enteritis usually is observed in young animals. It is a herd/flock problem with a wide variation in severity

of clinical signs. Poor body condition, depressed weight gain or weight loss and anorexia are general clinical signs. Haemonchosis is also characterized by anaemia and associated (submandibular) oedema. In most other forms of parasitic gastroenteritis, such as ostertagiosis, trichostrongylosis and nematodirosis, diarrhoea is a common feature. Factors in the history needed for the diagnosis of parasitic gastroenteritis include age (young animals), season, grazing history and anthelmintic usage. In outbreaks of parasitic gastroenteritis the history, in the authors' experience, will always be such that it explains why there is an outbreak. An exception to this may be when control has failed as a result of emerging anthelmintic resistance. Another exception may concern the occurrence of ostertagiosis type II caused by the resumed development of large numbers of inhibited larvae, because linking this to the grazing history is not immediately obvious. Laboratory confirmation of a clinical diagnosis, based on the combination of clinical signs, though not highly specific for parasitic gastroenteritis, and the history of the animals, is usually not necessary. In fact, when parasitic gastroenteritis is suspected, diseased animals should be treated with an anthelmintic before laboratory confirmation of the disease becomes available.

FAECAL EXAMINATION

Faecal egg counts

Faecal egg counts are the most widely used parameter in studies on gastrointestinal nematode infections of ruminants. The obvious reason is that it is an easily applicable low technology parameter. However, in most instances the faecal egg counts do not reflect nematode infection levels, and therefore should be used with great caution as an estimator of clinical disease. When calves are exposed to high primary nematode infections a stereotyped faecal egg count pattern is observed irrespective of whether experimental single or trickle infections or natural infections occur (Michel, 1969; Michel, Lancaster & Hong, 1973; Borgsteede, 1977) (Fig. 1). That typical pattern was thought to be independent from infection level (Michel, 1969). Indeed at the time clinical disease may develop, egg counts are not indicative of which animals or groups of animals develop disease and which do not. However, when initial infections are low at the beginning of the first grazing season, a gradual build-up of faecal egg counts occurs that is related to the level of the initial infection. As a consequence there is a high correlation between initial infection levels and the faecal egg counts approximately two months after turnout (Ploeger *et al.* 1994; Shaw *et al.* 1997). This could be used for herd health monitoring to assess in the middle of the grazing season whether 'low' or 'high' initial infections occurred and subsequently to decide on further worm control measures. Obviously, this is only possible if no anthelmintics are used early in the grazing season.

Another problem with the use of faecal egg counts as an infection parameter is that the most important nematode for the temperate world, *Ostertagia ostertagi* is a low egg producer and in first season grazing cattle *Cooperia oncophora* eggs usually dominate. In (sub)tropical environments *O. ostertagi* is replaced by *Haemonchus placei* and *C. oncophora* by *C. punctata*, *C. pectinata* and *C. spatulata*. Considering that perhaps faecal egg counts in *H. placei* infections may better reflect intensity of infection than in *O. ostertagi*, they may perform better for diagnostic purposes under (sub)tropical conditions than under temperate conditions. However, the main reason to use them in the (sub)tropics is that there are no suitable alternatives available.

A different situation exists for *Haemonchus contortus* infections in small ruminants, since a relationship exists between blood loss, the main pathological characteristic, and faecal egg counts (Allonby & Dargie, 1973). *H. contortus* is a high egg producer and faecal egg counts of several thousand eggs per gram faeces (EPG) in anaemic sheep form strong evidence that this nematode is causing the problem. A correlation between worm numbers and faecal egg counts has also been demonstrated in sheep infected with *H. contortus* (Roberts & Swann, 1981).

Faecal egg counts are very important for the detection of anthelmintic resistance. Even though alternative methods, such as a larval development

test (Lacey *et al.* 1990) have been developed, the Faecal Egg Count Reduction Test (FECRT) is still the most widely used tool for field diagnosis of this phenomenon (Waller, 1997).

The phenomenon that a wide variation in susceptibility against nematode infections exists within host populations has been known for a long time (Gregory, Miller & Stewart, 1940). Studies on the genetic background for this phenomenon are now highly fashionable (Gray, 1997; Stear, Strain & Bishop, 1999), since breeding against susceptibility to nematode infections is considered as one option for maintaining a sustainable (small) ruminant sector in the presence of widespread anthelmintic resistance (Waller, 1997). Faecal egg counts are a major parameter for distinguishing between susceptible and resistant phenotypes. In Australia (Anon, 1994) and New Zealand (McEwan *et al.* 1995) faecal egg counts are already used in selection programmes in sheep against susceptibility to nematode infections. Similarly it has been shown that there is a wide variation between calves in their response to *C. oncophora* infection which is partly sire-associated (Albers, 1981; Kloosterman & Ploeger, unpublished results).

Faecal larval cultures

Naturally infected ruminants always will have mixed infections of different species of gastrointestinal nematodes. Considering the differences in pathogenicity and in faecal egg output between species it is often necessary to further differentiate the egg counts. This is not a problem for eggs of some species such as *Strongyloides papillosus*, *Toxocara vitulorum*, *Nematodirus* spp., *Trichuris* spp. and *Capillaria* spp. because they can be differentiated to the species or at least the genus level. However, eggs of the most important trichostrongylid and strongylid species show considerable similarity and can only be differentiated with certainty as eggs of the 'strongyle' type. To differentiate these to species or genus level faecal larval cultures are needed. Identification of larvae can be made using keys by Keith (1953), Borgsteede & Hendriks (1974) and MAFF (1986). A disadvantage of these larval cultures is that yields are never 100% and, particularly when for some reason the yield is low, the proportion of larvae developing may differ between species. For normal diagnosis faecal larval cultures will not be required, but for parasitological field studies they are needed unless better molecular techniques to differentiate the eggs are available.

PASTURE LARVAL COUNTS

Pasture larval counts are extremely laborious and should therefore never be used as a routine method for the diagnosis of nematode infections. Another limitation for their use as a diagnostic tool on farms is that, depending on grazing history, a wide between-pasture variation in pasture infectivity occurs. Moreover, pasture infectivity on a single pasture will vary throughout the grazing season. On commercial farms it will be difficult to relate pasture infectivity to a clinical outbreak because it takes approximately 3–4 weeks after high pasture infections before the outbreak of clinical disease.

A relatively recent review on pasture larval counts has been written by Couvillon (1993). Virtually all laboratories use their own techniques to process herbage samples. There is a wide variation in the way herbage samples are collected, though the double W indicated by Taylor (1939) is used in many laboratories. Grass may be collected by picking small pieces by hand or by the use of scissors or knives. This implies that it is difficult to compare results between laboratories.

Pasture larval counts are particularly useful in longitudinal studies on the population dynamics of nematodes. They can also be used in epidemiological transversal studies with a well defined aim. For instance they might be used to assess the level of exposure of any age class of dairy or beef cows at a certain time, in particular the beginning and the end of the grazing season. It is important to perform differential pasture larval counts, using the same keys for larval differentiation as for faecal larval cultures, because there are significant differences between species in their bionomics. For instance *H. contortus* needs higher developmental temperatures than *Teladorsagia circumcincta* and hardly survives winter on pasture in Northern Europe (Eysker, 1980). Furthermore, the dynamics of *Nematodirus* spp. on pasture differ considerably from other trichostrongylids.

For both above-mentioned types of study pasture larval counts are the cheapest method to acquire meaningful data. However, the nature of the techniques involved implies that results are not very accurate. Herbage contaminated with faecal material should be avoided during sampling. Another point is that larvae are not randomly distributed on pasture (Bryan & Kerr, 1988). Cattle avoid grazing close to faecal pats and infective larvae concentrate in the resulting long grass surrounding the pats. Therefore, separate sampling of 'grazing areas' and 'long grass' or 'bushes' is sometimes performed, particularly in studies also involving *Dictyocaulus viviparus* larvae (Jørgensen, 1981). This also implies that pasture larval counts based on randomly collected samples may result in overestimation of the exposure of cattle to nematode infections, although a reasonable correlation has been observed between pasture larval counts and tracer worm counts (Shaw *et al.* 1998*a*).

POST-MORTEM EXAMINATION

Post-mortem examination can be performed on 'tracers' (parasite-naive fully susceptible animals that graze with the herd/flock for a short period) or by selecting sentinel animals from the herd/flock. In both instances it is obvious that, with perhaps the exception of sentinel sheep on large sheep farms, this is far too expensive for diagnostic purposes at farm level. Nevertheless, pathological examination of animals that died or were killed *in extremis* can demonstrate that parasitic gastroenteritis is a herd/flock problem on a certain farm. Necropsy studies are essential tools for determining the epidemiological pattern of infections and for identifying which species are present.

Tracer worm counts

Tracer worm counts are an alternative method to study pasture infectivity levels. Usually helminth-free young ruminants are used as tracer animals. They will be grazed with the herd/flocks for a certain period, usually 2–4 weeks, subsequently they will be housed to prevent further infection and necropsied 2–4 weeks later. It is important not to extend these periods because, in particular in calves exposed to high *C. oncophora* infections, large proportions of worms may then be expelled before necropsy (Albers, 1981).

The use of tracers is additive to pasture larval counts and should be linked to these. When done properly pasture larval counts will reflect ingestion of nematode larvae by grazing ruminants, whereas differential tracer worm counts will not only reflect ingestion but also subsequent establishment. These establishment rates may vary over the year, in particular when larvae on pasture are ageing. This implies that, in principle, tracer worm counts will better reflect what happens within a herd/flock than pasture larval counts. However, as mentioned earlier, there is a wide, genetically determined, variation in establishment of parasites between individuals. This implies that large enough groups of tracers should be used to get representative worm counts. Moreover, for results to be representative for the herd/flock, grazing behaviour must be similar for tracers and for 'residents'. This may not always be true because tracers have to adapt to grazing and may show deviant behaviour.

Obviously an advantage of tracer worm counts compared to pasture larval counts is that it is possible to differentiate the nematodes present to species level. This also depends on the length of the grazing and the housing periods as it will not be possible to perform proper worm counts for nematodes with long prepatent periods, such as *Bunostomum* spp. and *Chabertia ovina*. Another advantage of using tracer animals is that it is possible to assess whether inhibited development is an important factor in the epidemiology of the different nematode species (Michel, 1974; Schad, 1977; Armour, 1978; Gibbs, 1986; Eysker, 1997). Combined with results from sentinels, tracer worm counts will also reveal whether inhibition is primarily a seasonal phenomenon or primarily related to host resistance.

Sentinel worm counts

While tracer worm counts reflect pasture infectivity and recent exposure of a herd/flock, worm counts of sentinel animals rather reflect overall dynamics of nematode populations, including development of resistance. This implies that animals representative for the herd/flock should be selected and the group size should be large enough to cope with the wide within group variation of nematode burdens. Moreover, necropsy of sentinel animals will allow finding and enumeration of species with long prepatent periods that cannot be found easily using tracer animals.

MUCOUS MEMBRANES/RED BLOOD CELL VALUES

For infections with blood sucking nematodes like *H. contortus* the mucous membranes and the red blood cell values (haematocrit, packed cell volume) provide useful tools for diagnosis. In South Africa 'FAMACHA', a colour chart, is evaluated as a tool to decide which individuals in a flock should be selectively treated for haemonchosis (van Wyk, Malan & Bath, 1997). The colour of the mucous membranes of all sheep in a flock are regularly checked against the FAMACHA chart and those sheep with pale membranes will be treated with an anthelmintic. The rationale behind this selective treatment is that the faeces of non-treated sheep will remain to contaminate the pasture with nematode eggs, thus generating a refugia for maintaining genetic diversity of the nematode, thereby slowing the development of anthelmintic resistance (Martin, LeJambre & Claxton, 1981).

BLOOD PEPSINOGEN VALUES

Blood pepsinogen values have been used for 35 years as a diagnostic tool for ostertagiosis in cattle (Anderson *et al.* 1965; Jennings *et al.* 1966). In the first grazing season high pepsinogen values in cattle correlate with the occurrence of parasitic gastroenteritis. Different values have been given for the diagnosis of clinical ostertagiosis in young calves. According to Anderson *et al.* (1965) values above 3000 mU tyrosine are diagnostic for ostertagiosis, but Hilderson *et al.* (1989) indicated a threshold value of 5000 mU tyrosine. A problem in interpretation of these differences is that no standardized

method is used and between-laboratory comparisons are lacking.

Elevated blood pepsinogen values can also be found in clinically healthy cows (Chiejina 1977, 1978; Berghen *et al.* 1988), probably as a result of hypersensitivity following previous infections (McKellar, 1984–1985). This implies that for the diagnosis of ostertagiosis pepsinogen values always have to be used in conjunction with clinical and parasitological data. It has been demonstrated repeatedly that blood pepsinogen values correlate with infection levels of abomasal nematodes (Berghen *et al.* 1993; Ploeger *et al.* 1990*a–c*, 1994; Shaw *et al.* 1998*a, b*; Dorny, Shaw & Vercruysse, 1998). This renders blood pepsinogen values as a good candidate for herd health monitoring. A major disadvantage of the methods used was that they were very labour intensive and thus too expensive. However, recently a micro method was introduced that solved this problem (Dorny & Vercruysse, 1998). The potential of pepsinogen values for herd health monitoring compared to the other promising tool, serology, will be discussed later.

BLOOD GASTRIN VALUES

Hypergastrinaemia occurs around patency in *O. ostertagi* infections (Fox *et al.* 1989) and seems to be associated with the presence of adult worms (McKellar *et al.* 1987; Hilderson *et al.* 1992). Results of Ploeger *et al.* (1994) demonstrated a correlation between exposure levels and gastrin values in the second half of a simulated first grazing season. However, it was also clear that gastrin is far less sensitive than faecal egg counts, pepsinogen values and serology to detect low infections (Berghen *et al.* 1993; Ploeger *et al.* 1994). In fact gastrin is only elevated at levels which induce substantial growth depression. Levels inducing lower subclinical losses are not detected (Fig. 2). Furthermore, no gastrin elevations are observed in exposed immune cattle with low adult worm burdens. It can be concluded that gastrin is suitable for confirming an outbreak of clinical parasitic gastroenteritis, though it is expensive. For herd health monitoring it is not suitable.

SEROLOGY

An ELISA using crude worm antigens has been used for diagnosis of gastrointestinal nematode infections for many years (Keus, Kloosterman & van den Brink, 1981). Using the mean values of such ELISAs with crude *O. ostertagi* and *C. oncophora* proteins as antigens, Ploeger *et al.* (1989, 1990*a–c*) were able to demonstrate differences in infection levels between different commercial dairy farms in the Netherlands and between the different age classes on these farms. They could also link this to production traits like age-adjusted weight gain and even milk yield. The

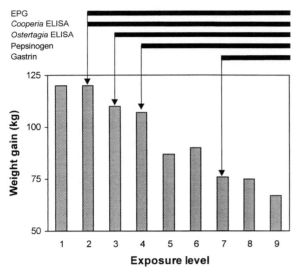

Fig. 2. The sensitivity of different parameters for the detection of gastrointestinal nematode infections in relation to production losses. The exposure levels are the group numbers from Ploeger & Kloosterman (1993) and Ploeger et al. (1994). Level 1 represents a non-infected control group. Horizontal bars indicate the range of exposure levels for which the parameter showed significantly elevated values. The vertical arrows indicate the lowest exposure at which the level of the parameter became significantly increased. In the ELISAs crude adult worm antigen was used.

obvious disadvantage of a crude worm ELISA is that antigenic epitopes are shared by many worms and cross reactions with other nematodes and even with *Fasciola hepatica* do occur. Moreover, it is generally recognized that a wide within-host variation exists in the ability to develop an antibody response against gastrointestinal nematode antigens, implying that an ELISA, like all previous parameters discussed, is not very appropriate for diagnosis of parasitic gastro-enteritis in individual animals. Finally it is very difficult to standardize a crude worm ELISA, considering that each batch of worms will differ. Nevertheless, the results of Ploeger et al. (1989, 1990a–c) clearly demonstrated the potential of serology with crude worm antigens during large-scale field surveys, as it was by far the most useful of their parameters. Later experimental studies of Ploeger et al. (1994) demonstrated a good correlation between exposure to nematodes and the results of a crude worm ELISA in the second half of an experimental 'first grazing season'. The *C. oncophora* ELISA demonstrated differences in exposure earlier than the *Ostertagia* ELISA or the pepsinogen values (Fig. 2). Nevertheless, it is also obvious that, though useful in research, crude worm antigen ELISAs never will satisfy all requirements to be suitable for herd health monitoring of nematode infections. Therefore, an ELISA was developed at Utrecht University that was based on a specific recombinant low molecular weight protein of *C. oncophora* (Poot et al. 1997). Sera from all calves with

a *C. oncophora* infection that were tested with this ELISA recognized the recombinant protein. Futhermore, a correlation between exposure and ELISA results was observed. Also, the protein proved to be highly specific for *Cooperia* in a competitive ELISA and cross reactions with helminths from other genera cannot be expected. For other *Cooperia* spp., such as *C. punctata*, *C. pectinata*, *C. spatulata* and *C. curticei* partial competition was observed in the competitive ELISA (Fig. 3; Kooyman & Yatsuda, unpublished results). Subsequently, the recombinant *C. oncophora* protein was recognized in sera of calves with mono infections with *C. punctata* (Yatsuda, unpublished results). This may imply that the ELISA with this recombinant protein is not only suitable for monitoring nematode infections under temperate conditions but also in the tropics. Currently the suitability of this test to monitor *C. punctata* infections is being evaluated.

Although initial studies using the ELISA with the recombinant *C. oncophora* protein showed a correlation with exposure to nematodes this had to be further validated. Data from experimental trickle infections (Ploeger et al. 1994) were used to link the level of cumulative exposure to *C. oncophora* infections to results with the recombinant ELISA. Subsequently another experimental infection trial (Eysker & Ploeger, unpublished results) and a series of field trials (Eysker et al. 1998a, b) were used to validate the test. This validation involved 'predicting' cumulative exposures based on observed IgG values in the recombinant ELISA, assuming a 4-week delay for an IgG response to infection. Generally, 'predicted' values correlated well to observed exposures (r values of 0·7–0·9). However, 'predicted' cumulative exposures underestimated observed exposures in the natural field studies. This was probably due to differences between parasite strains involved in the experimental studies and the field trials. Taken together, the results strongly indicate that the recombinant *C. oncophora* ELISA satisfies the basic requirements necessary for application in herd health monitoring programmes in young dairy cattle (Githiori et al. unpublished).

DNA TECHNOLOGY

DNA technology is becoming more important for diagnosis of infectious diseases. In particular the availability of the PCR allows the development of highly sensitive and specific tests for detecting the presence of infectious agents. Moreover, it allows measurement of within-species variation in populations of infectious agents. Development of such specific tests is possible for gastrointestinal nematodes of ruminants and specific probes have been developed for different species (Christensen, Zarlenga & Gasbarre, 1994; Hoste et al. 1995; Humbert & Cabaret, 1995; Zarlenga et al. 1994). However, in

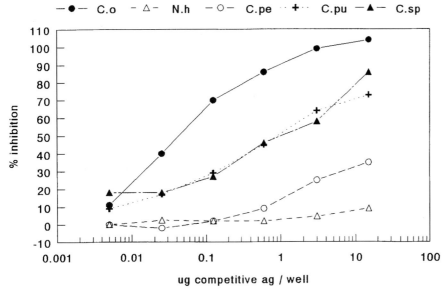

Fig. 3. Results of a competitive ELISA with the recombinant 14·2 kDa *Cooperia oncophora* protein with three other species of *Cooperia* spp. and with *C. oncophora* as positive and *Nematodirus helvetianus* as negative controls. Crude worm extracts of *Cooperia oncophora* (C.o), *C. pectinata* (C.pe), *C. punctata* (C.pu), *C. spatulata* (C.sp) and *Nematodirus helvetianus* (N.h) were used as competitive antigens.

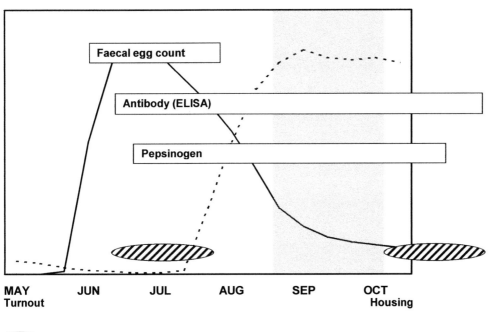

Fig. 4. Diagram indicating the epidemiological pattern of gastrointestinal nematode infections in first-season grazing calves and the periods when herd health monitoring of these infections could be applied using faecal egg counts, blood pepsinogen values and the recombinant ELISA. Solid and dotted lines see Fig. 1. The rectangles indicate the period in which the respective parameters reflect level of exposure to infection.

contrast with the value of DNA technology for the diagnosis of many viral, bacterial and protozoal infections, these specific molecular tests are not very useful for the diagnosis of gastrointestinal nematode infections. A qualitative test (one which indicates whether a parasite is present or not) does not generate additional information since grazing rumin-

ants will normally acquire infections during grazing. This implies that the DNA test should not only be specific but also quantitative. The only obvious sources for DNA are the eggs in the faeces. Thus, such quantitative DNA tests would only allow more accurate differential faecal egg counts. Considering that faecal egg counts are not a good parameter for

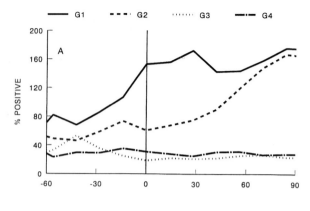

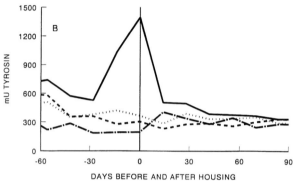

Fig. 5. Mean blood ELISA values with the recombinant *C. oncophora* protein as antigen (in mean percentage of OD value of a standard positive serum) (A) and pepsinogen values (in mU tyrosin) (B) in three groups of naturally infected and one permanently housed non-infected control group of calves in the last 2 months before and after housing. Infection levels, as measured from other parameters were moderate in G1, low in G2 and extremely low in G3.

diagnosis this automatically disqualifies a DNA-based test. Of course the DNA tests would be very useful for basic studies on the between-species and within-species variation of nematode parasites. Furthermore, molecular tools have been developed for the diagnosis of benzimidazole resistance of *Haemonchus contortus* (Kwa, Veenstra & Roos, 1993) and *Trichostrongylus* spp. (Grant & Mascord, 1996). These highly significant tools could have been used for early detection of a low prevalence of resistant genotypes within parasite populations. However, considering the current wide-spread occurrence of benzimidazole resistance it would have been more useful to have them available much earlier.

TOWARDS HERD HEALTH MONITORING OF GASTROINTESTINAL NEMATODE INFECTIONS

The requirements of parameters needed for herd health monitoring have been indicated earlier and the periods when sampling should occur in first-season grazing dairy replacement heifers have been shown in Fig. 1. From the parameters discussed only faecal egg counts, pepsinogen values and the ELISA, using recombinant proteins as antigens, are potential candidates.

The only period that faecal egg counts reflect exposure levels is between approximately 5 and 10 weeks after turnout in the first grazing season (Fig. 4). Therefore, faecal egg counts during this period can be used to assess the initial infection level and this may then result in an alteration of the control programme. Fig. 4 also indicates that the *Cooperia* ELISA, and to a lesser extent pepsinogen values, also reflect exposure levels during the latter half of this period. However, to use them in that period as a monitoring tool would necessitate taking blood samples from grazing calves, which imposes impractical demands on the farming system. Therefore, faecal examination is the parameter of choice for the middle of the first grazing season.

After housing no correlation between faecal egg counts and exposure exists anymore and only pepsinogen and the ELISA remain as potential candidate parameters for monitoring (Fig. 4). The number of animals sampled should be representative for the herd. Earlier studies indicated that 5 blood samples per age-class suffice (Boon *et al.* 1982). The ELISA has the obvious advantage that equipment to carry out the test is available in virtually any diagnostic laboratory. Furthermore, the test can be automated and thus can be relatively cheap. However, nowadays blood pepsinogen estimation is not very expensive since the development of a micro technique (Dorny & Vercruysse, 1998). It is difficult to judge at this moment which of those parameters should be used in preference for herd health monitoring. Results for both parameters on commercial farms correlate reasonably well (Ploeger *et al.* unpublished) and for both methods the predictive value needs to be further determined through large-scale epidemiological studies comparable to those performed on close to 100 farms in the Netherlands (Ploeger *et al.* 1989, 1990 *a–c*).

Comparison between blood pepsinogen values and results with the recombinant *C. oncophora* ELISA shows advantages and disadvantages for either technique. For both techniques there is a time delay of 3–4 weeks before 'positive' reactions are observed. For blood pepsinogen estimation it is essential that blood samples are taken at housing because values may decrease rapidly in the absence of continuous exposure whereas ELISA values may further increase (Fig. 5; Berghen *et al.* 1993). Moreover, this decrease in pepsinogen is further enhanced when animals are treated with an anthelmintic (Berghen *et al.* 1993). Fig. 4 also confirms that the *Cooperia* ELISA is more sensitive than pepsinogen values for the detection of low infections (Fig. 2). The further increase of ELISA values after termination of infection and even after anthelmintic treatment was also demonstrated using a crude worm ELISA for *Ostertagia* (Berghen *et al.* 1993). Thus, the possibility for flexibility of sampling dates after the first grazing season seems to favour the recombinant

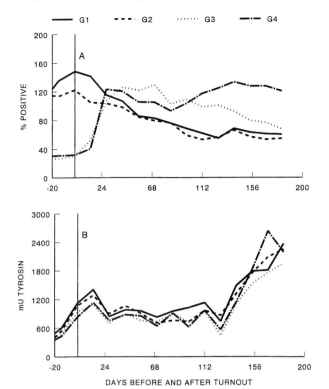

Fig. 6. Mean ELISA (A) and blood pepsinogen (B) values in the same groups as in Fig. 3 during the second grazing season when they were grazed as one herd. Exposure to nematode infections was low to moderate until the end of August and subsequently increased.

ELISA over the pepsinogen assay. On the other hand pepsinogen values probably perform better in the second grazing season, because recent data indicate that in groups of yearlings immune to *C. oncophora* infections recombinant ELISA values remained low or even decreased, whereas increasing pasture infectivity was reflected by elevations in pepsinogen values (Fig. 6).

CONCLUSIONS

Epidemiological studies have to be performed to determine the best type of monitoring and which diagnostic test has to be used. It is obvious that this can only be achieved when it is possible to link values of pepsinogen or ELISA to a prediction of risks for parasitic gastroenteritis or of subclinical production losses. The costs of a test should be low enough for farmers and veterinarians to accept it as a tool. Furthermore, veterinarians and farmers will need to be trained in the interpretation of the test to get it accepted.

ACKNOWLEDGEMENTS

Prof. Dr. A. W. C. A. Cornelissen is thanked for his support for this work and for critically reading the manuscript. Ing. F. N. J. Kooyman and A. P. Yatsuda DVM, MSc are thanked for their valuable contributions to the work.

REFERENCES

ALBERS, G. A. A. (1981). *Genetic Resistance to Experimental* Cooperia oncophora *Infection in Calves*. PhD thesis, Agricultural University, Wageningen.

ALLONBY, E. W. & DARGIE, J. D. (1973). Ovine haemonchosis: epidemiology, clinical signs and diagnosis; pathogenesis. In *Helminth Diseases of Cattle, Sheep and Horses* (ed. Urquhart, G. M. & Armour, J.), pp. 59–71. Glasgow: The University Press.

ANDERSON, N., ARMOUR, J., JARRETT, W. F. H., JENNINGS, F. W., RITCHIE, J. S. D. & URQUHART, G. M. (1965). A field study of parasitic gastroenteritis in cattle. *The Veterinary Record* **77**, 1196–1204.

ANKERS, P., ITTY, P., ZINNTAG, J., TRAWALLY, S. & PFISTER, K. (1998). Biannual anthelmintic treatment in village Djallonke sheep in The Gambia: effects on productivity and profitability. *Preventive Veterinary Medicine* **34**, 215–225.

ANON (1994). *Nemesis Breeders Worm Control Network – Recommendations for Breeding Sheep for Worm Resistance*. Armidale, CSIRO, p. 9.

ARMOUR, J. (1978). Arrested development in cattle nematodes with special reference to *Ostertagia ostertagi*. In *Facts and Reflections III* (ed. Borgsteede, F. H. M., Armour, J. & Jansen, J.), pp. 77–88. Lelystad: The Netherlands, Central Veterinary Institute.

BERGHEN, P., DORNY, P., VERCRUYSSE, J. & HILDERSON, H. (1988). The use of the serum pepsinogen assay in the epidemiological study of *Ostertagia ostertagi* (Dutch). *Vlaams Diergeneeskundig Tijdschrift* **57**, 157–173.

BERGHEN, P., HILDERSON, H., VERCRUYSSE, J. & DORNY, P. (1993). Evaluation of pepsinogen, gastrin and antibody response in diagnosing ostertagiosis. *Veterinary Parasitology* **46**, 175–195.

BOON, J. H., KLOOSTERMAN, A. & BRINK, R. VAN DEN (1982). The incidence of *Dictyocaulus viviparus* infections in cattle in The Netherlands. I. The Enzyme Linked Immunosorbent Assay as a diagnostic tool. *The Veterinary Quarterly* **4**, 155–160.

BORGSTEEDE, F. H. M. (1977). *The Epidemiology of Gastrointestinal Helminth-Infections in Young Cattle in The Netherlands*. PhD thesis, Utrecht University.

BORGSTEEDE, F. H. M. & HENDRIKS, J. (1974). Identification of infective larvae of gastrointestinal nematodes in cattle. *Tijdschrift voor Diergeneeskunde* **99**, 103–113.

BRYAN, P. & KERR, J. D. (1988). The grazing behaviour of cattle in relation to the sampling of infective nematode larvae on pasture. *Veterinary Parasitology* **30**, 73–82.

CHIEJINA, S. N. (1977). Plasma pepsinogen levels in relation to ostertagiasis in adult cattle. *The Veterinary Record* **100**, 120.

CHIEJINA, S. N. (1978). Field observations on the blood pepsinogen levels in clinically normal cows and calves and in diarrhoeic adult cattle. *The Veterinary Record* **103**, 278–281.

CHRISTENSEN, C. M., ZARLENGA, D. S. & GASBARRE, L. C. (1994). *Ostertagia, Haemonchus, Cooperia* and *Oesophagostomum*: construction and characterization of genus-specific DNA probes to differentiate important parasites of cattle. *Experimental Parasitology* **78**, 93–100.

COUVILLON, C. E. (1993). Estimation of numbers of trichostrongylid larvae on pasture. *Veterinary Parasitology* **46**, 197–203.

DORNY, P. & VERCRUYSSE, J. (1998). Evaluation of a micro method for the routine determination of serum pepsinogen in cattle. *Research in Veterinary Science* **65**, 259–262.

DORNY, P., SHAW, D. & VERCRUYSSE, J. (1998). The determination at housing of exposure to gastrointestinal nematode infections in first-season calves. *Veterinary Parasitology* **80**, 325–340.

EYSKER, M. (1997). Some aspects of inhibited development of trichostrongylids in ruminants. *Veterinary Parasitology* **72**, 265–283.

EYSKER, M. (1980). *Inhibited Development of Gastro-Intestinal Nematodes of Sheep*. PhD thesis, Utrecht University.

EYSKER, M., AAR, W. M. VAN DER, BOERSEMA, J. H., DOP, P. Y. & KOOYMAN, F. N. J. (1998a). The effect of Michel's dose and move system on gastrointestinal nematode infections in dairy calves. *Veterinary Parasitology* **75**, 99–114.

EYSKER, M., AAR, W. M. VAN DER, BOERSEMA, J. H., GITHIORI, J. B. & KOOYMAN, F. N. J. (1998b). The effect of repeated moves to clean pasture on the build up of gastrointestinal nematode infections in calves. *Veterinary Parasitology* **76**, 81–94.

FOX, M. T., GERELLI, D., PITT, S. R. & JACOBS, D. E. (1989). *Ostertagia ostertagi* infection in the calf: effects of a trickle challenge on the hormonal control of digestive and metabolic function. *Research in Veterinary Science* **47**, 299–304.

GIBBS, H. C. (1986). Hypobiosis in parasitic nematodes – an update. *Advances in Parasitology* **25**, 129–174.

GRANT, W. N. & MASCORD, L. J. (1996). Beta-tubulin gene polymorphism and benzimidazole resistance in *Trichostrongylus colubriformis*. *International Journal for Parasitology* **26**, 71–77.

GRAY, G. D. (1997). The use of genetically resistant sheep to control nematode parasitism. *Veterinary Parasitology* **72**, 345–366.

GREGORY, P. W., MILLER, R. F. & STEWART, M. A. (1940). An analysis of environmental and genetic factors influencing stomach-worm infestation in sheep. *Journal of Genetics* **39**, 391–400.

GROSS, S. J., RYAN, W. G. & PLOEGER, H. W. (1999). Anthelmintic treatment of dairy cows and its effect on milk production. *The Veterinary Record* **144**, 581–587.

HILDERSON, H., BERGHEN, P., VERCRUYSSE, J., DORNY, P. & BRAEM, L. (1989). The diagnostic value of pepsinogen for clinical diagnosis. *The Veterinary Record* **125**, 376–377.

HILDERSON, H., VERCRUYSSE, J., BERGHEN, P., DORNY, P. & MCKELLAR, Q. A. (1992). Diagnostic value of gastrin for clinical bovine ostertagiasis. *Journal of Veterinary Medicine Row B* **39**, 187–192.

HOSTE, H., CHILTON, N. B., GASSER, R. B. & BEVERIDGE, I. (1995). Differences in the second internal transcribed spacer (ribosomal DNA) between five species of *Trichostrongylus* (Nematoda: Trichostrongylidae). *International Journal for Parasitology* **25**, 75–80.

HUMBERT, J. F. & CABARET, J. (1995). Use of random amplified polymorphic DNA for identification of ruminant trichostrongylid nematodes. *Parasitology Research* **81**, 1–5.

JENNINGS, F. W., ARMOUR, J., LANSON, D. D. & ROBERTS, R. (1966). Experimental *Ostertagia ostertagi* infections in calves: studies with abomasal cannulas. *American Journal of Veterinary Research* **27**, 1249–1257.

JØRGENSEN, R. J. (1981). Monitoring pasture infectivity and pasture contamination with infective stages of *Dictyocaulus viviparus*. In *Epidemiology and Control of Nematodiasis in Cattle. An Animal Pathology Workshop in the CEC Programme of Coordination of Agricultural Research, Royal Veterinary and Agricultural University Copenhagen, Denmark, 4–6 February 1980* (ed. Nansen, P., Jørgensen, R. J. & Soulsby, E. J. L.), pp. 3–9. Martinus Nijhoff: The Hague, The Netherlands.

KEITH, R. K. (1953). The differentiation of infective larvae of some common nematode parasites of cattle. *Australian Journal of Zoology* **1**, 223–235.

KEUS, A., KLOOSTERMAN, A. & BRINK, R. VAN DEN (1981). Detection of antibodies to *Cooperia* spp. and *Ostertagia* spp. in calves with enzyme linked immunosorbent assay (ELISA). *Veterinary Parasitology* **8**, 229–236.

KWA, M. S., VEENSTRA, J. & ROOS, M. H. (1993). Molecular characterisation of beta-tubulin genes present in benzimidazole-resistant population of *Haemonchus contortus*. *Molecular and Biochemical Parasitology* **60**, 133–143.

LACEY, E., REDWIN, J. M., GILL, J. H., DEMARGHERITI, V. M. & WALLER, P. J. (1990). A larval development assay for the simultaneous detection of broad spectrum anthelmintic resistance. In *Resistance to Anthelmintic Drugs. Round Table Conference 7th International Congress of Parasitology, Paris. August 1990* (ed. Boray, J. C., Martin, P. J. & Roush, R. T.), pp. 177–183. MSD Agvet: Rahway, NJ.

MAFF (1986). *Manual of Veterinary Parasitological Laboratory Techniques*. 3rd edn. Ministry of Agriculture and Food, London.

MARTIN, P. J., LEJAMBRE, L. F. & CLAXTON, J. H. (1981). The impact of refugia in the development of thiabendazole resistance in *Haemonchus contortus*. *International Journal for Parasitology* **11**, 35–41.

MCEWAN, J. C., DODDS, K. G., WATSON, T. G., GREER, G. J., HOSKING, B. C. & DOUCH, P. G. C. (1995). Selection for host resistance to roundworms by the New Zealand sheep breeding industry: the WormFEC Service. *Proceedings of the Australian Association of Animal Breeding and Genetics* **11**, 70–73.

MCKELLAR, Q. A. (1984–1985). The role of the pepsinogen test in the diagnosis of ostertagiasis. *British Cattle Veterinary Association Proceedings* **1**, 11–14.

MCKELLAR, Q. A., DUNCAN, J. L., ARMOUR, J. & McWILLIAM, P. (1987). Further studies on the response to transplanted adult *Ostertagia ostertagi* in calves. *Research in Veterinary Science* **42**, 29–34.

MICHEL, J. F. (1969). The epidemiology and control of some nematode infections of grazing animals. *Advances in Parasitology* **7**, 211–282.

MICHEL, J. F. (1974). Arrested development of nematodes and some related phenomena. *Advances in Parasitology* **12**, 279–366.

MICHEL, J. F., LANCASTER, M. B. & HONG, C. (1973). *Ostertagia ostertagi*: protective immunity in calves. The development in calves of a protective immunity to infection with *Ostertagia ostertagi*. *Experimental Parasitology* **33**, 179–186.

OSAER, S., GOOSSENS, B., EYSKER, M. & GEERTS, S. (1999). The effects of prophylactic anthelmintic treatment on the productivity of traditionally managed Djallonke sheep and West African Dwarf goats kept under high trypanosomosis risk. *Acta Tropica*, in press.

PLOEGER, H. W., KLOOSTERMAN, A., BARGEMAN, G., WIJCKHUISE, L. V. & BRINK, R. VAN DEN (1990*a*). Milk yield increase after anthelmintic treatment of dairy cattle related to some parameters estimating worm infections. *Veterinary Parasitology* **35**, 103–106.

PLOEGER, H. W., KLOOSTERMAN, A., BORGSTEEDE, F. H. M. & EYSKER, M. (1990*b*). Effect of naturally occurring nematode infections in the first and the second grazing season on the growth performance of second-year grazing cattle. *Veterinary Parasitology* **36**, 57–70.

PLOEGER, H. W., KLOOSTERMAN, A., EYSKER, M., BORGSTEEDE, F. H. M., STRAALEN, W. VAN & VERHOEFF, J. (1990*c*). Effect of naturally occurring nematode infections on growth performance of first-season grazing cattle. *Veterinary Parasitology* **35**, 307–322.

PLOEGER, H. W. & KLOOSTERMAN, A. (1993). Gastrointestinal nematode infections and weight gain in dairy replacement stock: first-year calves. *Veterinary Parasitology* **46**, 223–241.

PLOEGER, H. W., KLOOSTERMAN, A., RIETVELD, F. W., BERGHEN, P., HILDERSON, H. & HOLLANDERS, W. (1994). Quantitative estimation of the level of exposure to gastrointestinal nematode infection in first-year calves. *Veterinary Parasitology* **55**, 287–315.

PLOEGER, H. W., SCHOENMAKER, G. J. W., KLOOSTERMAN, A. & BORGSTEEDE, F. H. M. (1989). Effect of anthelmintic treatment of dairy cattle on milk production related to some parameters estimating worm infection. *Veterinary Parasitology* **34**, 239–254.

POOT, J., KOOYMAN, F. N. J., DOP, P. Y., SCHALLIG, H. D. F. H., EYSKER, M. & CORNELISSEN, A. W. C. A. (1997). Use of two cloned excretory/secretory low molecular weight proteins of *Cooperia oncophora* in a serological assay. *Journal of Clinical Microbiology* **35**, 1728–1733.

ROBERTS, J. A. & SWAN, R. A. (1981). Quantitative studies of ovine egg counts and total worm counts. *Veterinary Parasitology* **8**, 165–171.

SCHAD, G. A. (1977). The role of arrested development in the regulation of nematode populations. In *Regulation of Parasite Populations* (ed. Esch, G. W.), pp. 111–179. New York: Academic Press.

SHAW, D., VERCRUYSSE, J., CLAEREBOUT, E., AGNEESSENS, J. & DORNY, P. (1997). Gastrointestinal nematode infections of first-grazing season calves in Belgium: General patterns and the effect of chemoprophylaxis. *Veterinary Parasitology* **69**, 103–116.

SHAW, D., VERCRUYSSE, J., CLAEREBOUT, E. & DORNY, P. (1998*a*). Gastrointestinal nematode infections of first-season grazing cattle in Western Europe: general patterns and the effect of chemoprophylaxis. *Veterinary Parasitology* **75**, 115–131.

SHAW, D., VERCRUYSSE, J., CLAEREBOUT, E. & DORNY, P. (1998*b*). Gastrointestinal nematode infections of first-grazing season calves in Western Europe: association between parasitological, physiological and physical factors. *Veterinary Parasitology* **75**, 133–151.

STEAR, M. J., STRAIN, S. & BISHOP, S. C. (1999). Mechanisms underlying resistance to nematode infection. *International Journal for Parasitology* **29**, 51–56.

TAYLOR, E. L. (1939). Technique for the estimation of pasture infestation by strongyloid larvae. *Parasitology* **31**, 473–478.

WALLER, P. J. (1997). Anthelmintic resistance. *Veterinary Parasitology* **72**, 391–412.

WYK, J. A. VAN, MALAN, F. S. & BATH, G. F. (1997). Rampant anthelmintic resistance in sheep in South Africa – What are the options? *Workshop: Managing anthelmintic resistance in endoparasites at 16th WAAVP Conference, Sun City, South Africa*, pp. 51–63.

ZARLENGA, D. S., STRINGFELLOW, F., NOBARY, M. & LICHTENFELS, J. R. (1994). Cloning and characterization of ribosomal RNA genes from three species of *Haemonchus* (Nematoda: Trichostrongyloidea) and identification of PCR primers for rapid differentiation. *Experimental Parasitology* **78**, 28–36.

Prospects for controlling animal parasitic nematodes by predacious micro fungi

M. LARSEN

Danish Centre for Experimental Parasitology, The Royal Veterinary and Agricultural University, 3 Ridebanevej, DK-1870 Frederiksberg C, Denmark

SUMMARY

Resistance against anthelmintics is widespread, particularly in parasitic nematode populations of small ruminants. Several new techniques or supplements have been developed or are under investigation. Biological control (BC) is one of these new methods. The net-trapping predacious fungus *Duddingtonia flagrans* produces thick walled resting spores, chlamydospores, which are able to survive passage through the gastrointestinal tract of cattle, horses, sheep and pigs. Under Danish climatic conditions it has been shown that the number of parasite larvae on pasture and the worm burden of the grazing animals is significantly reduced when animals are fed spores during the initial 2–3 months of the grazing season. Work with *D. flagrans* in France, Australia, USA, and Mexico has confirmed the strong BC potential of this fungus. Today much work is going into development of suitable delivery systems for grazing livestock worldwide. Ultimately, BC should be implemented in integrated parasite control strategies, both in conventional and organic livestock production.

Key words: Control, parasitic nematodes, predacious micro fungi.

INTRODUCTION

During the last 10–15 years there has been an increasing emphasis on the need for development of new alternatives to or supplements for chemical control of parasitic nematodes in grazing livestock. The background for this interest is multi-factorial but the major reason is the serious development of anthelmintic resistance (AR) in parasite populations. Other reasons include handling of parasite problems in organic livestock production, regulation of conventional drug use by legislation, plus political and consumer pressure for reduced chemical residues in products. The anthelmintic resistant gastrointestinal (GI) nematode populations constitute a major problem especially in small ruminants in the sub-tropics and tropics, but are also a serious threat to livestock in the rest of the world (Conder & Campbell, 1995; Waller, 1997; Sangster, 1999). Work to overcome these problems has been going on with increased intensity for more than a decade. Sutherst (1986) among others, has talked about the importance of implementing integrated parasite control based upon host resistance, immunisation, and non-living vaccines to reduce the use of chemotherapy. He has also mentioned that biological control should be used, with the focus on the dung as the environment for interaction between such biological control agents and various animal pest organisms including larvae of parasitic nematodes.

If we briefly look at the options available today or under investigation for control of parasitic nematodes, these can roughly be divided into two broad categories (Table 1): (a) elements with a rather narrow target spectrum; (b) elements able to affect a wide spectrum of parasitic nematodes. When we look at the narrow-spectrum elements, one striking feature is the strong focus on the important parasite, *Haemonchus contortus*. Substituted salicylanilide anthelmintics have a very narrow spectrum of target organisms. One such drug is closantel, which has been very efficient against *H. contortus*, but resistance is becoming an increasing problem.

Within the last decade, much effort has been directed at the development of a vaccine especially against *H. contortus*, either based upon naturally exposed or hidden antigens (Smith, 1999). Despite promising results and much appraisal over the years, we are still waiting for a commercial product to be released. Dictol, based on infective *Dictyocaulus viviparus* larvae attenuated by irradiation, is the only marketed vaccine against GI nematodes but it has only a very limited distribution.

Treating sheep with small amounts of copper-oxide wire particles (COWP) has been shown to be effective against both incoming and established *H. contortus* worms (Bang, Familton & Sykes, 1990). The activity is apparently insignificant against other GI nematodes. This principle is at present being tested out other countries.

Exploiting the fact that during an infection with *H. contortus* animals develop anaemia, a colour scoring card (FAMACHA) has been developed by South African researchers (van Wyk, Malan & Bath, 1997). This card shows the eye mucosal colour from healthy, unaffected animals to animals with severe anaemia in a stage where they will suffer severe clinical haemonchosis. Preliminary results have

Table 1. Elements, existing or under investigation, that could be implemented in integrated control of parasitic nematodes in livestock

Narrow-spectrum elements	Broad-spectrum elements
Some anthelmintics	Anthelmintics
Vaccines	Grazing management
Copper-oxide wire particles	Breeding for resistance
FAMACHA	Biological control

shown that this ingenious card is very easy to use and can be used by non-experts to group animals correctly into those needing no treatment at all and those with various degree of anaemia and therefore with various demands for treatment. On-going trials are testing the applicability of FAMACHA in other parts of the world.

With respect to the broad-spectrum measure, as it has also been emphasized by many other authors, broad-spectrum anthelmintics are still going to be a cornerstone for treatment of GI nematodes in livestock. But in the future farmers will have to adopt a more clever use of drugs (Hennessy, 1997a, b), which will be essential to slow down the development of resistance, and to extend the period for use of the still active anthelmintics for as long as possible. Apart from always being sure that the animals are treated according to right size and in the proper way with correctly adjusted tools, the following steps are suggested to optimize use of existing drugs: (a) treat animals after they have been taken off feed (8–24 hours); (b) when encountering problems with resistant worms, instead of one single treatment with many times the normal dose, animals should be dosed for several days in succession. The latter strategy prolongs the time of which the parasites in the host are exposed to active drug concentrations in the gut environment.

Grazing management strategies have been demonstrated to be useful to alleviate the impact of GI nematodes in livestock (Barger, 1999; Stromberg & Averbeck, 1999). Unfortunately, these strategies have not been adopted to their full extent, perhaps due to the ease for the farmer to use drugs and secondly, the increased demand for land, which makes this proposition less likely in many intensive livestock systems. Where it is used it is often in combination with drug treatment. In organic livestock production these strategies (minus drug treatment) are widely used, but are primarily based upon the availability of herbage rather than an active measure to control problems with GI nematodes (Thamsborg, Roepstorff & Larsen, 1999).

Breeding for resistance in host animals to GI nematodes has been attempted with some success (Kloosterman, Parmentier & Ploeger, 1992;

Woolaston & Baker, 1996; Gray, 1997), but although breeding programmes are promoted and adopted based upon these principles (e.g. Nemesis in Australia, Worm FEC in New Zealand), they are far from widely implemented.

The remaining part of this paper will focus on the last of the broad-spectrum measures, namely the possibility of using especially predacious micro fungi as biological control agents to control the free-living stages of GI nematodes in grazing livestock.

BIOLOGICAL CONTROL

This area of research has attracted increasing attention, especially within the last 5–7 years. Before going into details with biological control (BC) work, a few words on the principles for biological control and on the nematode-destroying fungi. The aim of a BC strategy is not to eliminate a given pest organism (in this case, parasitic nematodes in livestock), but to keep the population under a threshold, above which it would otherwise inflict harmful effects on the host population. Of the various definitions of BC which can be found in the literature, I find the definition by Papavizas & Lewis (1988) very appropriate. They defined BC as 'pest suppression with biotic agents, excluding the process of breeding for resistance to pests, sterility techniques, and chemical modifying pest behaviour'. A BC agent does not eliminate the target organism, but significantly reduces the pest population growth. As also emphasized by Sutherst (1986), when the pest population growth rate is under control, together with other methods (as those mentioned earlier) there are great prospects for long-term reduction of the pest.

The nematode-destroying fungi belong to a heterogeneous group of micro fungi that utilise nematodes either as the main source of nutrients or supplementary to a saprophytic existence (Barron, 1977). These fungi are found world wide in many different habitats, but are especially frequent in organically rich environments such as compost and faeces. In recent studies it was found that deposited fresh faeces are quickly colonized by various species of these fungi (Hay et al. 1997a, b; Bird et al. 1998). It has also been found that the fungi are picked up by grazing livestock (cattle, sheep and horses) and subsequently excreted in the voided faeces (Larsen, Faedo & Waller, 1994; Manueli et al. 1999; Saumell et al. 1999). Two main groups of fungi, with different ways of existence have been described: (a) predacious fungi that produce adhesive or non-adhesive nematode-trapping structures (hyphae, knobs, branches, nets and rings) on the growing mycelium, and (b) endoparasitic fungi that infect nematodes by means of either small, sticky spores adhering to the cuticle of the nematode, or by small, odd-shaped spores that are eaten by the nematode. While the first group of fungi is able to survive on dead organic matter as

Table 2. Predacious fungi (excluding *Duddingtonia flagrans*) and animal parasitic nematodes tested 1985–1999. The publications mentioned are not listed to match the fungus–parasite combination listed

Fungal species	Parasite	Publications
Endoparasitic fungi		
Drechmeria coniospora	*Ostertagia ostertagi*	Jansson *et al.* (1985)
	O. circumcincta	Santos & Charles (1995)
	Haemonchus contortus	
	Trichostrongylus colubriformis	
Harposporium anguillulae	*H. contortus*	Charles *et al.* (1996)
Predacious fungi		
Arthrobotrys oligospora	*Cooperia oncophora*	Grønvold (1989 *a*)
	C. curticei	Grønvold *et al.* (1985, 1987, 1988, 1989)
	O. ostertagi	Nansen *et al.* (1986, 1988)
	Dictyocaulus viviparus	Hashmi & Connan (1989)
	Strongyloides papillosus	Murray & Wharton (1990)
	H. contortus	Wharton & Murray (1990)
	O. circumcincta	Dackman & Nordbring-Hertz (1992)
	T. colubriformis	Waller & Faedo (1993)
	Oesophagostomum dentatum	Mendoza de Gives & Vasquez Prats (1994)
	O. quadrispinulatum	Mendoza de Gives *et al.* (1994)
	Cyathostomes	Charles *et al.* (1995)
		Chandrawathani *et al.* (1998)
A. robusta	*H. placei*	Araújo *et al.* (1993)
	H. contortus	Mendoza de Gives *et al.* (1992)
	S. papillosus	Mendoza de Gives & Vasquez Prats (1994)
		Gonzales-Cruz *et al.* (1998)
A. tortor	*H. contortus*	Grüner *et al.* (1985)
A. musiformis	*O. circumcincta*	Grüner *et al.* (1985)
Dactylaria candida	*O. circumcincta*	Grüner *et al.* (1985)
Monacrosporium eudermatum	*H. contortus*	Mendoza de Gives & Vasquez Prats (1994)
M. ellypsosporum	*H. placei*	Gonzales-Cruz *et al.* (1998)
M. gephyropagum	*S. papillosus*	Araújo *et al.* (1992)
Pleurotus pulmonarius	*O. ostertagi*	Larsen & Nansen (1990, 1991)
	C. oncophora	
	O. quadrispinulatum	
	Cyathostomes	

source of nutrients, the latter group is generally much more dependent on access to nematodes and is characterized as being obligate parasitic. Within the latter group, a subgroup of fungi has been characterized as being specialised in infecting eggs of various nematodes.

ENDOPARASITIC FUNGI AS BIOLOGICAL CONTROL AGENTS

As obligate parasites of nematodes, with very limited capacity to develop in the absence of or outside the prey, the effect of many of these fungi has been found to be density dependent (Jaffe, Tedford & Muldoon, 1993). This has led researchers to suggest that they might serve as better or stronger BC candidates against pest nematodes. So far, only a few species have been tested against animal parasitic nematodes (see Table 2). *Drechmeria coniospora* is a fungus producing sticky drops on very small conidia, which adhere to the cuticle of the nematode, penetrate it and subsequently destroy the victim. In

laboratory tests, on agar media or in faecal cultures, the effect of this fungus has been tested against free-living larvae of cattle and sheep GI nematodes. Only by applying a very high dose (10^8 conidia per gram of faeces) to faecal cultures, did this fungus show an effect against *H. contortus* larvae (Santos & Charles, 1995). This weak fungal response might be explained by the earlier findings by Jansson, Jayaprakash & Zuckerman (1985) who found in an agar test that only infective, third stage parasite larvae stripped of the protective extra (second stage) cuticle, became infected by the fungus. Another endoparasitic fungus, *Harposporium anguillulae* produce very small, half-moon shaped conidia which lodge in the digestive tract of the feeding nematode and after germination totally digest the victim before finally breaking through the cuticle to produce new conidia on short conidiophores. In a laboratory study involving faecal cultures, it was found that at a dose of 300000 conidia/g faeces, the number of *H. contortus* larvae recovered was significantly reduced (Charles, Roque & Santos, 1996). Despite these slightly optimistic results, it still has to be seen if and how spores of endoparasitic fungi will be able to perform as a true BC agent in infected animals, either housed or in field trials. It is my belief that the nature of these fungi, especially the need for spore dispersion or infection to be more or less directly from one infected individual to the next, severely limits or almost excludes the use of this group of fungi as practical BC agents.

BIOLOGICAL CONTROL OF NEMATODES EGGS BY FUNGI

As for other endoparasitic fungi, very few attempts with few species of fungi have been carried out to test the capability of these to attack and destroy long-lasting egg stages of parasitic nematodes. In the former Czechoslovakia (today the Republic of Slovakia) researchers studied a few species from this group of fungi. Eggs of what the authors refer to as *Ascaris lumbricoides* (anticipated to be *A. suum*), collected from naturally infected pigs were used to test the effect of mainly the fungus *Verticillium chlamydosporium* but also other *Verticillium* spp. The fungus was shown to be able to degrade the egg shell enzymatically and infect the eggs (Lýsek & Krajcí, 1987; Lýsek & Sterba, 1991; Kunert, 1992). Short exposure to high temperatures or UV-irradiation rendered the eggs more susceptible to fungal attack (Lýsek & Bacovský, 1979). Biological control of parasites like *Ascaris*, *Trichuris* and others with long-lasting egg stages in the environment is desirable, but unfortunately no new information has come from this group of researchers within the last 5–7 years. In the USA, Chien and colleagues have shown that *V. chlamydosporium* attacked and destroyed eggs of *Ascaridia galli* and *Parascaris equorum* but only rarely invaded *Trichuris suis* (L. Chien, unpublished observations). Unfortunately, details of this work have not been published yet. In Denmark, a range of different fungal species and isolates are at present under investigation as potential BC agents against *A. suum*, *T. suis*, *Nematodirus* spp. and *Fasciola hepatica*. Preliminary results clearly indicate good activity against the relatively thin-shelled eggs of *N. filicollis* and *N. battus*, and to a certain extent also to *A. suum*, while *T. suis* eggs seems to be much more resistant to attack (M. Larsen, unpublished observations). Araujo, Santos & Ferraz (1995) tested three fungal species (the netforming predacious fungi *Arthrobotrys conoides*, *A. robusta*, and the egg-parasitic fungus *Paecilomyces lilacinus*) for activity against eggs of *Toxocara canis*. As perhaps might be expected, based upon the species of fungi involved, neither species of the predacious fungi attacked the eggs while *P. lilacinus* showed some activity (16% eggs infected after 7 days).

PREDACIOUS FUNGI AS BIOLOGICAL CONTROL AGENTS

The idea of possibly using predacious micro fungi to control animal parasitic nematodes arose in the late 1930s, but for almost 50 years this subject was only sporadically investigated (for reviews of this early work, see Soprunov, 1958; Peloille, 1979; Grønvold, 1989a). It was not until the mid 1980s before thorough and systematic investigations were undertaken and since then, two lines of work can be clearly distinguished: (a) trials performed with mainly *Arthrobotrys* spp. (especially *A. oligospora*) and *Monacrosporium* spp. as BC agents (see Table 2); (b) the very fruitful line of trials, initiated by selection of the fungus *Duddingtonia flagrans* (see Tables 3 and 4). From 1985 until 1990, a series of trials was performed by a group of Danish researchers, testing the effect of the fungus *A. oligospora* primarily against parasitic nematodes in cattle but also in other livestock species (see Table 2). Testing different doses of spores mixed into faeces, 250 and 2500 conidia per gram of faeces was found to significantly lower the number (70 and 99% reduction, respectively) of developing *C. oncophora* larvae in faecal cultures (Grønvold et al. 1985). The trapping activity of the fungus was influenced by the motility of the infective larvae, but for most of the parasite species tested there seemed to be no difference in specificity of the fungus (Nansen et al. 1986, 1988). The significant nematode-reducing capacity of this fungus was also shown in plot studies (Grønvold et al. 1987, 1988). The aim of this line of research was to find a BC candidate that could be fed to grazing animals and be active in the excreted dung. Unfortunately various trials performed to test *A. oligospora* mycelium and conidia failed due to the destruction of these structures in the GI tract of the host animals (reported in Grønvold et al. 1993a, b). That it is experimentally possible to obtain an effect on the developing parasite larvae in the excreted faeces by feeding animals a very high number of spores was shown by Grüner et al. (1985). They fed a high dose (between 470 and 680 g of fungal material on millet) of 1 of 3 different fungal species (*A. musiformis*, *A. tortor*, *Dactylaria candida*) to housed lambs, harbouring a mono-infection of either *H. contortus* or *O. circumcincta*. This subsequently led to survival of *A. tortor* through the GI tract at a level high enough to significantly reduce the number of *H. contortus* larvae in faecal cultures. Hashmi & Connan (1989) reported that *A. oligospora* was naturally excreted by grazing cattle. The same authors stated that by feeding calves 16 million spores of fungus weekly for 3 months, a 50% reduction in the number of infective parasite larvae on herbage could be obtained.

Since 1990 more work has been published on *Arthrobotrys* spp. and other genera of predacious

Table 3. Test of the activity of *Duddingtonia flagrans* in either
laboratory or plot tests without spores first being fed to animals. L,
laboratory tests; P, plot trials; F, field trials

Host	Parasite	References
Cattle	*Ostertagia ostertagi*	Larsen *et al.* (1991) (L)
		Grønvold *et al.* (1996) (L)
		Fernandez *et al.* (1999*d*) (P)
	Cooperia oncophora	Grønvold *et al.* (1999) (L)
	Dictyocaulus viviparus	Henriksen *et al.* (1997) (L)
		Fernandez *et al.* (1999*c*) (L+P)
Sheep	*Haemonchus contortus*	Mendoza de Gives *et al.* (1999) (L)
	Ostertagia circumcincta	
	Trichostrongylus axei	
Pig	*Oesophagostomum dentatum*	Petkevicius *et al.* (1998) (L)
Horse	Cyathostomes	Bird & Herd (1994) (L), (1995) (L)

Table 4. Experiments testing the activity of *Duddingtonia flagrans after* passage through the gastrointestinal tract of different animals species. L, laboratory tests; P, plot trials; F, field trials

Host	Parasite	References
Cattle	*Ostertagia ostertagi*	Larsen *et al.* (1992) (L), (1995*b*) (F)
		Grønvold *et al.* (1993*a*) (P)
		Wolstrup *et al.* (1994) (F)
		Nansen *et al.* (1995) (F)
		Fernandez *et al.* (1999*b*) (F)
	Cooperia oncophora	Nansen *et al.* (1995) (F)
		Larsen *et al.* (1995) (F)
Horse	*Strongylus vulgaris*	Larsen *et al.* (1996) (F)
	Strongylus edentatus	
	Cyathostomes	Larsen *et al.* (1995*a*) (L), (1996) (F)
		Baudena *et al.* (2000) (P)
		Fernandez *et al.* (1997) (P), (1999*a*) (P)
Sheep	*Haemonchus contortus*	Larsen, Faedo & Waller (1994) (L)
		Larsen *et al.* (1998) (L)
		Llerandi-Juarez & Mendoza de Gives (1998) (L)
		Mendoza de Gives *et al.* (1998) (L)
	Trichostrongylus colubriformis	Faedo, Larsen & Waller (1997) (L)
		Faedo *et al.* (1998) (P)
	Nematodirus spp.	Githigia *et al.* (1997) (F)
	O. circumcincta	Peloille (1991) (L)
		Githigia *et al.* (1997) (F)
Pigs	*Oesophagostomum dentatum*	Nansen *et al.* (1996) (F)
	Hyostrongylus rubidus	

fungi, mainly focusing on their activity on agar media or in faecal cultures. As it can be seen from Table 2, most of the important parasitic nematodes with free-living larval stages on pasture have been tested in combination with many different predacious fungi. In the tests performed on agar media or in faecal cultures many of these fungi were able to cause significant reduction in the number of parasite larvae. But very few of the fungi tested have been taken to the crucial next step of orally dosing infected animals with the candidate fungus to see whether or not it will stand as a good BC agent. One exception is the work by Walker *et al.* (1994). They examined the activity of initially eight species of

fungi, out of which three (*A. oligospora, A. oviformis, Geniculifera eudermata*) were tested for their nematode-trapping activity after passage through the GI tract of sheep. The fungi were recovered from the faecal samples and also caused some reduction in the number of *H. contortus* larvae, but the fungi were never tested in a field situation.

In conclusion it can be stated that if species of known predacious fungi are tested on agar or in faecal cultures in laboratory experiments, the researchers are more or less guaranteed some sort of positive response, perhaps depending on dose size and number of nematodes involved. It can also be concluded that what appears to be a good nematode-trapping fungus monitored by these tests does not necessarily automatically prove also to be a good BC candidate. It is highly likely that spores of most of these fungi will be destroyed by passage through the GI tract of livestock and only if fed in very high doses or otherwise protected will we achieve the desired high reduction in developing parasite larvae. This being said, it can not totally be excluded that there might exist species, not yet tested, with spores able to sustain the stress through the GI tract and which are excellent BC agents. The extensive screening performed by Waller & Faedo (1993) suggests however that the likelihood of this is low. They tested 94 different species/isolates of known nematode-trapping fungi from the CBS (Centraalbureau voor Schimmelcultures, Barn, The Netherlands) collection, but eventually only three species made it to survival through the GI tract of sheep (see Waller *et al.* 1994), and this at a level which I would not consider sufficient to be of necessary efficacy also in field trials. As mentioned earlier, these fungi have never been tested in a natural field situation.

Since 1991 we do have at least one very promising BC candidate has caused a major break-through in this area of research, namely the fungus *D. flagrans*. This predacious fungus produces three dimensional, sticky networks on its growing hyphae, but as a unique feature, it also produces an abundance of intercalary (= inside the growing hyphae) thick-walled resting spores, chlamydospores. It is probably this special feature which ultimately led to the successful result in the initial Danish search for a better candidate through a two-step stress selection procedure (Larsen *et al.* 1991, 1992), and also later in Australia with screening a huge number of fresh faecal samples from sheep, cattle and horses (Larsen *et al.* 1994). Interestingly, in parallel with the Danish studies, work in France also led to the conclusion that *D. flagrans* was a very promising BC agent (Peloille, 1991). Since then, the interest in predacious fungi as BC agents has grown fast. If we look at the number of articles published in refereed international journals, there is a 50:50 balance (26 and 25) between articles on *D. flagrans* (1991–9) and articles with other predacious fungi alone (1985–99). This is excluding the relatively high number (10) of reviews published on this subject during the same period (1985–99). The growing interest can also be illustrated by the fact that before 1990, work had only been published from the former USSR, France, Australia, Belgium, Sweden, Denmark and UK, but after 1990 we have seen results from Denmark, Australia, USA, Brazil, New Zealand, UK, Mexico, Malaysia and Fiji. Although not yet published in international journals, work is also on-going in Lithuania, Sweden, Portugal, Switzerland, Poland, Ukraine, India, South Africa, Sri Lanka, Thailand and Indonesia. In most of these countries researchers

are working with *D. flagrans*. Like the other predacious fungi mentioned previously, *D. flagrans* has been tested against many of the most important nematodes in cattle, sheep, horses and pigs (see Tables 3 and 4). A distinct feature that characterizes this research is that a majority of the published articles (almost 80%) represents work where the spores have been fed orally to the animals. Of the work published for the rest of the predacious fungi since 1980, only a few articles (approximately 20%) deal with survival of fungal elements through the GI tract (Grüner *et al.* 1985; Grønvold *et al.* 1993 *a*, *b*; Waller *et al.* 1994; Manueli *et al.* 1999; Saumell *et al.* 1999). These are very often fungi found as result of a survey, being naturally excreted in faeces.

In comparison with many other species of predacious fungi, *D. flagrans* is a relatively slow growing fungus (Grønvold *et al.* 1996) and as with other predacious fungi growth is strongly influenced by temperature (Fernandez *et al.* 1999 *e*). Since traps are formed on the growing hyphae it might be expected that faster-growing fungi in laboratory tests on agar or in faecal cultures perhaps would appear as better trappers. But when such potential fungal BC candidates have either been exposed to conditions simulating passage through ruminants (Larsen *et al.* 1991; Waller *et al.* 1994) or have been fed directly to an animal (Larsen *et al.* 1992; Faedo, Larsen & Waller, 1997; Grønvold *et al.* 1993 *a*, *b*; Waller *et al.* 1994), it has been clearly demonstrated that spores of these fungi are much more sensitive to the stress of the GI tract than the chlamydospores of *D. flagrans*. Plot trials involving faecal material from animals fed spores of *D. flagrans* have shown a good reduction of free living larval stages of parasitic nematodes in cattle (Grønvold *et al.* 1993 *a*, *b*), sheep (Peloille, 1991), and horses (Fernandez *et al.* 1997, 1999 *a*; Baudena *et al.* 1999). The BC potential of *D. flagrans* has also been clearly demonstrated in field trials involving calves (Wolstrup *et al.* 1994; Larsen *et al.* 1995 *a*, *b*; Nansen *et al.* 1995; Fernandez *et al.* 1999 *b*), horses (Larsen *et al.* 1996), sheep (Githigia *et al.* 1997) and pigs (Nansen *et al.* 1996). These field tests have shown that daily feeding of fungal spores (chlamydospores) to grazing animals for 2–3 months prevents build-up of dangerous levels of infective larvae on pasture. Daily doses of 1 million chlamydospores per kg live weight or higher have been used in the published field tests, but very recent data seem to imply that this dose level can be lowered for the small ruminants (M. T. Pena, unpublished observations). Only few publications or little information exist beyond the publications covering the full-scale field trials performed under Danish conditions (turn out of animals in May, housing again in late September or early October) with animals fed spores in a feed supplement. In an Australian study Knox and Faedo found that sheep fed supplement containing *D. flagrans* chlamydospores had lower egg counts and improved liveweight gains compared to un-treated control animals (M. Knox & M. Faedo, unpublished observations). Manueli (1998) mentioned that a field trial with sheep was on-going in Fiji. Good parasite control has been found in a field trial with calves in Lithuania, where feeding with *D. flagrans* spores for 2 months led to a reduced number of parasite larvae on herbage, lower worm burdens and prevented the outbreak of clinical parasitic gastroenteritis (M. Sarkunas, unpublished observations).

POTENTIAL FOR IMPLEMENTATION OF BIOLOGICAL CONTROL

The plot and field trials mentioned above involving *D. flagrans* were all performed with spores given in a supplement, but perhaps soon also feed blocks as well as slow-release devices will be available (Waller, 1998).

If we consider the potential for implementation of BC by fungi in various production systems, it seems that supplement feeding should be possible in intensive systems, in particular where animals are handled on a daily basis e.g. dairy cattle, sheep or goats. It also seems to be suited to the control of large and small strongyles in horses. Perhaps to become a practical option for farmers, development of automatic feeders with animal-released pulses of the supplement-fungus mix needs to be considered. Feed blocks where fungal spores are incorporated into attractive, palatable blocks that should last for many weeks, are another relatively simple way of delivery. The blocks should perhaps be placed in a protected rack to reduce the influence of moisture as much as possible. Moisture is one of the major constraints for feed block shelf-life. Spores exposed to a moist environment will germinate and subsequently become extremely sensitive to the stress of the GI tract of the animals. An other option could be to incorporate fungal spores in hard mineral blocks, but then it might be difficult to guarantee the necessary daily consumption. At least under North European conditions, in the beginning of a grazing season when animals have access to plenty of lush pasture, they are not consuming much of the supplied mineral blocks. Later in the season when the amount of grass is dwindling, frequent use of the blocks can be observed. This behaviour is not compatible with the aim to prevent a mid-summer rise in pasture larval numbers by supplying fungal spores to the grazing young stock in the beginning of the season. If a slow-release device (SRD) with incorporated fungal spores was to be developed, it would make BC available as an element in sustainable control strategies also for producers where animals are raised in extensive grazing systems. Whether only one fungus SRD would be sufficient to supply spores during the critical period on pasture would depend on both the technology (how many spores could be incorporated in a single SRD) and the local epidemiological situation. In the conventional livestock industry, a combined SRD system containing fungal spores, released at a constant rate, and an anthelmintic drug, released once or perhaps twice during the season, could be an option during first year for introduction of BC on permanent grazed pastures. Then for the following season or production cycle, a pure fungus bolus could be used. Whether this principle will work in practice will of course need to be tested out thoroughly, and managed according to the local overall epidemiological situation.

CONCLUSION

The focus of this presentation has primarily been on the potential of *D. flagrans* as a BC agent against parasitic nematodes, but again it should be emphasised that to successfully manage the control parasitic nematodes in different livestock production systems, farmers will have to rely on an integrated parasite control strategy involving a combination of several elements. Organic farmers will have to use integrated strategies based on preventive measures such as grazing, biological control and perhaps in some places FARMACHA. Whether or not the different national or farmers association regulations will accept the use of vaccines will have to be determined, but if acceptable they should of course also be incorporated. In conventional production systems the same elements could be integrated with a reduced and smarter use of anthelmintics. A very useful tool in the development and testing of integrated control strategies could also be mathematical models. Barnes, Dobson & Barger (1995) have already used their *T. colubriformis* model to predict the outcome of use of either vaccines or predacious fungi. The predicted activity of a novel vaccine or nematode-trapping fungus was calculated based on a standard regime of 30 dead lambs out of 2000 under 'normal' conditions, which means 3 drug treatments per year and move of lambs to safe pasture at weaning. It was predicted that by using a novel vaccine lamb deaths could be reduced by 63 and 77 % (80 % of the flock responding) plus 83 and 97 % (90 % responding) for a vaccine with 70 and 80 % efficacy, respectively. In the simulation of fungal treatment it was predicted that the reduction in lamb deaths would be 73 % if the fungus caused 75 % reduction in infective larvae during a 90 days period. If the fungus caused 90 % reduction in larvae, the reduction in deaths would be 87 and 43 % for a treatment over 90 or 60 days, respectively.

In conclusion there seem to be several either readily available or very promising elements to supplement anthelmintic treatment for worm control. It is necessary that these elements are integrated in practical control strategies and thoroughly tested under field conditions in the near future. Although BC could become an important element in such strategies, there still exist several issues which need to be attended to. Manufacturing of fungal material as well as further technological refinement of delivery systems should be strengthened. The interest in testing the principle is growing fast, but it requires a product to test. Comparative studies on various *D. flagrans* isolates should be undertaken to examine the possible genetic and physiological differences or similarities between isolates from all over the world. If it is not possible to find essential differences between isolates, this might perhaps make it easier to manufacture and distribute a fungal product to various markets worldwide. The potential environmental impact by continuous use of nematode-trapping fungi also needs to be thoroughly investigated. Yates, Waller & King (1997) could not find any differences in soil nematode populations between samples from pastures where sheep were given *D. flagrans* spores and pastures where sheep were treated according to the normal anthelmintic regime. This could indicate that there will not be negative environmental effects of the use of nematode-trapping fungi in a preventive strategy, but of course further studies are needed to fully answer the question. On of the things we need to look at closely is what happens after years of fungal treatment. Several of these issues are being investigated closely at present.

REFERENCES

ARAUJO, J. V., SANTOS, M. A. & FERRAZ, S. (1995). Efeito ovicida de fungos nematófagos sobre ovos embrionados de *Toxocara canis* (Ovicidal effect of nematophagous fungi on embryonated eggs of *Toxocara canis*). *Arquivo Brasileiro de Medicina Veterinaria et Zootecnia* **47**, 37–42.

ARAUJO, J. V., SANTOS, M. A., FERRAZ, S. & MAIA, A. S. (1993). Antagonistic effect of predacious *Arthrobotrys* fungi on infective *Haemonchus placei* larvae. *Journal of Helminthology* **67**, 136–138.

ARAUJO, J. V., SANTOS, M. A., FERRAZ, S., MAIA, A. S. & MAGELHÃES, A. C. M. (1992). Controle de larvas infectantes de *Haemonchus placei* por fungos predadores da espécie *Monacrosporium ellypsosporum* em condicões de laboratório (Control of infective *Haemonchus placei* larvae by the predacious fungus *Monacrosporium ellypsosporum*). *Arquivo Brasileiro de Medicina Veterinaria et Zootecnia* **44**, 521–526.

BANG, K. S., FAMILTON, A. S. & SYKES, A. R. (1990). Effect of copper oxide wire particle treatment on establishment of major gastrointestinal nematodes in lambs. *Research in Veterinary Sciences* **49**, 132–137.

BARGER, I. A. (1999). The role of epidemiological knowledge and grazing management for helminth control in small ruminants. *International Journal for Parasitology* **29**, 41–48.

BARNES, E., DOBSON, R. & BARGER, I. (1995). Worm control and anthelmintic resistance : adventures with a model. *Parasitology Today* **11**, 56–63.

BARRON, G. L. (1977). The nematode-destroying fungi. *Topics in Mycobiology* **1**, 140 pp.

BAUDENA, M. A., CHAPMAN, M. E., LARSEN, M. & KLEI, T. R. (2000). Seasonal efficacy of the nematophagous fungus *Duddingtonia flagrans* in reducing numbers of equine cyathostome larvae on pasture in southern Louisiana. *Veterinary Parasitology* (In press).

BIRD, J. & HERD, R. P. (1994). Nematophagous fungi for the control of equine cyathostomes. *Compendium on Continuing Education for the Practising Veterinarian* **16**, 658–665.

BIRD, J. & HERD, R. P. (1995). *In vitro* assessment of two species of nematophagous fungi (*Arthrobotrys oligospora* and *A. flagrans*) to control the development of infective cyathostome larvae from naturally infected horses. *Veterinary Parasitology* **56**, 181–187.

BIRD, J., LARSEN, M., NANSEN, P., KRAGELUND, H. O., GRØNVOLD, J., HENRIKSEN, S. A. & WOLSTRUP, J. (1998). Dung-derived biological agents associated with reduced numbers of infective larvae of equine strongyles in faecal cultures. *Journal of Helminthology* **72**, 21–26.

CHANDRAWATHANI, P., OMAR, J. & WALLER, P. J. (1998). The control of the free-living stages of *Strongyloides papillosus* by the nematophagous fungus, *Arthrobotrys oligospora*. *Veterinary Parasitology* **76**, 321–325.

CHARLES, T. P., RODRIGUES, M. L. A. & SANTOS, C. P. (1995). Reducão do número de larvas de *Cyathostominae* em fezes de eqüinos tratadas com conídios de *Arthrobotrys oligospora* (Predacious activity of the nematode-destroying fungus *Arthrobotrys oligospora*, on *Cyathostominae* nematodes). *Arquivo Brasileiro de Medicina Veterinaria et Zootecnia* **47**, 87–89.

CHARLES, T. P., ROQUE, M. V. C. & SANTOS, C. DE P. (1996). Reduction of *Haemonchus contortus* infective larvae by *Harposporium anguillulae* in sheep faecal cultures. *International Journal for Parasitology* **26**, 509–510.

CONDER, G. A. & CAMPBELL, W. C. (1995). Chemotherapy of nematode infections of veterinary importance, with special reference to drug resistance. *Advances in Parasitology* **35**, 1–84.

DACKMAN, C. & NORDBRING-HERTZ, B. (1992). Conidial traps – a new survival structure of the nematode-trapping fungus *Arthrobotrys oligospora*. *Mycological Research* **96**, 194–198.

FAEDO, M., LARSEN, M. & WALLER, P. J. (1997). The potential of nematophagous fungi to control the free-living stages of nematode parasites of sheep: comparison between Australian isolates of *Arthrobotrys* spp. and *Duddingtonia flagrans*. *Veterinary Parasitology* **72**, 149–155.

FAEDO, M., BARNES, E. H., DOBSON, R. J. & WALLER, P. J. (1998). The potential of nematophagous fungi to control the free-living stages of nematode parasites of sheep: pasture plot study with *Duddingtonia flagrans*. *Veterinary Parasitology* **76**, 129–135.

FERNÁNDEZ, A. S., HENNINGSEN, E., LARSEN, M., NANSEN, P., GRØNVOLD, J. & SØNDERGAARD, J. (1999*a*). A new isolate of the nematode-trapping fungus *Duddingtonia* *flagrans* as biological control agent against free-living larvae of horse strongyles. *Equine Veterinary Journal* **31**, 488–491.

FERNÁNDEZ, A. S., LARSEN, M., HENNINGSEN, E., NANSEN, P., GRØNVOLD, J., BJØRN, H. & WOLSTRUP, J. (1999*b*). Effect of *Duddingtonia flagrans* against *Ostertagia ostertagi* in cattle grazing under different stocking rate. *Parasitology* **119**, 105–111.

FERNÁNDEZ, A. S., LARSEN, M., NANSEN, P., GRØNVOLD, J., HENRIKSEN, S. A., BJØRN, H. & WOLSTRUP, J. (1999*c*). The efficacy of two isolates of the nematode-trapping fungus *Duddingtonia flagrans* against *Dictyocaulus viviparus* larvae in faeces. *Veterinary Parasitology* **85**, 287–302.

FERNÁNDEZ, A. S., LARSEN, M., NANSEN, P., GRØNVOLD, J., HENRIKSEN, S. A. & WOLSTRUP, J. (1997). Effect of the nematode-trapping fungus *Duddingtonia flagrans* on the free-living stages of horse parasitic nematodes: a plot study. *Veterinary Parasitology* **73**, 257–266.

FERNÁNDEZ, A. S., LARSEN, M., NANSEN, P., HENNINGSEN, E., GRØNVOLD, J., WOLSTRUP, J., HENRIKSEN, S. A. & BJØRN, H. (1999*d*). The ability of *Duddingtonia flagrans* to reduce the transmission of infective *Ostertagia ostertagi* larvae from faeces to herbage. *Journal of Helminthology* **73**, 115–122.

FERNÁNDEZ, A. S., LARSEN, M., WOLSTRUP, J., NANSEN, P., GRØNVOLD, J. & BJØRN, H. (1999*e*). Growth rate and trapping efficacy of nematode-trapping fungi under constant and fluctuating temperatures. *Parasitology Research* **85**, 661–668.

GITHIGIA, S. M., THAMSBORG, S. M., LARSEN, M., KYVSGAARD, N. & NANSEN, P. (1997). The preventive effect of the fungus *Duddingtonia flagrans* on trichostrongyle infections of lambs on pasture. *International Journal for Parasitology* **27**, 931–939.

GONZALEZ-CRUZ, M. E., MENDOZA DE GIVES, P. & QUIROZ ROMERO, H. (1998). Comparison of the trapping ability of *Arthrobotrys robusta* and *Monacrosporium gephyropagum* on infective larvae of *Strongyloides papillosus*. *Journal of Helminthology* **72**, 209–213.

GRAY, G. D. (1997). The use of genetically resistant sheep to control nematode parasitism. *Veterinary Parasitology* **72**, 345–366.

GRÜNER, L., PELOILLE, M., SAUVÉ, C. & CORTET, J. (1985). Parasitologie animale – survie et conservation de l'activité prédatrice vis-à-vis de nématodes trichostrongylides après ingestion par des ovins de trois hyphomycètes prédateurs. *Comptes Rendus de l'Académie des Sciences* **300**, 525–528.

GRØNVOLD, J. (1989*a*). Induction of nematode-trapping organs in the predacious fungus *Arthrobotrys oligospora* (Hyphomycetales) by infective larvae of *Ostertagia ostertagi* (Trichostrongylidae). *Acta Veterinaria Scandinavia* **30**, 77–87.

GRØNVOLD, J. (1989*b*). *Transmission of infective larvae of Ostertagia ostertagi and Cooperia oncophora (Trichostrongylidae: Nematoda).* Dissertation, The Royal Veterinary and Agricultural University, Copenhagen, 93 pp.

GRØNVOLD, J., HENRIKSEN, S. A., NANSEN, P., WOLSTRUP, J. & THYLIN, J. (1989). Attempts to control infection with *Ostertagia ostertagi* (Trichostrongylidae) in grazing calves by adding mycelium of the nematode-trapping fungus *Arthrobotrys oligospora*

(Hyphomycetales) to cow pats. *Journal of Helminthology* **63**, 115–126.

GRØNVOLD, J., WOLSTRUP, J., HENRIKSEN, S. A. & NANSEN, P. (1987). Field experiments on the ability of *Arthrobotrys oligospora* (Hyphomycetales) to reduce the number of larvae of *Cooperia oncophora* (Trichostrongylidae) in cow pats and surrounding grass. *Journal of Helminthology* **61**, 65–71.

GRØNVOLD, J., NANSEN, P., HENRIKSEN, S. A., LARSEN, M., WOLSTRUP, J. & FRIBERT, L. (1996). Induction of traps by *Ostertagia ostertagi* larvae, chlamydospore production and growth rates in the nematode-trapping fungus *Duddingtonia flagrans*. *Journal of Helminthology* **70**, 291–297.

GRØNVOLD, J., KORSHOLM, H., WOLSTRUP, J., NANSEN, P. & HENRIKSEN, S. A. (1985). Laboratory experiments to evaluate the ability of *Arthrobotrys oligospora* to destroy infective larvae of *Cooperia* species, and to investigate the effect of physical factors on the growth of the fungus. *Journal of Helminthology* **59**, 119–126.

GRØNVOLD, J., NANSEN, P., HENRIKSEN, S. A., THYLIN, J. & WOLSTRUP, J. (1988). The capability of the predacious fungus *Arthrobotrys oligospora* (Hyphomycetales) to reduce numbers of infective larvae of *Ostertagia ostertagi* (Trichostrongylidae) in cow pats and herbage during the grazing season in Denmark. *Journal of Helminthology* **62**, 271–280.

GRØNVOLD, J., WOLSTRUP, J., LARSEN, M., HENRIKSEN, S A. & NANSEN, P. (1993*a*). Biological control of *Ostertagia ostertagi* by feeding selected nematode-trapping fungi to calves. *Journal of Helminthology* **67**, 31–36.

GRØNVOLD, J., WOLSTRUP, J., NANSEN, P., HENRIKSEN, S. A., LARSEN, M. & BRESCIANI, J. (1993*b*). Biological control of nematode parasites in cattle with nematode-trapping fungi – a survey of Danish studies. *Veterinary Parasitology* **48**, 311–325.

GRØNVOLD, J., WOLSTRUP, J., NANSEN, P., LARSEN, M., HENRIKSEN, S. A., BJØRN, H., KIRCHHEINER, K., LASSEN, K., RAWAT, H. & KRISTIANSEN, H. L. (1999). Biotic and abiotic factors influencing growth rate and production of traps by the nematode-killing fungus *Duddingtonia flagrans* when induced by *Cooperia oncophora* larvae. *Journal of Helminthology* **73**, 129–136.

HASHMI, H. A. & CONNAN, R. M. (1989). Biological control of ruminant trichostrongylids by *Arthrobotrys oligospora*, a predacious fungus. *Parasitology Today* **5**, 28–30.

HAY, F. S., NIEZEN, J. H., MILLER, C., BATESON, L. & ROBERTSON, H. (1997*a*). Infestation of sheep dung by nematophagous fungi and implications for the control of free-living stages of gastro-intestinal nematodes. *Veterinary Parasitology* **70**, 247–254.

HAY, F. S., NIEZEN, J. H., RIDLEY, G. S., BATESON, L., MILLER, C. & ROBERTSON, H. (1997*b*). The influence of pasture, species and time of deposition of sheep dung on infestation by nematophagous fungi. *Applied Soil Ecology* **6**, 181–186.

HENNESSY, D. R. (1997*a*). Modifying the formulation or delivery mechanism to increase the activity of anthelmintic compounds. *Veterinary Parasitology* **72**, 367–390.

HENNESSY, D. R. (1997*b*). *Practical Aspects of Parasite Treatment.* 16th WAAVP, Sun City, South Africa:

Workshop on managing anthelmintic resistance in endoparasites (ed. van Wyk, J. A. & van Schalkwyk, P. C.), pp. 40–49.

HENRIKSEN, S. A., LARSEN, M., GRØNVOLD, J., NANSEN, P. & WOLSTRUP, J. (1997). Nematode-trapping fungi in biological control of *Dictyocaulus viviparus*. *Acta Veterinaria Scandinavia* **38**, 175–179.

JANSSON, H.-B., JEYAPRAKASH, A. & ZUCKERMAN, B. M. (1985). Differential adhesion and infection of nematodes by the endoparasitic fungus *Meria coniospora* (Deuteromycetes). *Applied and Environmental Microbiology* **49**, 552–555.

JAFFEE, B. A., TEDFORD, E. C. & MULDOON, A. E. (1993). Tests for density-dependent parasitism of nematodes by nematode-trapping and endoparasitic fungi. *Biological Control* **3**, 329–336.

KLOOSTERMAN, A., PARMENTIER, H. K. & PLOEGER, H. W. (1992). Breeding cattle and sheep for resistance to gastrointestinal nematodes. *Parasitology Today* **8**, 330–335.

KUNERT, J. (1992). On the mechanism of penetration of ovicidal fungi through egg shells of parasitic nematodes. Decomposition of chitinous and ascaroside layers. *Folia Parasitologica* **39**, 61–66.

LARSEN, M. & NANSEN, P. (1990). Effects of the oyster mushroom *Pleurotus pulmonarius* on preparasitic larvae of bovine trichostrongyles. *Acta Veterinaria Scandinavia* **31**, 509–510.

LARSEN, M. & NANSEN, P. (1991). The ability of the fungus *Pleurotus pulmonarius* to immobilize preparasitic nematode larvae. *Research in Veterinary Science* **51**, 246–249.

LARSEN, M., FAEDO, M. & WALLER, P. J. (1994). The potential of nematophagous fungi to control the free-living stages of nematode parasites of sheep: survey for the presence of fungi in fresh faeces of grazing livestock in Australia. *Veterinary Parasitology* **53**, 275–281.

LARSEN, M., FAEDO, M., WALLER, P. J. & HENNESSY, D. R. (1998). The potential of nematophagous fungi to control the free-living stages of nematode parasites of sheep: studies with *Duddingtonia flagrans*. *Veterinary Parasitology* **76**, 121–128.

LARSEN, M., NANSEN, P., GRØNDAHL, C., THAMSBORG, S. M., GRØNVOLD, J., WOLSTRUP, J., HENRIKSEN, S. A. & MONRAD, J. (1996). The capacity of the fungus *Duddingtonia flagrans* to prevent stongyle infections in foals on pasture. *Parasitology* **113**, 1–6.

LARSEN, M., NANSEN, P., GRØNVOLD, J., WOLSTRUP, J. & HENRIKSEN, S. A. (1992). *In vivo* passage of nematophagous fungi selected for biocontrol of parasitic nematodes in ruminants. *Journal of Helminthology* **66**, 137–141.

LARSEN, M., WOLSTRUP, J., HENRIKSEN, S. A., DACKMAN, C., GRØNVOLD, J. & NANSEN, P. (1991). *In vitro* stress selection of nematophagous fungi for biocontrol of pre-parasitic nematode larvae of ruminants. *Journal of Helminthology* **65**, 193–200.

LARSEN, M., NANSEN, P., HENRIKSEN, S. A., WOLSTRUP, J., GRØNVOLD, J., ZORN, A. & WEDØ, E. (1995*a*). Predacious activity of the nematode-trapping fungus *Duddingtonia flagrans* against cyathostome larvae in faeces after passage through the gastrointestinal tract of horses. *Veterinary Parasitology* **60**, 315–320.

LARSEN, M., NANSEN, P., WOLSTRUP, J., GRØNVOLD, J., HENRIKSEN, S. A. & ZORN, A. (1995*b*). Biological control of trichostrongylosis in grazing calves by means of the fungus *Duddingtonia flagrans*. *Veterinary Parasitology* **60**, 321–330.

LLERANDI-JUÁREZ, R. D. & MENDOZA DE GIVES, P. (1998). Resistance of chlamydospores of nematophagous fungi to the digestive processes of sheep in Mexico. *Journal of Helminthology* **72**, 155–158.

LÝSEK, H. & BACOVSKÝ, J. (1979). Penetration of ovicidal fungi into altered eggs of *Ascaris lumbricoides*. *Folia Parasitologica* **26**, 139–142.

LÝSEK, H. & KRAJCÍ, D. (1987). Penetration of ovicidal fungus *Verticillium chlamydosporium* through the *Ascaris lumbricoides* egg-shells. *Folia Parasitologica* **34**, 57–60.

LÝSEK, H. & STERBA, J. (1991). Colonization of *Ascaris lumbricoides* eggs by the fungus *Verticillium chlamydosporium* Goddard. *Folia Parasitologica* **38**, 255–259.

MANUELI, P. R. (1998). Livestock production, effects of helminth parasites and prospects for their biological control in Fiji. In *Biological Control of Gastrointestinal Nematodes of Ruminants Using Predacious Fungi*, FAO Animal Production and Health Paper 141, FAO Rome 1998, pp. 47–53.

MANUELI, P. R., WALLER, P. J., FAEDO, M. & MAHOMMED, F. (1999). Biological control of nematode parasites of livestock in Fiji: screening of fresh dung of small ruminants for the presence of nematophagous fungi. *Veterinary Parasitology* **81**, 39–45.

MENDOZA DE GIVES, P., DAVIES, K. G., CLARK, S. J. & BEHNKE, J. M. (1999). Predatory behaviour of trapping fungi against *sfr* mutants of *Caenorhabditis elegans* and different plant and animal parasitic nematodes. *Parasitology* **119**, 95–104.

MENDOZA DE GIVES, P., FLORES CRESPO, J., HERRERA-RODRIGUEZ, D., VAZQUEZ-PRATS, V. M., LIEBANO HERNANDEZ, E. & ONTIVEROS FERNANDEZ, G. E. (1998). Biological control of *Haemonchus contortus* infective larvae in ovine faeces by administering an oral suspension of *Duddingtonia flagrans* chlamydospores to sheep. *Journal of Helminthology* **72**, 343–347.

MENDOZA DE GIVES, P. & VAZQUEZ-PRATS, V. M. (1994). Reduction of *Haemonchus contortus* infective larvae by three nematophagous fungi in sheep faecal cultures. *Veterinary Parasitology* **55**, 197–203.

MENDOZA DE GIVES, P., ZAVALETA-MEJIA, E., HERRERA-RODRIGUES, D. & QUIROZ-ROMERO, H. (1994). *In vitro* trapping capability of *Arthrobotrys* spp. on infective larvae of *Haemonchus contortus* and *Nacobbus aberrans*. *Journal of Helminthology* **68**, 223–229.

MENDOZA DE GIVES, P., ZAVALETA-MEJIA, E., QUIROZ-ROMERO, H., HERRERA-RODRIGUEZ, D. & PERDOMO-ROLDAN, F. (1992). Interaction between the nematode-destroying fungus *Arthrobotrys robusta* (Hyphomycetales) and *Haemonchus contortus* infective larvae *in vitro*. *Veterinary Parasitology* **41**, 101–107.

MURRAY, D. S. & WHARTON, D. A. (1990). Capture and penetration processes of the free-living juveniles of *Trichostrongylus colubriformis* (Nematoda) by the nematophagous fungus, *Arthrobotrys oligospora*. *Parasitology* **101**, 93–100.

NANSEN, P., GRØNVOLD, J., HENRIKSEN, S. A. & WOLSTRUP, J. (1986). Predacious activity of the nematode-destroying fungus *Arthrobotrys oligospora* on preparasitic larvae of *Cooperia oncophora* and on soil nematodes. *Proceedings of the Helminthological Society of Washington* **53**, 237–243.

NANSEN, P., GRØNVOLD, J., HENRIKSEN, S. A. & WOLSTRUP, J. (1988). Interaction between the predacious fungus *Arthrobotrys oligospora* and third-stage larvae of a series of animal-parasitic nematodes. *Veterinary Parasitology* **26**, 329–337.

NANSEN, P., LARSEN, M., GRØNVOLD, J., WOLSTRUP, J., ZORN, A. & HENRIKSEN, S. A. (1995). Prevention of clinical trichostrongylidosis in calves by strategic feeding with the predacious fungus *Duddingtonia flagrans*. *Parasitological Research* **81**, 371–374.

NANSEN, P., LARSEN, M., ROEPSTORFF, A., GRØNVOLD, J., WOLSTRUP, J. & HENRIKSEN, S. A. (1996). Control of *Oesophagostomum dentatum* and *Hyostrongylus rubidus* in outdoor-reared pigs through daily feeding with the microfungus *Duddingtonia flagrans*. *Parasitology Research* **82**, 580–584.

PAPAVIZAS, G. C. & LEWIS, J. A. (1988). The use of fungi in integrated control of plant diseases. In *Fungi in Biological Control Systems* (ed. Burge, M. N.), pp. 235–253. Manchester University Press.

PELOILLE, M. (1979). *Etude des Hyphomycetes predateurs de nematodes rencontres sur une praire du Limousin : morphologie-physiologie, frequence et distribution.* Thesis, l'Université de Rennes, pp. 106.

PELOILLE, M. (1991). Selection of nematode-trapping fungi for use in biological control. In *IOBC/WPRS Bulletin 1991/XIV/2, Working Group 'Integrated Control of Soil Pests', 'Methods for Studying Nematophagous Fungi'* (ed. Kerry, B. R. & Crump, D. H.), pp. 13–17.

PETKEVČIUS, S., LARSEN, M., BACH KNUDSEN, K. E., NANSEN, P., GRØNVOLD, J., HENRIKSEN, S. A. & WOLSTRUP, J. (1998). The effect of the nematode-destroying fungus *Duddingtonia flagrans* against *Oesophagostomum dentatum* larvae in faeces from pigs fed different diets. *Helminthologia* **35**, 111–116.

SANGSTER, N. C. (1999). Anthelmintic resistance: past, present and future. *International Journal for Parasitology* **29**, 115–124.

SANTOS, C. P. & CHARLES, T. P. (1995). Efeito da aplicacão de conidios de *Drechmeria coniospora* em cultivos de fezes contendo ovos de *Haemonchus contortus* (Effect of an endoparasitic fungus, *Drechmeria coniospora*, in fecal cultures containing eggs of *Haemonchus contortus*). *Arquivo Brasileiro de Medicina Veterinaria et Zootecnia* **47**, 123–128.

SAUMELL, C. A., PADIILHA, T., SANTOS, C. P. DE & ROQUE, M. V. C. (1999). Nematophagous fungi in fresh feces of cattle in the Mata region of Minas Gerais state, Brazil. *Veterinary Parasitology* **82**, 217–220.

SMITH, W. D. (1999). Prospects for vaccines of helminth parasites of grazing ruminants. *International Journal for Parasitology* **29**, 17–24.

SOPRUNOV, S. S. (1958). Predacious hyphomycetes and their application in the control of pathogenic nematodes. Ashkahabad, 292 pp. (Translated from Russian by Nemchonok, S., Israel Program for Scientific Translations Ltd., Jerusalem, 1966).

STROMBERG, B. E. & AVERBECK, G. A. (1999). The role of parasite epidemiology in the management of grazing cattle. *International Journal for Parasitology* **29**, 33–40.

SUTHERST, R. W. (1986). Epidemiological concepts and strategies for parasite control: What changes are likely to occur? In *Parasitology – Quo Vadit? Proceedings of the 6th International Congress of Parasitology* (ed. Howell, M. J.), pp. 721–729. Australian Academy of Sciences.

THAMSBORG, S. M., ROEPSTORFF, A. & LARSEN, M. (1999). Integrated and biological control of parasites in organic and conventional production systems. *Veterinary Parasitology* **84**, 169–186.

VAN WYK, J. A., MALAN, F. S. & BATH, G. F. (1997). Rampant anthelmintic resistance in sheep in South Africa – what are the options? In *Managing Anthelmintic Resistance in Endoparasites*. Workshop proceedings at 16th WAAVP Conference, Sun City, South Africa, pp. 51–63.

WALLER, P. J. (1997). Anthelmintic resistance. *Veterinary Parasitology* **72**, 391–412.

WALLER, P. J. (1998). Possible means of using nematophagous fungi to control nematode parasites of livestock. In *Biological Control of Gastro-intestinal Nematodes of Ruminants Using Predacious Fungi*, FAO Animal Production and Health Paper 141, FAO Rome 1998, pp. 11–14.

WALLER, P. J. & FAEDO, M. (1993). The potential of nematophagous fungi to control the free-living stages of nematode parasites of sheep: Screening studies. *Veterinary Parasitology* **49**, 285–297.

WALLER, P. J., LARSEN, M., FAEDO, M. & HENNESSY, D. R. (1994). The potential of nematophagous fungi to control the free-living stages of nematode parasites of sheep: *in vitro* and *in vivo* studies. *Veterinary Parasitology* **51**, 289–299.

WHARTON, D. A. & MURRAY, D. S. (1990). Carbohydrate/lectin interactions between the nematophagous fungus, *Arthrobotrys oligospora*, and the infective juveniles of *Trichostrongylus colubriformis*. *Parasitology* **101**, 101–106.

WOOLASTON, R. R. & BAKER, R. L. (1996). Prospects for breeding for parasite resistance. *International Journal for Parasitology* **26**, 845–855.

WOLSTRUP, J., GRØNVOLD, J., HENRIKSEN, S. A., NANSEN, P., LARSEN, M., BØGH, H. O. & ILSØE, B. (1994). An attempt to implement the nematode-trapping fungus *Duddingtonia flagrans* in biological control of trichostrongyle infections of first year grazing calves. *Journal of Helminthology* **68**, 175–180.

YATES, G. W., WALLER, P. J. & KING, K. L. (1997). Soil nematodes as indicators of the effect of management on grasslands in the New England Tablelands (NSW): effect of measures for control of parasites of sheep. *Pedobiologia* **41**, 537–548.

Onchocerca ochengi infections in cattle as a model for human onchocerciasis: recent developments

A. J. TREES[1]*, S. P. GRAHAM[2], A. RENZ[3], A. E. BIANCO[2] *and* V. TANYA[4]

[1] *Veterinary Parasitology, Liverpool School of Tropical Medicine/Faculty of Veterinary Science*
[2] *Division of Molecular Biology and Immunology, Liverpool School of Tropical Medicine, University of Liverpool, Pembroke Place, Liverpool L3 5QA, UK*
[3] *Fachgebiet Parasitologie, Universität Hohenheim, Stuttgart 70599, Germany*
[4] *Institut de Recherche Agricole pour le Developpement, Wakwa, BP 65, Ngaoundere, Cameroon*

SUMMARY

The bovine parasite *Onchocerca ochengi* is a nodule-dwelling filarial nematode, closely related to *O. volvulus*, the causal agent of human River Blindness, and, sharing with it, the same vector. This brief review, based on a presentation at the BSP Autumn Symposium 1999, describes recent work supported by the WHO Drug Development Research Macrofil programme and the Edna McConnell Clark Foundation vaccine development programme, to research the chemotherapy and immunology of onchocerciasis utilising this model system, with experimental infections in Liverpool and field infections in northern Cameroon. In a series of chemotherapeutic trials involving 10 compounds in 20 treatment regimes, the comparability of drug efficacy against *O. ochengi* with that described against *O. volvulus* has been demonstrated. Repeated, long-term treatment with oxytetracycline has been shown to be macrofilaricidal and the effect is hypothesized to be related to action on *Wolbachia* endobacteria, abundant in *O. ochengi*. Avermectins/milbemycins are not macro-filaricidal (even in high and repeated long-term treatments) but induce sustained abrogation of embryogenesis. In prospective, field exposure experiments with naive calves, prophylactic treatments with ivermectin and moxidectin prevented the development of adult worm infection, raising the possibility that drug-attenuated larval challenge infections may induce immunity. Putatively immune adult cattle exist in endemically exposed populations, and these have been shown to be significantly less susceptible to challenge than age-matched naive controls, whereas radically drug-cured, previously patently-infected cattle were not. Experimental infections with *O. ochengi* have revealed the kinetics of the immune response in relation to parasite development and demonstrate analogous responses to those reported in *O. volvulus* infection in humans and chimpanzees. In an immunization experiment with irradiated L_3 larvae, cattle were significantly protected against experimental challenge – the first such demonstration of the experimental induction of immunity in a natural *Onchocerca* host–parasite system. Taken collectively, these studies not only demonstrate the similarity between the host–parasite relationships of *O. ochengi* in cattle and *O. volvulus* in humans, but promise to advance options for the control of human onchocerciasis.

Key words: onchocerciasis, *Onchocerca ochengi*.

INTRODUCTION

The aim of this paper, which is based on a presentation at the BSP Autumn Symposium 1999, is to provide a brief review of aspects of *Onchocerca ochengi* relevant to its experimental use as a model system for *O. volvulus* and to outline some of the recent research results relating to its immunology, immunoprophylaxis and chemotherapy generated in drug and vaccine development projects supported, respectively, by the WHO and the Edna McConnell Clark Foundation. It does not aim to provide a comprehensive review on *O. ochengi*, and much recent, excellent research, notably on the epidemiology of the infection and its consequences for human onchocerciasis by Renz, Wahl, Achu-Kwi

and Harnett, is not covered. Many of the results are currently the subject of papers submitted or in preparation and will be only briefly described as unpublished observations, although some of the findings are of significance. For earlier reviews readers are referred to Trees (1992) and Trees *et al.* (1998).

BIOLOGY, PARASITOLOGY AND INFECTION

Experimental infections

Experimental infections of cattle have been achieved using either fresh L_3s, generated in surrogate British simuliids (McCall & Trees, 1989) by intrathoracic infection of dermal microfilariae (mf) collected and cryopreserved as previously described (Bianco *et al.* 1980; Ham *et al.* 1981), or cryopreserved L_3s, dissected from wild-caught *S. damnosum*, after engorgement on an *O. ochengi*-infected bait cow and cryopreserved with DMSO using a protocol

* Corresponding author: Veterinary Parasitology, Liverpool School of Tropical Medicine, Pembroke Place, Liverpool L3 5QA, UK. Tel: 44 151 708 9393. Fax: 44 151 709 3681. E-mail: trees@liverpool.ac.uk

Parasitology (2000), **120**, S133–S142.   Printed in the United Kingdom   © 2000 Cambridge University Press

Table 1. *Summary of parasitological observations following experimental infections with* Onchocerca
ochengi

Calf	No. of L_3 inoculated	Earliest detection of nodules (days p.i.)	No. of nodules detected	Earliest detection of microfilariae (days p.i.)	Peak microfilarial density (per g skin)
704	350	ND*	ND	ND	ND
706	350	600	1	ND	ND
707	350	110	7	279	3330
708	350	209	4	404	7750
709	350	ND	ND	539†	ND
712	350	322	1	348	140
751	700	172	5	294	4000
752	700	187	4	368	2350
C2	1000	210	3	322	167
C9	1000	210	8	294	1230
C11	1000	154	9	294	290
C12	1000	210	2	ND	ND
C13	1000	210	4	382	167
C17	1000	154	16	322	100
C18	1000	182	3	ND	ND
C22	1000	182	4	382	32

* ND, not detected.
† Single biopsy identified positive by PCR.
Animals observed to 600 days post-infection (p.i.) except those inoculated with 1000 L_3s which were observed until
382 days p.i.

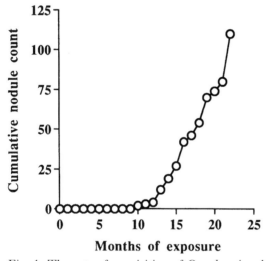

Fig. 1. The rate of acquisition of *O. ochengi* nodules in
calves exposed to field challenge. The cumulative group
total ($n = 14$) of nodules in naive calves exposed from
birth to 22 months is shown.

modified from Lowrie (1983) (Lustigman, personal
communication). The infections were produced in
the course of experiments to study the kinetics of the
immune response and to determine the immunity
conferred by immunization with irradiated L_3s
(Graham, Trees, Collins, Taylor, Moore, Guy &
Bianco, unpublished observations; Graham, Trees,
Lustigman & Bianco, unpublished observations).
The parasite intensity, time to appearance of nodules
and time to first mf detection from a series of
infections following a single subcutaneous inoculation at the umbilicus in 2–3 month old calves are
shown in Table 1. Infections with 700 or 1000 L_3s

led to adult infections in all cattle with a median (and
range) time to nodule detection of 182 days (154–210,
$n = 10$). Of six cattle inoculated with 350 L_3s, four
developed detectable nodules, and in a fifth skin mf
were detected by PCR (method of Meredith *et al.*
1991) but not by parasitological means. For all
infected cattle, the minimum mf prepatent period
was 279 days. There was a trend towards increased
nodule number with increased L_3 dose and cryopreserved L_3s were not obviously less infective than
freshly dissected L_3s. For 10 animals each inoculated
with 1000 cryopreserved L_3s, the worms recovered
at 382 days represented 0·83% of the number
inoculated. Nodules were distributed over the ven-
tral abdomen and upper hind leg. All of 50 nodules
dissected from cattle inoculated with 1000 L_3s
contained only one female worm with numbers of
males varying from 0 to 3. Of 50 female worms, 17
were fecund at the time of examination and in all
cases but one, male worms were present in their
nodules. All nodules were intradermal and, after
careful post-mortem dissection, no nodules were
found in deeper sites. The anatomical distribution of
nodules and of mf, and the numbers of worms in
nodules was in accord with the properties in naturally
infected cattle, except the latter may often carry large
numbers of males in some nodules (Trees *et al.* 1992;
Wahl *et al.* 1994).

O. ochengi *infections do not cause disease*

Whilst *O. ochengi* infections in cattle serve as an
excellent parasitological analogue of *O. volvulus* in
humans, the dermal and ocular pathology associated

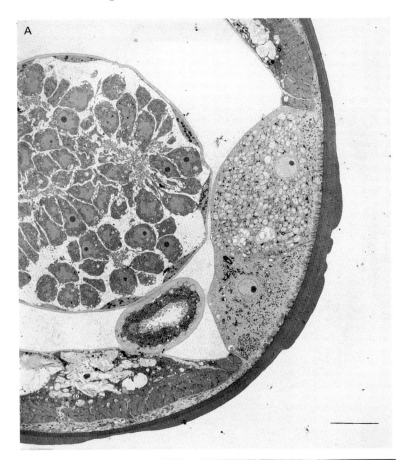

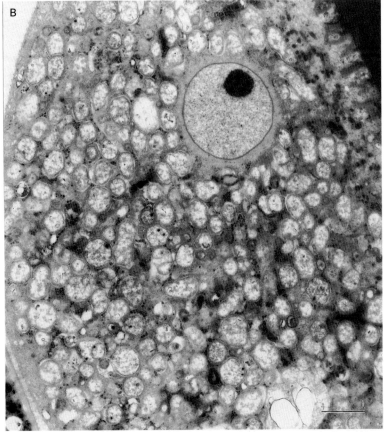

Fig. 2. (A) *Wolbachia* bacteria are abundant in one hypodermal cell in this transverse section of a male *O. ochengi*. The other hypodermal cell appears uninfected (bar, 10 μm). (B) Higher power of a male *O. ochengi* showing packed bacteria in the hypodermal cells, the nucleus of which is prominent (bar, 2 μm).

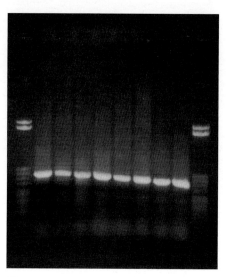

Fig. 3. *Wolbachia*-specific amplification product in each of four male (left) and four female *O. ochengi*. PCR using nested eubacterial and *Wolbachia*-specific 16S rDNA (see text). The amplified product has an estimated size of 766 bp.

with the latter is not observed in *O. ochengi* infections of cattle. Not only has dermatitis not been observed, but we have failed to provoke Mazzotti-like reactions in infected cattle treated with diethylcarbamazine (Renz, Bronsvoort & Trees, unpublished observations). Microfilariae have been found in bovine eyes (Daiber & Renz, unpublished observations) but no pathology has been associated with their presence. The lack of pathology and clinical signs may be a result of an inherently reduced pathogenicity, a lower parasite intensity (mf densities in bovine skin are one or two $\log_{10}$ increments lower than those seen in human onchocerciasis), a difference in the host's immune response, all of which may indicate a better adapted host–parasite relationship. Perhaps an added factor is that cattle do not live long enough to develop clinical signs.

Experimental field exposure reveals the kinetics of nodule acquisition and insusceptible individuals

The experimental infection data are consistent with field data, where, under conditions of high but seasonal challenge, calves become patently infected with nodules as early as 7 months and with mf as early as 10 months of age (Achu-Kwi *et al.* 1994; Tchakouté *et al.* 1999). The rate of acquisition of nodules under ongoing field exposure is shown in Fig. 1 (Gilbert, Langworthy, Tanya, Renz & Trees, unpublished observations). We have not continued experiments for long enough to determine if the rate of increase in nodule density slows, although an earlier cross-sectional study found that there was a progressive increase in group nodule density with age in three age cohorts up to > 8 years of age (Trees *et al.* 1992). In experiments involving exposure of

calves from birth to natural challenge, individuals exhibit highly variable parasite loads, and there are some apparently insusceptible individuals (one of six and two of 14 after 21 and 22 months of exposure respectively; Tchakouté *et al.* 1999; Gilbert, Langworthy, Renz, Tanya & Trees, unpublished observations). These proportions are consistent with 3/30 cattle of > 8 years old maintained for life in an endemic area and in which infection was not detectable (Trees *et al.* 1992) and similar proportions of apparently uninfected cattle from the endemic area of the Adamawa reported by Wahl *et al.* (1994 and 1998). These individuals conform parasitologically and in terms of history of exposure with the endemic normals or putative immunes (PIs) recognized in human populations exposed to *O. volvulus*. Unlike human PIs, we have been able to test the susceptibility of such cattle to exposure – see later.

Drugs freely penetrate the nodule and adult male worms

The intradermal nodule of *O. ochengi* comprises a well developed capsule enclosing a matrix in which coiled worms lie. Detailed studies on nodule vascularization, such as done for *O. volvulus* (George, Palmieri & Connor, 1985; Smith *et al.* 1988) have not been conducted but extensive vascularization is obvious in nodule sections. *In vivo* studies have shown that drugs (suramin and ivermectin) readily penetrate into the matrix and nodule wall and achieve concentrations equivalent to or greater than those in plasma and skin (Cross *et al.* 1997). *In vitro* experiments with isolated male worms have demonstrated that ivermectin is readily taken up, predominantly across the cuticle (Cross, Renz & Trees, 1998). Transmission and scanning EM studies show the cuticle of female worms to be invested with an electron-dense coat (of unknown origin) into which cuticle lamellae project (Determann *et al.* 1997; Cross, Renz & Trees, 1998). The male cuticle exhibits a pitted and honey-combed surface appearance (Cross, Renz & Trees, 1998). These structures presumably serve to increase the interface with the surrounding medium and may facilitate molecular exchange across the cuticle.

O. ochengi contains Wolbachia-*like bacteria*

Intracellular rickettsia-like bacteria in filariae are currently the subject of much interest and molecular phylogeny has shown them to be closely related to the *Wolbachia* of insects (see review of Taylor & Hoerauf, 1999). In *O. ochengi*, *Wolbachia* are abundant (Determann *et al.* 1997) and can be especially concentrated in hypodermal cells of adults (see Fig. 2). Partial sequencing of the 16S rDNA of the *O. ochengi Wolbachia* (AF 172401) showed that in 858 nucleotides only one base differed from the homologous sequence of the bacteria in *O. volvulus* (AF 069069) (Langworthy *et al.* 2000).We have investigated the prevalence of *Wolbachia* infection *O. ochengi* collected from cattle in northern Cameroon by polymerase chain reaction (PCR) analysis of DNA from individual worms using a nested PCR with primers for eubacterial and *Wolbachia*-specific 16S rDNA (Taylor, Bilo & Cross, unpublished observations). All of 89 male worms and 105 female worms were shown to harbour the infection – Fig. 3 (Whittle, McGarry, Cross, Taylor and Trees, unpublished observations). This prevalence study should be extended to populations elsewhere, but the results are consistent with a relationship of mutual dependence between worm and bacterium.

ATTRIBUTES AS A MODEL SYSTEM

High degree of genetic and antigenic similarity with O. volvulus

Some of the attributes of *O. ochengi* as a model have been previously described (Trees, 1992). The close morphological and biological similarities between *O. ochengi* and *O. volvulus* (Bain, 1981), which include sharing the same vector (Wahl, Ekale & Schmitz, 1998), are confirmed by molecular phylogenetic studies (Xie, Bain & Williams, 1994). There is extensive conservation between the two species such that there is a high degree of antibody cross-reactivity to native or recombinant antigens (Hoch *et al.* 1994; Graham *et al.* 1999). This cross-reactivity has potentially important epidemiological consequences and field studies have provided evidence that exposure to *O. ochengi* infective stages cross-protects humans against *O. volvulus* (Wahl *et al.* 1998).

Multiple infection allows sequential evaluations of worm status

An important practical aspect of *O. ochengi* infection in cattle as an experimental system is that multiple infection is common in adult cattle exposed to natural infection in endemic areas. Thus in chemotherapeutic trials, for example, sequential nodulectomies from the same individual are possible over a period of months or years. This, in turn, requires that nodules are identified at the commencement of experiments involving chemotherapy, since it is impractical to prevent exposure to continuing infection and, hence, nodule acquisition, in experiments of long duration under field conditions. Nodules to be examined during the course of therapeutic experiments are pre-designated by mapping their location, marking the location with tattoo ink (three spots at the angles of a triangle enclosing each nodule), and subcutaneously implanting adjacent to each nodule a microchip transponder which

carries a unique, ten digit signal. The location of each designated nodule can be subsequently determined by reference to a hide map and percutaneous scanning with the transponder detector to confirm the nodule identity and position, followed by shaving to reveal the tattoo marks.

CHEMOTHERAPY AND CHEMOPROPHYLAXIS

Methods for drug efficacy studies

Efficacy trials of potential macrofilaricides are conducted to a standard design. Naturally infected cattle of 3–5 years of age, each of which has > 20 palpable *O. ochengi* nodules, are recruited and assigned to treatment groups (usually three per group) to match for weight and nodule number. Nodules for subsequent examination are pre-designated (as described above) and, at each sampling, four are excised under local anaesthesia from each animal providing 12 nodules per treatment group at each time point. Skin biopsies are taken simultaneously and occasionally at other times and mf densities determined as previously described (Renz *et al.* 1995). Experiments usually last 9 months and may continue for 2 years.

Following nodulectomy under local anaesthesia, worms can be expressed from cut nodules and immediately obtained clean enough for viability estimates without recourse to collagenase digestion. In therapeutic studies the viability of worms is determined in three ways (triple assay) as described by Renz *et al.* (1995); by motility assessment, ability to reduce tetrazolium (MTT) to formazan (Comley *et al.* 1989 *a, b*) and a quantitative and qualitative assessment of embryogenesis (Schulz-Key, 1988). MTT reduction is standardized per unit length (10 mm) of worm.

The search for macrofilaricidal agents

There is currently no compound safe enough for mass application to radically cure human infections with adult *O. volvulus* and the development of such a drug is an objective of the MACROFIL programme of the Drug Development Research division of WHO/TDR. *O. ochengi* infections in naturally infected cattle are being used as a tertiary screen for the development of a human macrofilaricide (Renz *et al.* 1995; Trees *et al.* 1998). This project, a collaboration between the Universities of Liverpool and Hohenheim and the Institut de Recherche Agricole pour le Developpement (IRAD), Cameroon, is based near Ngaoundere at the IRAD station at Wakwa. The activities of a number of drugs have been studied, some to validate the model system and others as lead research compounds. Results are summarized in Table 2, updated and adapted from Trees *et al.* (1998). Melarsomine and parenteral UMF-078 (a flubendazole derivative)

Table 2. *Summary of compounds tested for activity against* O. ochengi *in cattle*

| Compound | Dose | Route* | Frequency | Activity† | | Note |
				Mf	Adult	
Ivermectin	200 μg/kg	SC	×1	+	−	Renz *et al.* (1995)
	500 μg/kg/m‡	SC	×7	+	−	By 24 months; Bronsvoort *et al.* (unpublished)
Doramectin	500 μg/kg/d‡	SC	×7	+	−	By 24 months; Bronsvoort *et al.* (unpublished)
Moxidectin	500 μg/kg	SC	×1	+	−	In progress
	500 μg/kg/m	SC	×6	+	−	
Suramin	10 mg/kg/d	IV	×6	±	−	By 137 days; Renz *et al.* (1995)
	17 mg/kg/w‡	IV	×7	±	−	By 6 months; Tchakouté *et al.* (unpublished)
Melarsomine	4 mg/kg alternate days	IV	×3	+	+	Tchakouté *et al.* (unpublished)
CG 20376	20 mg/kg	PO	×1	+	−	Renz *et al.* (1995)
WR 251993	140 mg/kg	SC	×1	−	−	
	35 mg/kg	SC	×1	−	−	
UMF 078	150 mg/kg	IM	×1	+	+	Bronsvoort *et al.*
	50 mg/kg	IM	×1	±	±	(unpublished)
	150 mg/kg	IA	×1	+	−	
	50 mg/kg	IA	×1	±	−	
	150 mg/kg/d	PO	×2	+	−	Toxic
Oxytetracycline	10 or 20 mg/kg	IM	×13	+	+	Treatment intermittent over 6 months

* SC, subcutaneous; IV, intravenous; PO, *per os*; IA, intra-abomasal; IM, intramuscular.
† + for adults denotes effect characterized by reduction in nodule size, calcification or disappearance of worms or elimination of worm motility and ability to reduce MTT and for mf complete or almost complete absence from skin biopsies; ± means partial effects or total effect on only some worms; −, none of the foregoing effects.
‡ m, month; d, day; w, week.

were macrofilaricidal (Tchakouté, Renz, Tanya & Trees, unpublished observations; Bronsvoort, Renz, Tanya, Ekale & Trees, unpublished observations). Both compounds are potentially too toxic for mass human application and the development of UMF-078 has been abandoned. Nonetheless, the results showed that death of worms was followed by the resolution of intradermal nodules within months. As an effective macrofilaricide against *O. ochengi*, which is adequately tolerated by cattle, melarsomine provides a useful research tool to investigate immunity in drug-cured cattle (see later).

Avermectins/milbemycins and their action against adult Onchocerca

It has been difficult in human studies to determine the macrofilaricidal activity, if any, of ivermectin, since assessments depend on once-only, retrospective assays of inherently aged worm populations. Most studies on *O. volvulus* have concluded that ivermectin is not macrofilaricidal (Chavasse *et al.* 1992; Kläger, Whitworth & Downham, 1996),

although there is a profound abrogation of embryogenesis which is potentially reversible but in a proportion of worms, after repeated treatments, may be irreversible (see Plaisier *et al.* 1995). We have been able, in *O. ochengi*-infected cattle, to study the activity of ivermectin and some newer avermectins, in a protocol which permits pre- as well as post-treatment worm evaluation, in addition to comparisons with concurrent controls. Single or repeated high dose treatment with ivermectin (Table 2) caused sustained abrogation of embryogenesis and accumulation of dead mf *in utero*, analogous to findings with *O. volvulus*, but worms were not killed up to 2 years post-treatment (Renz *et al.* 1995; Bronsvoort, Renz, Tanya, Ekale & Trees, unpublished observations).

The milbemycin compound moxidectin ('Cydectin', Fort Dodge) also exhibits a potent effect on adult embryogenesis and recent results indicate that 2 years after six monthly treatments at 500 μg/kg normal embryogenesis was not restored. Following a single treatment at 500 μg/kg, effects were less marked. However, neither regime was macrofilaricidal.

Oxytetracycline is macrofilaricidal

Our interest in the effect of tetracycline against *O. ochengi* was initially aroused following the resolution of all nodules in a heavily infected animal which was undergoing a prolonged course of antibiotic treatment for the incidental dermatophytic infection, streptothricosis. A growing awareness of the existence of intracellular bacteria in filarial nematodes (which had first been described in *O. volvulus* in 1977 by Kozek & Marroquin) prompted us to mount a controlled trial. This has demonstrated that intermittent treatment with oxytetracycline over a 6-month period completely eliminated *O. ochengi* infections and led to resolution of all nodules by 9 months post treatment (Langworthy *et al.* 2000). Death of worms occurred after the destruction of intracellular, *Wolbachia*-like bacteria, visualized by EM. It is not possible to rule out direct drug action against the worm but the sequence of events in which bacteria were eliminated before lethal effects on worms were observed suggests that bacterial death was causally related to worm death. Whether this is due to toxic products of bacterial degradation or due to the elimination of some essential function provided by the bacteria is not clear, but the 100% prevalence of bacterial infection in worms (see above) suggests a relationship of mutual dependency. Other accumulating evidence of the effects of antibiotics on filarial nematodes, indicates that these effects are due to indirect effects on the bacteria rather than direct efficacy against the worms (Hoerauf *et al.* 1999). In other filarial nematodes sublethal effects have been described and a macrofilaricidal effect has not been observed (Bosshardt *et al.* 1993; Bandi *et al.* 1999; Hoerauf *et al.* 1999). This may be partly related to the much shorter period of treatment used in other host–parasite systems. Much research is needed to confirm and extend these findings and to determine whether an efficacious regime using antibiotics is practicable, but these results provide the most promising lead yet in the search for a safe macrofilaricide against *Onchocerca* spp.

Avermectins are prophylactic against O. ochengi

Administration of ivermectin to whole communities on an annual basis is the principal tool of the African Programme for Onchocerciasis Control, APOC (Molyneux & Davies, 1997) which, by suppressing mf, aims to eliminate onchocerciasis as a public health problem. Since humans will be treated under conditions of continuing exposure to infection, we were interested to determine whether ivermectin was efficacious against L_3/L_4 stages. An experiment with *O. volvulus* in chimpanzees had suggested a partial effect against L_3 larvae but not against later stages (Taylor *et al.* 1988). Accordingly a prospective controlled study was conducted against natural infection in Cameroon. It showed conclusively that monthly treatment of growing calves with ivermectin at 200 μg/kg or at 500 μg/kg completely prevented infection, as determined by nodule development, up to 21 months of exposure, at which time 5/6 untreated calves had acquired a total of 54 *O. ochengi* nodules (Tchakouté *et al.* 1999). A second experiment has confirmed and extended these results with both ivermectin and moxidectin (at 150 μg/kg and 200 μg/kg doses respectively) and at 1 month and 3 month treatment intervals (Gilbert, Langworthy, Bronsvoort, Tanya, Renz & Trees, unpublished observations). These studies are not designed to provide a chemoprophylactic regime for mass human use. Their significance is twofold. Firstly, they indicate that if the annual ivermectin treatments carried out under APOC are timed to coincide with times of maximum transmission, a double benefit will accrue – namely prophylaxis as well as clearance of existing mf burdens. Secondly, they raise the possibility that chemoprophylactic attenuation of L_3/L_4 larvae will immunize individuals and protect against subsequent challenge (see later).

IMMUNOLOGY AND IMMUNITY

These studies have been carried out as part of the Edna McConnell Clark Foundation programme to develop a vaccine for human onchocerciasis. They have involved collaboration between Liverpool and IRAD, together with many laboratories involved in the EMCF programme. With respect to *O. ochengi* the objectives have been to (1) generate experimental infections and study the immune responses involved; (2) to determine if immunity to *O. ochengi* exists naturally; (3) to determine if immunization is possible with irradiated L_3s; (4) and finally to determine if recombinant antigens can immunize against field challenge. The first three objectives have been achieved and a major experiment to address the fourth is currently underway.

Immune responses are down-regulated at patency in experimental infections

Cattle were infected with single-pulse infections with 350 or 700 L_3 larvae as briefly described above. This infection permitted the serial observation of synchronous infections and the determination of immunological events in relation to the parasite life-cycle (Graham, Trees, Collins, Taylor, Moore, Guy & Bianco, unpublished observations). Antigen-specific proliferation and *ex vivo* cytokine production by peripheral blood mononuclear cells (PBMCs) followed infection and was maintained throughout the prepatent period but both were down-regulated following the onset of microfilarial patency. Antigen-specific antibody production remained elevated throughout the post-infection (p.i.) period to

600 days p.i. The results are consistent with evidence from human infections of cellular hyporesponsiveness in association with patent infections (Soboslay *et al.* 1997) and provide evidence of analogous immune responses in *O. ochengi* infections of cattle and *O. volvulus* infections of humans.

Putative-immune (PI) cattle are significantly protected against heavy challenge infections but radically drug-cured cattle are not

The existence of aparasitotic individual cattle reared in endemic areas has already been referred to above and the analogy made with PI humans. We sought to test whether such cattle were indeed resistant to parasite challenge by re-exposing them, together with age-matched, non-infected controls imported from a non-endemic area. Additionally, we sought to determine if previously patently-infected cattle are resistant to re-infection after radical drug treatment with a macrofilaricide (melarsomine). These experiments are difficult to undertake in humans, but would enhance understanding of protective immunity in onchocerciasis. Cattle were exposed for 19 months in an area of high natural transmission (estimated Annual Transmission Potential of between 2800 and 25300 depending on proximity to simuliid breeding site). PI cattle did develop palpable nodules but far fewer ($P = 0.0002$) than naive or treated cattle; previously infected, melarsomine-treated cattle were fully susceptible to challenge infection and nodule loads were not statistically different from those in naive controls (Tchakouté, Bianco & Trees, unpublished observations). These results are consistent with observations on re-infection dynamics of *O. volvulus* in humans cured with suramin or stibocaptate (Duke, 1968).

Immunization with irradiated L₃ larvae induces significant protection

The use of chamber models has previously indicated that protection can be induced against L_3 larvae of *O. volvulus* implanted in mice (Lange *et al.* 1993; Taylor *et al.* 1994) and in calves, immunization by live *O. volvulus* L_3 larvae and subsequent challenge with *O. ochengi* L_3s showed a reduction in the numbers of *O. ochengi* L_3s developing to adults in vaccinated calves compared with naive controls (Achu-Kwi, Renz & Enyong, unpublished observations). The latter experiment was designed to be analogous to natural conditions, where cross-transmission of *O. ochengi* L_3s to humans is associated with reduced susceptibility to *O. volvulus* infection (Renz *et al.* 1994; Wahl *et al.* 1998). By the classical method of using irradiated larvae, attempted immunization of chimpanzees had given equivocal results due to limitations of the experimental system

(Prince *et al.* 1992). Finding PI cattle in endemic areas and the results of the exposure experiment involving PI cattle (above) suggested that reduced susceptibility to *O. ochengi*, possibly immune-mediated, existed naturally and hence might be induced experimentally. Using irradiated *O. ochengi* L_3 larvae cattle were immunized then challenged under controlled conditions in Liverpool (Graham, Trees, Lustigman & Bianco, unpublished observations). Immunized cattle were significantly protected against challenge with 1000 *O. ochengi* L_3 larvae and adult worm burdens were very significantly reduced ($P < 0.01$). This experiment illustrated that it was possible to induce and measure immunity to *O. ochengi* and has led to a current field trial to evaluate the efficacy of selected recombinant antigens to induce protection.

CONCLUSIONS AND FUTURE RESEARCH

As research continues it is clear that the biological and molecular similarities between *O. ochengi* and *O. volvulus* extend to functional aspects of their response to drugs and the hosts' immune responses to them. The possibility of inducing experimental infections of *O. ochengi* in cattle, although logistically challenging, has provided opportunities to study the kinetics of immune responses and relate them to stages in parasite development in the host, and to conduct precise infection and challenge experiments. The prevalence of naturally infected animals harbouring multiple infections is enabling studies in Africa to identify chemotherapeutic agents, and the high rate of endemic transmission is enabling field challenge experiments to assess chemo- or immuno-prophylactic regimes within practicable time-frames. Current and future research will seek to examine the potential of antibiotics as macrofilaricides, the relationship of *Wolbachia* and *Onchocerca*, and the efficacy of other compounds arising from the DDR programme of WHO. A major field challenge trial is underway to determine the protection conferred by eight recombinant antigens, identified and selected by several laboratories participating in the EMCF vaccine development programme. The results of that will be known in 2001, as will those of a second field challenge experiment in which the immunity of calves reared under ivermectin and moxidectin chemoprophylaxis will be assessed. Concurrent with the advances in our knowledge of the parasitology of *O. ochengi*, developments in bovine immunology and the availability of reagents to assay antibody isotypes, cytokines and T cell subsets means that this host–parasite system can provide a highly comparable and experimentally malleable model for *O. volvulus* which can contribute to the fundamental understanding of onchocerciasis in humans.

ACKNOWLEDGEMENTS

The work described has been supported by the MACROFIL programme of the WHO/TDR/OCP/APOC, and the Edna McConnell Clark Foundation, to whom we are extremely grateful. Additional support for postgraduate students has been given by WHO, the University of Liverpool, the Liverpool School of Tropical Medicine and the University of Hohenheim. AR has been supported by the commission of the European Communities (STD programme). We are indebted to numerous collaborators and helpers, especially in Cameroon who have provided essential help, notably David Ekale and other members of our team, to the staff of IRAD, Wakwa, and to numerous farmers and ranchers who have helped our field work. We thank Daniel Achu-Kwi, Mark Bronsvoort, Helen Cross, Wolfgang Daiber, Jeff Gilbert, Nigel Langworthy, Mark Taylor, Virginia Tchakouté (nee Wood) and Richard Whittle for their permission to quote unpublished research. Finally a particular thank you to Goetz Wahl.

REFERENCES

ACHU-KWI, D., DAIBER, W. H., RENZ, A., WAHL, G. & WANJI, S. (1994). Prepatency period and some aspects of the epizootiology of *Onchocerca ochengi* infestation in cattle in the Adamawa Plateau, Cameroon. *Parasite* 1, 10–12.

BAIN, O. (1981). Le genre *Onchocerca*: hypothèses sur son évolution et clé dichotomique des espèces. *Annales de Parasitologie* 56, 503–526.

BANDI, C., McCALL, J. W., GENCHI, C., CORONA, S., VENCO, L. & SACCHI, L. (1999). Effects of tetracycline on the filarial worms *Brugia pahangi* and *Dirofilaria immitis* and their bacterial endosymbionts *Wolbachia*. *International Journal for Parasitology* 29, 357–364.

BIANCO, A. E., HAM, P. J., EL SINNARY, K. & NELSON, G. S. (1980). Large-scale recovery of *Onchocerca* microfilariae from naturally infected cattle and horses. *Transactions of the Royal Society of Tropical Medicine and Hygiene* 74, 109–110.

BOSSHARDT, S. C., McCALL, J. W., COLEMAN, S. U., JONES, K. L., PETIT, T. A. & KLEI, T. R. (1993). Prophylactic activity of tetracycline against *Brugia pahangi* infection in jirds (*Meriones unguiculatus*). *Journal of Parasitology* 79, 775–777.

CHAVASSE, D. C., POST, R. J., LEMOH, P. A. & WHITWORTH, J. A. G. (1992). The effect of repeated doses of ivermectin on adult female *Onchocerca volvulus* in Sierra Leone. *Tropical Medicine and Parasitology* 43, 256–262.

COMLEY, J. C. W., REES, M. J., TURNER, C. H. & JENKINS, D. C. (1989a). Colorimetric quantitation of filarial viability. *International Journal for Parasitology* 19, 77–83.

COMLEY, J. C. W., TOWNSON, S., REES, M. J. & DOBINSON, A. (1989b). The further application of MTT-formazan colorimetry to studies on filarial worm viability. *Tropical Medicine and Parasitology* 40, 311–316.

CROSS, H. F., BRONSVOORT, B. M., WAHL, G., RENZ, A., ACHU-KWI, D. & TREES, A. J. (1997). The entry of ivermectin and suramin into *Onchocerca ochengi* nodules. *Annals of Tropical Medicine and Parasitology* 4, 393–401.

CROSS, H. F., RENZ, A. & TREES, A. J. (1998). *In vitro*

uptake of ivermectin by adult male *Onchocerca ochengi*. *Annals of Tropical Medicine and Parasitology* 6, 711–720.

DETERMANN, A., MEHLHORN, H. & GHAFFAR, F. A. (1997). Electron microscope observations on *Onchocerca ochengi* and *O. fasciata* (Nematoda: Filaroidea). *Parasitology Research* 83, 591–603.

DUKE, B. O. L. (1968). Reinfections with *Onchocerca volvulus* in cured patients exposed to continuing transmission. *Bulletin of the World Health Organization* 39, 307–309.

GEORGE, G. H., PALMIERI, J. R. & CONNOR, D. H. (1985). The onchocercal nodule: interrelationship of adult worms and blood vessels. *American Journal of Tropical Medicine and Hygiene* 34, 1144–1148.

GRAHAM, S. P., WU, Y. K., HENKLE DUEHRSEN, K., LUSTIGMAN, S., UNNASCH, T. R., BRAUN, G., WILLIAMS, S. A., McCARTHY, J., TREES, A. J. & BIANCO, A. E. (1999). Patterns of *Onchocerca volvulus* recombinant antigen recognition in a bovine model of onchocerciasis. *Parasitology* 119, 603–612.

HAM, P. J., TOWNSON, S., JAMES, E. R. & BIANCO, A. E. (1981). An improved technique for the cryopreservation of *Onchocerca* microfilariae. *Parasitology* 83, 139–146.

HOCH, B., WAHL, G., ENYONG, P., LUDER, C. G., HARNETT, W. & RENZ, A. (1994). Onchocerciasis of cattle and man: serological recognition of parasite specific and cross-reactive antigens. *Parasite* 1, 14.

HOERAUF, A., NISSEN-PÄHLE, K., SCHMETZ, C., HENKLE-DÜHRSEN, K., BLAXTER, M. L., BÜTTNER, D. W., GALLIN, M. Y., AL-QAOUD, K. M., LUCIUS, R. & FLEISCHER, B. (1999). Tetracycline therapy targets intracellular bacteria in the filarial nematode *Litomosoides sigmodontis* and results in filarial infertility. *Journal of Clinical Investigation* 103, 11–18.

KLÄGER, S., WHITWORTH, J. A. G. & DOWNHAM, M. D. (1996). Viability and fertility of adult *Onchocerca volvulus* after 6 years of treatment with ivermectin. *Tropical Medicine and Health* 1, 581–589.

KOZEK, W. J. & MARROQUIN, H. F. (1977). Intracytoplasmic bacteria in *Onchocerca volvulus*. *American Journal of Tropical Medicine and Hygiene* 26, 663–678.

LANGE, A. M., YUTANAWIBOONCHAI, W., LOK, J. B., TRPIS, M. & ABRAHAM, D. (1993). Induction of protective immunity against larval *Onchocerca volvulus* in a mouse model. *American Journal of Tropical Medicine and Hygiene* 49, 783–788.

LANGWORTHY, N. G., RENZ, A., MECKENSTEDT, U., HENKLE-DÜHRSEN, K., BRONSVOORT, B. M. deC., TANYA, V. N., DONNELLY, M. J. & TREES, A. J. (2000). Macrofilaricidal activity of tetracycline against the filarial nematode, *Onchocerca ochengi*: elimination of *Wolbachia* precedes worm death and suggests a dependent relationship. *Proceedings of the Royal Society London B* (in press).

LOWRIE, R. C. (1983). Cryopreservation of third-stage larvae of *Brugia malayi* and *Dipetalonema viteae*. *American Journal of Tropical Medicine and Hygiene* 32, 767–771.

McCALL, P. J. & TREES, A. J. (1989). The development of *Onchocerca ochengi* in surrogate temperate Simuliidae, with a note on the infective larva. *Tropical Medicine and Parasitology* 40, 295–298.

MEREDITH, S. E. O., LANDO, G., GBAKIMA, A. A., ZIMMERMAN, P. A. & UNNASCH, T. R. (1991). *Onchocerca volvulus*: application of the polymerase chain reaction to identification and strain differentiation of the parasite. *Experimental Parsitology* **73**, 335–344.

MOLYNEUX, D. H. & DAVIES, J. B. (1997). Onchocerciasis control: moving towards the millenium. *Parasitology Today* **13**, 418–425.

PLAISIER, A. P., ALLEY, E. S., BOATIN, B. A., VAN OORTMARSSEN, G. J., REMME, H., DE VLAS, S. J., BONNEUX, L. & HABBEMA, J. D. F. (1995). Irreversible effects of ivermectin on adult parasites in onchocerciasis patients in the Onchocerciasis Control Programme in West Africa. *Journal of Infectious Diseases* **172**, 204–210.

PRINCE, A. M., BROTMAN, B., JOHNSON, E. H., SMITH, A., PASCUAL, D. & LUSTIGMAN, S. (1992). *Onchocerca volvulus*: immunization of chimpanzees with X-irradiated third stage (L_3) larvae. *Experimental Parasitology* **74**, 239–250.

RENZ, A., ENYONG, P. & WAHL, G. (1994). Cattle, worms and zooprophylaxis. *Parasite* **1**, 4–6.

RENZ, A., TREES, A. J., ACHU-KWI, D., EDWARDS, G. & WAHL, G. (1995). Evaluation of suramin, ivermectin and CGP 20376 in a new macrofilaricidal drug screen, *Onchocerca ochengi* in African cattle. *Tropical Medicine and Parasitology* **46**, 31–37.

SCHULZ-KEY, H. (1988). The collagenase technique: how to isolate and examine adult *Onchocerca volvulus* for the evaluation of drug effects. *Tropical Medicine and Parasitology* **39**, 423–440.

SMITH, R. J., COTTER, T. P., WILLIAMS, J. F. & GUDERIAN, R. H. (1988). Vascular perfusion of *Onchocerca volvulus* nodules. *Tropical Medicine and Parasitology* **39**, 418–421.

SOBOSLAY, P. T., GEIGER, S. M., WEISS, N., BANLA, M., LUDER, C. G. K., DREWECH, C. M., BATCHASSI, E., BOATIN, B. A., STADLER, A. & SCHULTZ-KEY, H. (1997). The diverse expression of immunity in humans at distinct states of *Onchocerca volvulus* infection. *Immunology* **90**, 592–599.

TAYLOR, H. R., TRPIS, M., CUPP, E. W., BROTMAN, B., NEWLAND, H. S., SOBOSLAY, P. T. & GREENE, B. M. (1988). Ivermectin prophylaxis against experimental *Onchocerca volvulus* infection in chimpanzees. *American Journal of Tropical Medicine and Hygiene* **39**, 86–90.

TAYLOR, M. J., VAN ES, R. P., SHAY, K., FOLKAR, S. G., TOWNSON, S. & BIANCO, A. E. (1994). Protective immunity against *Onchocerca volvulus* and *O. lienalis* infective larvae in mice. *Tropical Medicine and Parasitology* **45**, 17–23.

TAYLOR, M. J. & HOERAUF, A. (1999). *Wolbachia* bacteria of filarial nematodes. *Parasitology Today* **15**, 437–442.

TCHAKOUTÉ, V. L., BRONSVOORT, M., TANYA, V., RENZ, A. & TREES, A. J. (1999). Chemoprophylaxis of *Onchocera* infections: in a controlled, prospective study ivermectin prevents calves becoming infected with *O. ochengi*. *Parasitology* **118**, 195–199.

TREES, A. J. (1992). *Onchocerca ochengi*: mimic, model or modulator of *O. volvulus*? *Parasitology Today* **8**, 337–339.

TREES, A. J., WAHL, G., KLÄGER, S. & RENZ, A. (1992). Age-related differences in parasitosis may indicate acquired immunity against microfilariae in cattle naturally infected with *Onchocerca ochengi*. *Parasitology* **104**, 247–252.

TREES, A. J., WOOD, V. L., BRONSVOORT, M., RENZ, A. & TANYA, V. N. (1998). Animal models – *Onchocerca ochengi* and the development of chemotherapeutic and chemoprophylactic agents for onchocerciasis. *Annals of Tropical Medicine and Parasitology* **1**, S175–S179.

WAHL, G., ACHU-KWI, M. D., MBAH, D., DAWA, O. & RENZ, A. (1994). Bovine onchocercosis in North Cameroon. *Veterinary Parasitology* **52**, 297–311.

WAHL, G., EKALE, D. & SCHMITZ, A. (1998). *Onchocerca ochengi*: assessment of the *Simulium* vectors in North Cameroon. *Parasitology* **116**, 327–336.

WAHL, G., ENYONG, P., NGOSSO, A., SCHIBEL, J. M., MOYOU, R., TUBBESING, H., EKALE, D. & RENZ, A. (1998). *Onchocerca ochengi*: epidemiological evidence of cross-protection against *Onchocerca volvulus* in man. *Parasitology* **116**, 349–362.

XIE, H., BAIN, O. & WILLIAMS, S. A. (1994). Molecular phylogenetic studies on filarial parasites based on 5s ribosomal spacer sequences. *Parasite* **1**, 141–151.